Alfred P. (Alfred Payson) Gage

The Elements of Physics

A Text-book for High Schools and Academie

Alfred P. (Alfred Payson) Gage

The Elements of Physics
A Text-book for High Schools and Academie

ISBN/EAN: 9783744763868

Printed in Europe, USA, Canada, Australia, Japan

Cover: Foto ©Paul-Georg Meister /pixelio.de

More available books at **www.hansebooks.com**

THE
ELEMENTS OF PHYSICS

A TEXT-BOOK FOR HIGH SCHOOLS
AND ACADEMIES

BY

ALFRED PAYSON GAGE, Ph.D.

AUTHOR OF "PRINCIPLES OF PHYSICS," "INTRODUCTION TO
PHYSICAL SCIENCE," ETC.

REVISED EDITION

BOSTON, U.S.A.
GINN & COMPANY, PUBLISHERS
The Athenæum Press
1898

PREFACE.

SEVENTEEN years ago, in the preface to the first edition of this work, the author urged the importance of experiments to be performed by the pupil, as an aid to his mastery of the science of Physics, in opposition to the then universal and exclusive text-book-memorizing or illustrated-lecture system. In these seventeen years he has seen the value of the "objective method of study" demonstrated in the case of more than two thousand pupils in his own classes, and has witnessed the spread of the "laboratory method" of teaching science through the larger portion of the high schools of the United States. But with this great improvement in educational method, he has observed the development of a tendency which threatens seriously to impair its usefulness.

This is the tendency to allow enthusiasm for experimentation, for mere manipulation of apparatus, to obscure the importance of an intellectual mastery of the facts and their underlying principles. In their zeal for "training the mental powers" of their pupils, teachers may easily reduce the study of physical science to a *mere gymnastic*, forgetting that the object of training the mind is to render possible the greater acquisition of knowledge. If, then, the pupil has not at the end of his course a reasonably comprehensive and accurate grasp of the subject-matter of Physics, the instructor may be sure that he has failed to preserve a due proportion between study and experiment.

The author is convinced that both mental discipline and the acquisition of knowledge will be promoted if theory and experiment be somewhat sharply divided. If on the same page the pupil reads a statement of a scientific principle, the facts on which it is based, the reasoning by which it is deduced, together with elaborate detailed directions for preparing apparatus for exhibiting the facts in a series of experiments, and minute directions for the manipulation, he is almost certain to fail to discriminate in regard to the relative importance of each of these matters, and the relation which each sustains to the purposes of his study; he is likely to bestow equal attention upon each — even (if not wholly emancipated from childish methods) to memorize all alike. To obviate the danger of this enormous misdirection of effort, the author has been led to construct a separate Manual of "Physical Experiments," by the use of which the attention of the pupil is concentrated upon apparatus, process, result, and inference, and to treat matter in the text-book principally as a body of systematized knowledge to be mastered — a science in the strictest application of that term.

In the laboratory, then, the pupil should *do*, should *observe*, should *reflect*, should *infer*, and should *record* the results of these processes in a suitable manner. In the class-room he should concentrate his attention upon the *theory* and *principles* of the science as laid down in the text-book, striving to bring these and the results recorded in his manual into their proper relations in his mind as a portion of a unified whole, the possession of which shall make him master of one department of human knowledge.

Certain topics that come under the head of molecular physics, *e.g.* absorption, osmosis, and crystallization, have been omitted in this revision, since their processes are deemed too obscure and imperfectly understood to be classified under the title of scientific *knowledge*. That department of ether dynamics known as double refraction and polarization of light has been omitted, as it is generally conceded to be too difficult of apprehension for the average high-school pupil.

The author wishes to make special acknowledgments for valuable suggestions and aid in correcting the proofs to his friend, Dr. Arthur W. Goodspeed, of the University of Pennsylvania, and to Mr. Albert P. Walker, his colleague in the English High School, Boston, who has read the work in manuscript and in proof sheets.

CONTENTS.

CHAPTER I.

Introduction.

PAGES

Domain of Physics. Some properties of matter. Physical measurements. Kinematics. Laws of accelerated motion. Composition of motions and velocities. Kinds of motion 1-24

CHAPTER II.

Molar Dynamics.

Force. Newton's Laws of Motion. Momentum. Measurement of force. Composition of forces. Moments of force. Center of mass. Curvilinear motion. The pendulum. Gravitation. Work, energy, and power. Machines. Properties of matter due to molecular forces 25-97

CHAPTER III.

Dynamics of Fluids.

Law of transmission of pressure. Pascal's Principle. Atmospheric pressure. Boyle's Law. Barometer. Principle of Archimedes. Specific gravity 98-123

CHAPTER IV.

Molecular Dynamics. Heat.

Theory of heat. Sources of heat. Thermometry. Calorimetry. Specific heat. Laws of gaseous bodies. Fusion. Latent heat. Artificial cold. Hygrometry. Diffusion of heat. Thermo-dynamics. Steam engine 124-165

CHAPTER V.

Energy of Mass Vibration.

Simple harmonic motion. Wave-motion. Sound-waves. Reënforcement of sound-waves. Pitch. Composition of wave-forms. Discord and harmony. Musical instruments. Vocal organs. The ear 166–205

CHAPTER VI.

Ether Dynamics. Radiant Energy.

The ether. Radiation. Light. Intensity of illumination. Mirrors. Refraction. Prisms and lenses. Prismatic analysis of light. Spectroscopy. Color. Optical instruments. The eye. Thermal effects of radiation 206–267

CHAPTER VII.

Ether Dynamics. Electrostatics.

Electrification. Conduction. Induction. Potential. Atmospheric electricity 268–278

CHAPTER VIII.

Electrokinetics.

Voltaic cells. Effects producible by electric current. Electrical quantities. Ohm's law. Instruments for measurement of electric current. Resistance and its measurement. Divided circuits. Methods of combining cells. Magnets and magnetism. Magnetic relations of the current. Mutual action of currents. Electromagnetic induction. Dynamos. Electric motors. Storage batteries. Thermo-electric currents. Electric light. Electroplating and electrotyping. Telegraph. Telephone and microphone. Electro-magnetic theory of light. Radiography. Tesla's investigations. 279–365

Appendix 367–375

Index 377–381

LIST OF PLATES AND PORTRAITS.

	PAGE
SPECTRUMS, PLATE I	*Frontispiece.*
SIR ISAAC NEWTON	28
GALILEO	38
ARCHIMEDES	118
BENJAMIN FRANKLIN	278
LORD KELVIN	306
MICHAEL FARADAY	330
TELEGRAPH, PLATE II	355
RADIOGRAPH, PLATE III	362

Read Nature in the Language of Experiment.

ELEMENTS OF PHYSICS.

CHAPTER I.
INTRODUCTION.

SECTION I.
DOMAIN OF PHYSICS.

1. The perception of changes constantly taking place in the external world is a universal human experience. The manifestations of these changes to the senses are called *phenomena;* that which undergoes change is called *matter*. Since most of the changes which come within the scope of our present study are due to the *motions*[1] of portions of matter, we may adopt the following provisional definition:[2]

Physics is the science which treats of the phenomena of matter and motion.

2. **Some Properties of Matter or of Material Bodies:**
(1) *Extension.* Every portion of matter, however small, occupies space, *i.e.* it has length, breadth, and thickness. The property thus manifested is called *extension*.

---

[1] Sound, heat, and light, considered as distinct from the sensations to which they give rise, are nothing but motions.

"I do not believe that there exists in external bodies anything for exciting tastes, smells, and sounds, but motion, swift or slow; and if tongues, noses, and ears were removed, I am of the opinion that motion would remain, but there would be an end of tastes, smells, and sounds." — GALILEO.

[2] A provisional definition is one that answers present needs.

(2) *Impenetrability.* While matter occupies any portion of space it excludes all other matter from that portion. It is an axiom, exemplified in everyday experience, that *no two bodies (e.g.* our two hands) *can occupy the same space at the same time.*[1] The property in virtue of which a body occupies space to the exclusion of all other bodies is called *impenetrability.*

FIG. 1.

Air and other gases are invisible, and hence are not readily recognized as matter. If, however, we show that air possesses impenetrability, we have reason to believe that air is matter, since it possesses one of the characteristic properties of matter.

Experiment 1. — Float a cork on a surface of water, cover it with a tumbler as in Fig. 1, and force the tumbler, mouth downward, deep into the water. State what evidence the experiment furnishes that air is matter.

(3) *Divisibility.* Bodies of matter with which we are acquainted are divisible beyond any appreciable limits, but scientists assume that there are ultimate portions which cannot undergo further division. It is certain that every kind of material substance known has a minimum limit of division which *is never exceeded* in that substance. *A molecule of any substance is one of the ultimate homogeneous portions of which the substance is composed.*

A grain of musk will scent a room for years by constantly discharging into the air particles of musk. These particles are so small that the original grain does not perceptibly diminish in weight. Yet the smallest particle of musk dust, or the smallest particle of any substance that can be obtained by mechanical division, is very large in comparison with the molecules of which the particle is composed.[2]

---

[1] This is quite as axiomatic as the fact that "a body cannot be in two places at the same time."

[2] "In a cubic inch of any gas at atmospheric pressure and at ordinary temperatures there are about 3×10^{20} detached particles absolutely similar and equal to one another." — TAIT.

(4) *Compressibility and expansibility.* All bodies of matter are compressible and expansible, though in very different degrees. A proof of the existence of molecules and the molecular constitution of bodies (though by no means the most conclusive proof) may be found in this fact. Matter (*e.g.* gold or water) is either continuous as it appears to the eye, or it is discontinuous, granular, composed of distinct particles (molecules) somewhat as represented in Fig. 2. *But bodies are compressible and expansible.* On the supposition that matter is continuous, these phenomena are unexplainable; but on the supposition that matter is granular or molecular they are easily explainable. According to the latter supposition, a change of volume by contraction or expansion means simply *a coming together or a separation of the molecules composing the body*, as represented in Fig. 2.

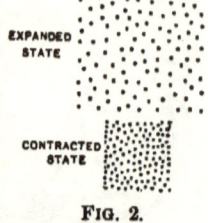

Fig. 2.

3. Theory of the Constitution of Matter. Porosity. — For reasons which will appear as our knowledge of matter is extended, physicists have generally adopted the following theory of the constitution of matter: *Every body of matter except the molecule is composed of exceedingly small disconnected particles, called molecules. No two molecules of matter in the universe are in permanent contact with each other. Every molecule is in rapid motion, moving back and forth among its neighbors, hitting and rebounding from them. When we heat a body we simply cause the molecules to move more rapidly through their spaces, so they strike harder blows on their neighbors, and usually push them away a very little; hence the body expands.*

If the molecules of a body are never in contact except at the instants of collision, it follows that there are spaces between them. These spaces are called *pores*.

It is estimated that even in dense solids the average distance between molecules is many times the diameter of a molecule.

- All matter is porous; thus water may be forced through the pores of cast iron; and gold, one of the densest of substances, absorbs liquid mercury.

Impenetrability may be affirmed of molecules, but not necessarily of masses. The term *pores*, in physics, is restricted to the *invisible spaces* that separate molecules, and does not include such cavities as may be seen with the naked eye in sponges, and with a microscope in wood, etc.

4. Physical Measurements. — Physics is essentially a science of measurements. Measuring consists in finding how many times a definite quantity, called *a unit*, is contained in the quantity to be measured. For example, should we wish to measure the length of a table, we may choose arbitrarily for a unit of measurement the length of a certain pencil and proceed to find how many times this pencil may be laid along the length of the table. If ten times, we say the table is ten pencil-lengths long.

The unit of measurement must be a definite quantity of the same kind as the thing to be measured. Thus a unit for measuring length must be a certain length, a unit for measuring surface must be a certain quantity of surface, and a unit for measuring volume must be a definite volume. A unit which has become legalized either by statute or by common usage is called a *standard unit*. The *expression* of a physical quantity consists of a statement of the *concrete unit* employed, *e.g.* pound, foot, quart, etc., with the *number of those units* prefixed. The numerical part, called the *numeric*, is obtained by measurement.

5. Metric System of Measures. — [In this connection the Tables of Metric Measures, in the Appendix, should be studied.] The term *metric* is derived from the word *meter*, which is the name of the fundamental unit employed in this system for measuring length. The international standard meter is defined by law to be the shortest distance between two lines engraved on a given platinum bar (carefully preserved by the French government) at the temperature of

0° C. (32° F.).[1] The metric system is now generally employed in scientific work.[2]

6. Volume, Mass, and Density. — The quantity of space a body of matter occupies is its *volume,* and is expressed in cubic inches, cubic centimeters, etc. By the *mass* of a body is meant the *quantity of matter* in the body.

The unit of mass generally employed in science is *the gram,* or, in the British system, *the pound.* The gram is the one-thousandth part of the standard *kilogram.* This standard is a piece of platinum carefully preserved by the French government at Paris. Originally it was intended to represent the mass of a cubic decimeter of pure water at the temperature of 4° C. *A kilogram of any substance is that quantity of the substance which, placed on a scale pan, would just balance in a vacuum the standard kilogram placed on the other pan.*

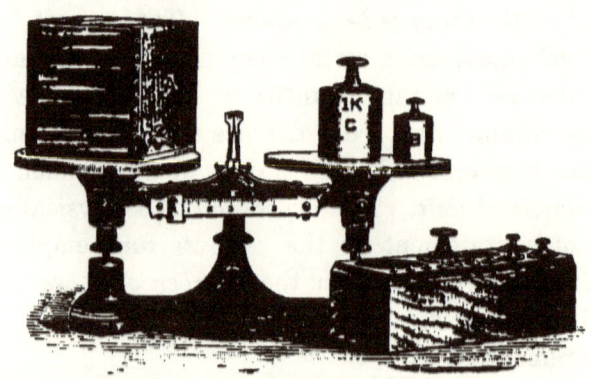

FIG. 3.

Experiment 2. — Place on one pan of a balance (Fig. 3) a vessel, A, whose capacity is one liter, *i.e.* one cubic decimeter (see Appendix). Place upon the other pan some body, B, which will just counterbalance

---

[1] The United States Government carefully preserves in Washington copies of the international meter. These are declared by Congress to be the standard units of length for this country.

[2] "The British measurements are infinitely inconvenient and wasteful of brain energy." — TAIT. "I look upon our English system as a wickedly brain-destroying piece of bondage under which we suffer." — LORD KELVIN.

the empty vessel. Then place upon the same pan a kilogram mass, C. Now pour water slowly into the vessel until the water and kilogram mass counterbalance each other. What mass of water does the vessel contain? How does the mass of water in A compare with the mass of the body C? How does the volume of water compare with the volume of the body C?

Mass is quite distinct from weight. The *weight* of a body is the measure of the attraction between it and the earth, and may vary with change of position, because the earth's attraction varies with the distance of the body from the earth, while the mass of the body remains constant. Mass does not depend upon weight, but weight depends upon mass.

When we open a heavy iron gate, it is its mass with which we have to deal; if it were lying on the ground and we should try to raise it, we should have to deal simultaneously with both its mass and its weight.

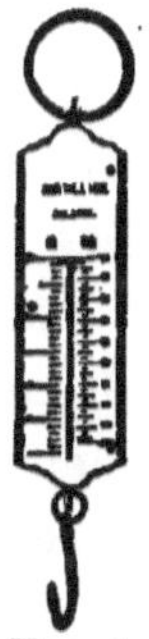

The process of measuring the mass of a body must not be confounded with the process of finding *how heavy* a body is, although both processes are, in common usage, called weighing. Weighing a body to ascertain its *mass* consists in balancing it with a body or bodies of known mass, and is performed with a scale balance (Fig. 3) and a set of masses (commonly called a set of weights). Weighing to ascertain *weight* should be performed with an instrument adapted to measuring *force*, e.g. a spring balance (Fig. 4). For most practical purposes, however, the latter instrument may be used to measure mass, inasmuch as *at the same place mass is proportional to weight*.[1]

FIG. 4.

Equal volumes of different substances (*e.g.* cork, cheese, lead) *contain unequal quantities of matter*. Of two substances, that which contains the greater quantity of matter in the same volume is said to be the *denser*. By the *density* of a substance

---

[1] This is one of many instances in physics in which one quantity is indirectly measured by measuring another proportional to it.

is meant the *mass in a unit of volume* of that substance. The density of water (at 4° C.) is one gram per cubic centimeter, and the density of cast iron is about 7.12 grams per cubic centimeter.

The mean (or average) density of a body is found by dividing its mass by its volume. Thus, if the mass of a body be 32 g. and its volume be 5 cc., its mean density is (32 ÷ 5 =) 6.4 g. per cc.

7. Three States of Matter. Fluidity. We recognize three states or conditions of matter, *viz. solid, liquid,* and *gaseous,* distinctively represented by earth, water, and air. Everyday observation teaches us that *solids tend to preserve a definite volume and shape; liquids tend to preserve a definite volume only, while their shape conforms to that of the containing vessel; gases tend to preserve neither definite volume nor shape, but to expand indefinitely.*

In consequence of their manifest tendency to flow, liquids and gases are called *fluids*.

Susceptibility of motion of the molecules of a body around and among one another is called *fluidity*. All bodies of matter, including solids, possess this property, but in very different degrees. It is due to this property that solids can be bent, stretched, and compressed, and that most metals can be drawn into wires and rolled or hammered into sheets.

EXERCISES.

1. (*a*) Give names of at least three substances. (*b*) Give a name of a *body* of each substance named.

2. What additional idea does the term "impenetrability" imply besides extension?

3. How may it be shown that air is matter?

4. (*a*) What is an air bubble (*e.g.* in water)? (*b*) What property does it show that air possesses?

5. Why is matter compressible?

6. How may bodies of matter be expanded?

7. Heating a body is attended with what molecular changes?

8. (a) Distinguish between the mass and the weight of a body. (b) Which may change, and why, while the other remains constant?

9. Which instrument represented in Figs. 3 and 4 will not show a change of weight of a body occasioned by a change of distance from the earth?

10. (a) Which is more difficult, to roll a cannon ball along a smooth floor or to raise the same from the floor? (b) In rolling the ball, with which do we deal, its mass or its weight? (c) In raising the ball, with what do we deal?

11. In weighing articles of groceries, such as tea, sugar, etc., is the purpose to measure their mass or the force of gravity?

12. If the mass of 45 cc. of cork be 10.8 g., what is the density of cork?

SECTION II.

KINEMATICS.

8. Kinematics. Kinematics treats of motions without reference to their causes. Nearly all physical phenomena are attributable to motions (§ 1), hence the great importance of this subject. Motions of bodies of sensible size are called *molar or mechanical motions* in distinction from molecular motions. The motions of the molecules of a body constitute its *heat*, and will therefore be treated under that head.

9. Motion. *Motion is a continuous change of position.* That which moves is known as matter. The position of a particle of matter is determined by its direction and distance from another particle, or from some point of reference. A particle moves relatively to a given point while an imaginary straight line connecting it with the point changes either in *direction* or in *length*. A particle is at rest relative to a given point while a straight line joining them changes *neither* in direction *nor* in length.

MOTION. 9

When you open or shut the legs of a pair of dividers (A, Fig. 5), a straight line, a' b', connecting the points at the ends of the legs changes *in length;* hence there is relative motion between these points. If (B, Fig. 5) you open the legs a little way, and, fixing the end of one of the legs upon a plane surface, trace a circle with the end of the other leg around the former as a center, there will be relative motion between the two points, since a line joining them, a b, a b', etc., changes *in direction.*

If (C, Fig. 5) you trace with the points of the open dividers two straight parallel lines on a plane surface, the two points will be relatively at rest, just as surely as if the dividers were lying upon the table, since in both cases a straight line connecting the points a b, a' b', etc., changes neither in length nor in direction.

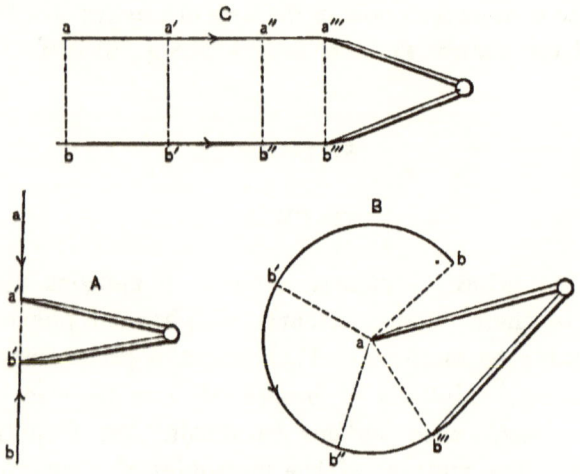

FIG. 5.

A point may be at the same instant at rest with reference to certain points and in motion with reference to certain other points. For example, while the points of the dividers are tracing straight lines on the plane surface (C, Fig. 5), and are relatively at rest, they are in motion with reference to every point in the plane surface. A passenger in a railway car may be at rest relative to the car and the other passengers, but in rapid motion relative to objects by the roadside.

In ordinary language the phrase "a body at rest" means a body that does not change its position with reference to that on which it stands, as, for instance, the surface of the earth or the deck of a ship. It can mean nothing else, for both the body said to be "at rest" and all points on the earth's surface are in rapid motion with reference to the sun and other heavenly bodies, and also with reference to the earth's axis.

10. Velocity. *Velocity is rate of change of position.* It involves units of distance (or length) and units of time, and is commonly expressed in *units of distance per unit of time*, *e.g.* feet per second, miles per hour, etc.

If a particle move through equal distances in equal periods of time, the velocity is said to be *uniform*. If the distances traversed in equal intervals of time continually increase or continually decrease, the velocity is said to be *accelerated*.

Velocity is determined by dividing the distance traversed by the time consumed. If a body move s feet in t seconds, its velocity, v, is $\frac{s}{t}$ feet per second, or $v = \frac{s}{t}$. In case the velocity be accelerated, this result is to be regarded as the *average* velocity for that distance; and in the case of uniform motion the average velocity is the same as the *actual* velocity at every instant. It is evident that the actual velocity of a body whose rate of motion changes can be given only at some definite instant or point in its journey. It denotes the space which *would be traversed in a unit of time*, if at the given instant the velocity should become uniform.

When a particle experiences equal changes of velocity in equal units of time, its motion is said to be *uniformly accelerated*, and its change of velocity per unit of time is called its *rate of acceleration*, or simply its *acceleration*, and is represented by a. When the velocity increases, as in the case of a falling stone, its acceleration is said to be positive ($+a$); when the velocity decreases, as in the case of a stone thrown upward, its acceleration is said to be negative ($-a$).

The acceleration of a body falling in a vacuum and of a body projected vertically upward in a vacuum is practically uniform; in the former case it is about 32.2 feet (or 9.8 m.) per second; in the latter case it is a negative acceleration of about 32.2 feet per second.

ACCELERATED MOTION.

The rate of acceleration of a particle in traversing a certain distance in a given time is found by dividing the entire change in velocity, v, by the units of time, t, consumed in making the change; i.e. $a = \dfrac{v}{t}$, whence $v = a\,t$.

Thus, if the velocity of a railroad train at a certain instant be 25 miles per hour, and half an hour hence it be 15 miles per hour, then the entire change of velocity, v, is -10 miles per hour; hence the average acceleration, i.e. the acceleration if it were uniformly distributed throughout the 30 minutes, is $\dfrac{-10}{30} = \left(-\dfrac{1}{3}\right)$ of a mile per minute. Again, if a stone falling with a uniformly accelerated velocity acquire in 4 seconds a velocity of 128.8 feet per second, its acceleration is $\dfrac{128.8}{4} = 32.2$ feet per second.

SECTION III.

LAWS OF UNIFORMLY ACCELERATED MOTION.

11. First Law. If a particle starting from a state of rest move with uniform acceleration, a, its velocity, v, at the end of any given unit of time, t, is found by the equation (1). $v = a\,t$, as given in § 10. From this equation we derive the following law:[1]

Change of velocity due to uniform acceleration is equal to the product of the acceleration multiplied by the units of time; or, the change of velocity is proportional to the rate of acceleration and to the time occupied.

But if a particle be in motion, and at a certain instant have a velocity, V, and its acceleration be a, then its velocity at any subsequent instant is expressed as follows:

After a lapse of one unit of time, $\quad v = V \pm (a \times 1)$.
" " " " two units " " $\quad v = V \pm (a \times 2)$.
" " " " t " " " (2) $v = V \pm a\,t$.

---

[1] A physical law is an expression of the relation which has been discovered to exist between certain physical quantities.

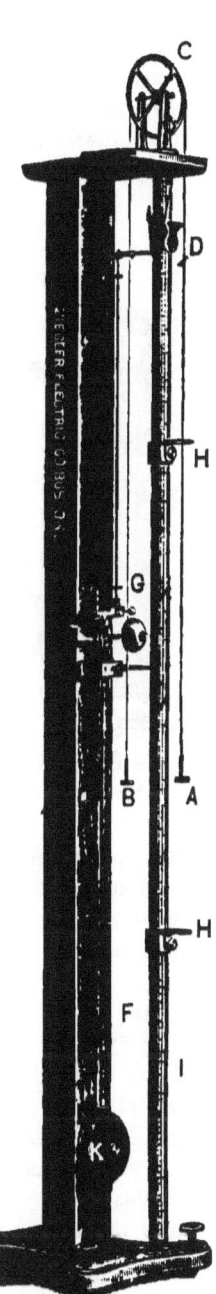

FIG. 6.

12. Second Law. Since the velocity of a particle starting from a state of rest increases from zero to at, the average velocity must be $\dfrac{0+at}{2} = \tfrac{1}{2}\,at$. At this rate in the same time, t, it would traverse a distance, S, equal to $\tfrac{1}{2}\,at \times t = \tfrac{1}{2}\,at^2$ units; hence, (3) $S = \tfrac{1}{2}\,at^2$. From this we derive the law:

The distance traversed in a given time by a particle starting from a state of rest and having uniformly accelerated velocity is one half the product of the acceleration and the square of the units of time; or, *the entire distance traversed is proportional to the rate of acceleration and to the square of the time occupied.*

If a particle, instead of starting from a state of rest, have an initial velocity, V, it would move in t units of time without acceleration a distance $V \times t$; to this distance must be added the distance it moves in consequence of acceleration, in order to obtain the entire distance traversed in t units, and our formula becomes (4)

$$S = Vt + \tfrac{1}{2}\,at^2.$$

If it be required to find the distance passed over during any specified unit of time we may subtract the distance traversed in $t-1$ units from the distance traversed in t units. Thus, representing the required distance traversed during a specified unit of time t by s, we have (5)

$$s = \tfrac{1}{2}\,at^2 - \tfrac{1}{2}\,a(t-1)^2 = \tfrac{1}{2}\,a(2t-1).$$

ACCELERATED MOTION.

13. Verification. The laws given above are verified approximately and conveniently by the use of the venerable Atwood's machine.[1] The equal weights A and B (Fig. 6) are suspended by a thread passing over the wheel C. Inasmuch as the weights are equal, they counterbalance each other and remain at rest. Raise the weight A and place it on the platform D as shown in Fig. 7. Place on this weight a small additional one, E, called a "rider," the weight of which sets the system in motion. Set the pendulum F swinging. At each swing it causes a stroke of the hammer on the bell G. At the instant of the first stroke the pendulum causes the platform D to drop so as to allow the weights to move. When they reach the ring H, the rider, not being able to pass through, is caught off by the ring. Raise and lower the ring on the graduated pillar I, and ascertain by repeated trials the average distance the weights move between the first two strokes of the bell, *i.e.* during one swing of the pendulum. Inasmuch as all swings of the pendulum are made in equal intervals of time, we may take the time of one swing as a *unit of time*. We will also, for convenience, take for a

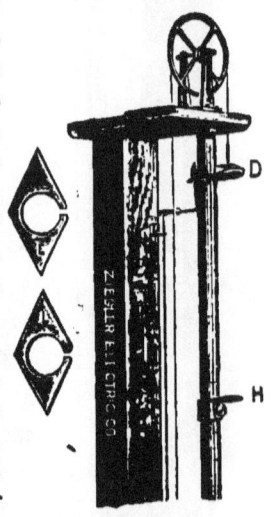

FIG. 7.

unit of distance the distance the weights move during the first unit of time, call this unit a *space*, and represent the unit graphically by the line a b (Fig. 8).

Next ascertain how far the weights move from the starting point during two units of time, *i.e.* in the interval of time between the first and third strokes of the bell. The distance will be found to be four spaces, or four times the distance that they moved during the first unit of time. This distance is represented by the line a c.

Now ascertain the velocity which the weights have at the end of the first unit of time. Place the ring H at the point (b, Fig. 8) which the weights have been found by trial to reach at the end of the first unit of time. Allow the weights to descend as before. At the end of the first unit of time the rider is caught off. At this instant acceleration ceases, and the motion becomes uniform. Ascertain how far the weights move with uniform velocity during the second unit of time; this velocity is evidently the velocity which the weights have at the end of the first unit

---

[1] This machine is a contrivance which enables us to increase the mass to be moved without increasing the force which moves it, thus so decreasing the acceleration as to render approximate measurements feasible.

14 INTRODUCTION.

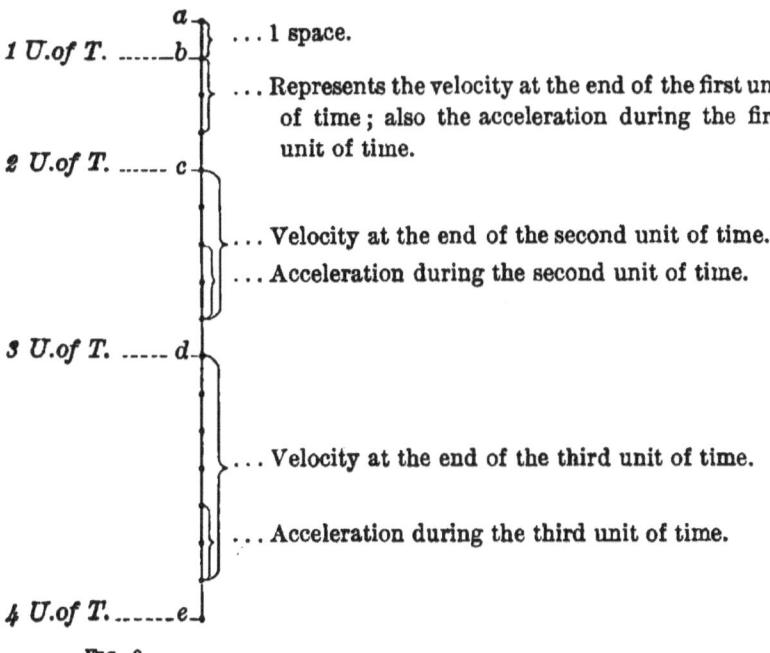

FIG. 8.

of time. This distance will be found to be (approximately [1]) two spaces; hence the velocity at the end of the first unit of time is *two spaces per unit of time*. But the velocity at the beginning of the first unit of time was zero; hence the acceleration during the first unit of time is *two spaces per unit of time*.

In like manner determine the velocity at the end of the second unit of time. It will be found to be four spaces per unit of time. And as the velocity at the end of the first unit of time was two spaces per unit of time, the acceleration during the second unit of time is two spaces per unit of time. Hence the acceleration during the first two units of time is *uniform*, and the change of velocity during the first two units of time, as stated in the First Law, $= at = 2 \times 2 = 4$ spaces per unit of time.

[1] Approximately, since they are retarded by the resistance of the air and the friction of the wheel.

EXERCISES.

1. (a) What is the meaning of the statement that "the velocity of a falling body at the end of the first second of its fall is 32.2 feet per second"? (b) Has the body the same velocity at any other point?

2. What is the relation between the velocity of a freely falling body at the end of the first unit (of time) of its fall and its acceleration?

3. The velocity of a particle at a certain instant is V; its acceleration is a. What will be its velocity, v, in t units of time afterward?

4. If the initial velocity of a body be V, its acceleration a, and its final velocity v, how long, t, was it in acquiring its final velocity?

5. If a body having an initial velocity V acquire in t seconds a velocity v, what is its acceleration?

6. If a body move from a state of rest with a uniform acceleration a, what space, S, will it traverse in t units of time?

7. If a body move from a state of rest with an acceleration a, in what time, t, will it traverse the space S?

8. The velocity of a particle at a certain instant is 20 feet per second; its acceleration is 3 feet per second. What will be its velocity 10 seconds hence?

9. Suppose that the acceleration of the particle mentioned above be — 2 feet per second, what will be its velocity 5 seconds after the instant named?

10. (a) A body falls from a state of rest; its velocity increases (if we disregard the resistance of the air) 32.2 feet per second. What is its velocity at the end of the first second? (b) What at the end of the tenth second? (c) What at the end of half a second?

11. If the initial velocity of a body be 5 feet per second, its final velocity 25 feet per second, and its acceleration 2 feet per second, what was the time consumed in acquiring the final velocity?

12. A bullet is projected vertically upward with an initial velocity of 161 feet per second. What will be its velocity at the end of the third second ($a = -33.2$ feet per second)?

13. How long will the bullet named in the last problem rise?

14. What velocity will the bullet have at the end of the sixth second, and in what direction will it be moving?

15. A person throws a stone vertically upward to a distance of 78.4 meters. With what velocity does the stone leave his hand?

INTRODUCTION.

16. A stone thrown vertically downward is given an initial velocity of 40 feet per second. How far will it descend in 10 seconds?

17. (a) A bullet is projected vertically upward with an initial velocity of 225.4 feet per second. How long will it rise? (b) How far will it rise?

18. How long will it take a body to fall from a state of rest 1030.4 feet?

19. (a) A body falls during $1\frac{1}{2}$ seconds. What is its final velocity? (b) How far does it fall?

20. A body falls 297.6 feet in 4 seconds. What was its initial velocity? *Ans.* 10 feet per second.

21. What initial velocity must be given a body that it may rise 6 seconds?

22. A falling body acquires a velocity of 68.6 meters per second. How long does it fall ($a = 9.8$ meters)?

23. A body acquires in falling a velocity of 98 meters per second. From what hight has it fallen?

24. A body is projected vertically upward with a velocity of 128.8 feet per second. Where will it be at the end of 8 seconds?

25. (a) A body at rest receives a uniform acceleration of 10 meters per minute. How far will it move in half an hour? (b) What will be its velocity at the end of the half-hour?

26. A stone is thrown vertically upward with a velocity of 100 meters per second. In how many seconds will it return to its original position?

27. (a) In what time would a stone fall to the earth from a balloon 3 miles high? (b) What velocity would it acquire?

28. A body starts from a state of rest and moves with a uniform acceleration of 18 feet per second. Find the time required to traverse the first foot.

29. A stone dropped from a hight of 4 feet will reach the ground in what time?

30. Find the depth of a well in which a stone, if dropped, takes $1\frac{1}{2}$ seconds to reach the bottom.

31. A body falls from a state of rest. (a) How many feet does it fall during the fifth second? (b) How many meters does it fall during the fourth second?

32. How far does the stone referred to in Exercise 16 descend during the tenth second?

SECTION IV.

COMPOSITION AND RESOLUTION OF MOTIONS AND VELOCITIES.

14. Graphical Representation of Motion and of Velocity. A person who would describe to you the motion of a ball struck by a bat must tell you three things:[1] (1) *where it starts*, (2) *in what direction it moves*, and (3) *how far it goes*. These three essential elements may be represented graphically by a straight line. Thus, suppose balls at A and D (Fig. 9) to be struck by bats, and to move respectively to B and E in one second. Then the points A and D are their starting points, the lines A B and D E represent the direction of their motions, and the lengths of the lines represent the distances traversed. The lengths of these lines are not equal to the distances traversed by the two balls, but represent these distances drawn to some convenient arbitrary scale; thus on a scale of 1 cm. = 10 m., these lines represent distances of 32 m. and 20 m., respectively.

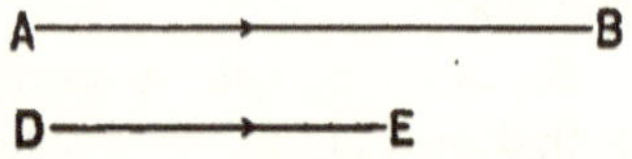

FIG. 9.

The velocity of a moving body is described by giving (1) *the direction of its motion*, and (2) *the units of distance traversed per unit of time*. Since the lines A B and D E represent the distances traversed by the two balls during the same unit of time, these lines likewise represent their *average velocities* during this time; *i.e.* A B may represent an average velocity of 32 m. per second, and D E an average velocity of 20 m. per second.

15. Composition of Simultaneous Motions and Velocities. If by any means a particle have two or more separate and independent motions communicated to it simultaneously, and if the motions imparted be themselves constant in velocity

---
[1] It is assumed that the motion is rectilinear.

and direction, the result of the two motions is a single motion in a straight line with a single velocity and direction.

This is illustrated in the following manner :

With the handle A in the position shown in Fig. 10, push it forward, carrying the frame B C to the right. This frame carries a pencil, D, whose point presses the paper below, and as the frame advances the line a b is traced upon the paper, graphically representing the motion of the pencil. If, when the pencil point is at a and the frame is at rest, the string G be pulled, the pencil will trace the line a c at right angles to a b. Now these two independent motions may be communicated to

FIG. 10.

the pencil *simultaneously* by fastening the string E to the binding screw F and pushing forward the handle A. The pencil point will not move in either of the lines a b or a c, but its motion will be intermediate between the two, and it will trace the line a d. This single motion, which is the result of the concurrence of two motions, is called their *resultant;* and they, with regard to the resultant, are called its *components*.

The distance a d is traversed in exactly the same time that the distance a b would be traversed if the pencil had no other motion, the handle A being pushed forward with the same speed in both cases; likewise the distance a d is traversed in the same time that the distance a c is accomplished when the string is simply pulled over the pulley G with the same speed, and has no other motion. The lines a b, a c, and a d represent not only the distances traversed in the several directions, but also the magnitudes and directions of the respective velocities. For example, if the velocity be constant and the pencil reach successively at the end of equal intervals of time the points m″, n″, and d (Fig. 11), then a m″, m″ n″, and n″ d represent its velocities in the successive intervals,

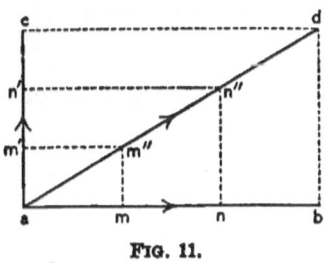

FIG. 11.

and a m, m n, and n b represent the velocities for the same intervals in the direction a b; and a m′, m′ n′, and n′ c the velocities in the direction a c. If points c and d, and d and b be joined by (dotted) lines, we have a parallelogram of which the line a d, representing the resultant, is a diagonal.

Hence, to find the resultant of two simultaneous velocities when they make an angle with each other, the rule is: *Construct a parallelogram of which the adjacent sides represent the two velocities; then the diagonal which lies between these adjacent sides represents their resultant.*

When more than two components are given, find the resultant of any two of them, then of this resultant and a third, and so on until every component has been used. For example, let the several velocities imparted to a particle be represented by the lines A B, A C, A D, and A E (Fig. 12). The resultant of A B and A C is A F; the resultant of A F and A D is A G; that of A G and A E is A H, which represents the resultant of the four velocities.

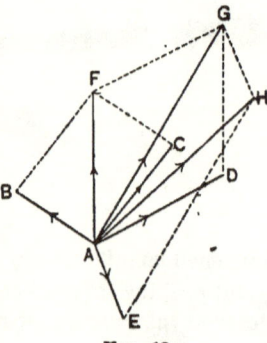

FIG. 12.

When two components are at right angles to each other, it is evident that we may obtain the magnitude of the resultant by finding the square root of the sum of the squares of the two components.

In case a particle has several velocities imparted to it, all in the same direction, their resultant is the sum of all. If some are opposite to others, one of the two directions is considered as positive and the opposite direction as negative, and these signs being prefixed to the numerical values, their algebraic sum is the resultant.

16. Resolution of a Motion or a Velocity into Components. Any motion or velocity may be resolved into two or any given number of motions or velocities. Let A B (Fig. 13) represent the velocity and direction of motion of a particle. Draw a line, A C, to represent, either arbitrarily or according to the conditions of the problem,

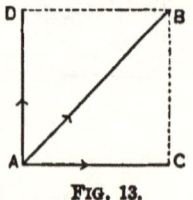

FIG. 13.

one of the required components. Connect B and C, draw A D parallel to B C, and D B to A C, and thus complete a parallelogram of which A B is a diagonal. The two adjacent sides A C and A D represent two component velocities of the particle; in other words, a particle having a velocity represented by the line A B has at the same time velocities represented in magnitude and direction by the lines A C and A D.

17. Composition of Constant with Accelerated Velocity. Experience teaches that a body, *e.g.* a stone, projected in a horizontal direction moves, not in a horizontal path, but in a path intermediate between a horizontal and a vertical one, showing that its velocity is composed of a horizontal and a

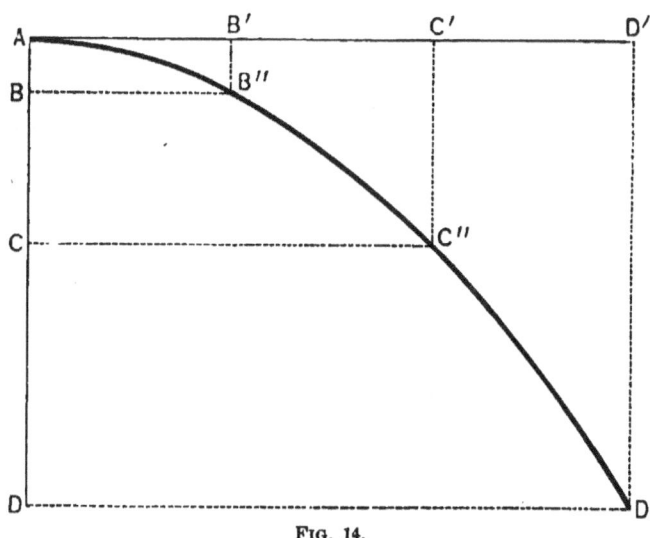

FIG. 14.

vertical component. Its horizontal velocity (if the resistance of the air be disregarded) is constant and its vertical velocity is uniformly accelerated. Let A B (Fig. 14) represent the vertical component of the motion during the first second; then B C and C D will represent its vertical motion during the second and third seconds, respectively. Let A B', B' C', and C' D' represent successive horizontal motions during the same three periods. Then by the law of the composition of motions it is evident that the body will pass from A to B″ during the first second, from B″ to C″ during the second second, and from C″ to D″ during the third second. The body traverses a curvilinear path called a *parabola,* as shown in the figure. In practice, the resistance of the air would modify the nature of the curve somewhat, so that its real path is a peculiar curve known in the science of gunnery as a *ballistic curve* or *trajectory.*[1]

It should be borne in mind that *one of the component velocities of a particle moving in a curvilinear path is always accelerated.*

EXERCISES.

1. (*a*) If a ship move east at the rate of 10 miles an hour, and a person on deck walk towards the bow at the rate of 2 miles an hour, what is the resultant of these two velocities? (*b*) With reference to what has he this velocity?

2. (*a*) Suppose the person on the ship mentioned above to walk aft at the rate of 2 miles an hour, what will be the resultant of the two velocities? (*b*) Prefix suitable signs to the numbers given and represent the addition which gives the resultant.

3. Suppose the person to walk directly north across the deck at the rate of 4 miles an hour, what will be the resultant of the two velocities?

4. Suppose the person to walk northeast at the rate of 4 miles an hour, what will be his resultant velocity? [In drawing the parallelogram of velocities, represent the component velocities to some scale, *e.g.* ¼ of 1 inch or 1 cm. = 1 mile; then, having completed the parallelogram and having drawn the diagonal which represents the resultant, measure the latter, and the result will express, on the scale chosen, the resultant velocity required.]

5. Suppose an attempt to be made to row a boat at the rate of 6 miles an hour directly across a stream flowing at the rate of 10 miles an hour, determine the direction and velocity of the boat.

---

[1] If the velocity of a projectile be very great, the trajectory at first will be very flat. For example, if the initial velocity of a rifle bullet be 2550 feet a second, the vertical component of the trajectory for the first 1500 feet may not be more than 2½ feet.

INTRODUCTION.

6. A vessel sails south-southeast (*i.e.* 22.5° east of south) at the rate of 14 miles an hour. Determine its southerly and its easterly velocity.

7. Represent graphically, to scale, a velocity of 100 feet per second, and resolve this velocity into two components which shall have between them an angle of 45°.

8. Imagine a body to be projected obliquely upward at an angle of 45°. Represent arbitrarily its vertically downward accelerated motion, and its obliquely upward constant motion for 3 seconds, and determine the actual path traversed by the body during this time.

9. When a ship is sailing northeast at the rate of 10 miles per hour, with what speed is it approaching a north and south coast lying to the east? *Ans.* 7.071 miles per hour.

SECTION V.

KINDS OF MOTION.

18. Motion of Translation and of Rotation. In pure motion of *translation* all the points of a body move with the same velocity and in the same direction (Fig. 15). Example: the

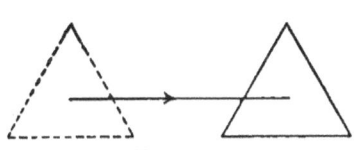

Fig. 15.
Rectilinear motion of translation.

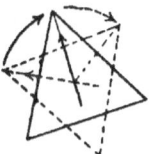

Fig. 16.
Motion of rotation.

motion of an elevator, or of a piston in the cylinder of a stationary steam engine. When the points of a body describe arcs of circles, the motion is one of *rotation* (Fig. 16). Example: the motion of a top, or of a wheel in a watch. When a body rotates, every particle in the body describes a circle round some point in a straight line which forms the *axis* of rotation.

The velocity of a point far from the axis is greater than that of a point nearer the axis, and, generally, the velocity of a point is proportional to its distance from the axis; hence the expression "velocity of a rotating body" is meaningless.

We may, however, speak of the *angular* velocity of the rotating body, which is the same for all points in the body. Angular velocity is *rate of rotation*, and is measured by the angle through which the rotating body turns in any given unit of time.

19. Rectilinear and Curvilinear Motion. Besides change in velocity, there may be a change in *direction* of motion. When a particle moves in a constant direction, *i.e.* in a straight line, as in the case of a freely falling bullet, its motion is said to be *rectilinear*. But if its motion constantly changes in direction, *i.e.* at every point, as is the case of every particle in a rotating wheel (except points on its axis), its motion is said to be *curvilinear*. It is evident that the direction of a motion in a curvilinear path can be given only for some specified point; and, furthermore, that the direction can be represented only by a straight line, for a curved line is a line having an infinite number of directions.

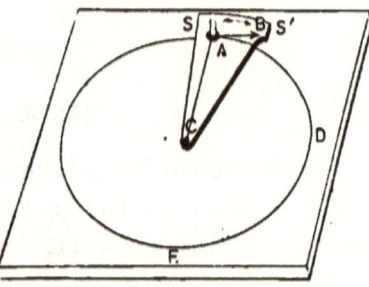

Fig. 17.

Let A (Fig. 17) represent a body mounted on a cardboard sector, S S', which is rotated about the axis C in the direction indicated by the arrow. The body will move in the circular path A D E F. The straight line A B will indicate the direction of the motion at every point, but it will be seen that this line changes its direction constantly. At whatever point the body may be at any instant, the line A B, which shows the direction of the motion, is tangent to the curve at that point.

20. Analysis of Circular Motion. If a particle move in a circular path, *e.g.* a stone whirled in a sling, its motion every instant is the resultant of a tangential motion and a centripetal (toward the center) motion. If when it passes point A (Fig. 18) its tangential velocity be represented by A B, its

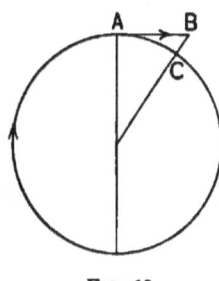

FIG. 18.

centripetal velocity may be represented by B C, because at the end of the unit of time in which it would reach B if it were moving in a straight line, it is found to be not at B, but at some other point, C, nearer the center by the distance B C. The centripetal motion is an accelerated motion (§ 17).

EXERCISES.

1. What kind of motion is that of the earth in its orbit?

2. Why is it meaningless to speak of the velocity of rotation of a body?

3. (a) What motions have the wheels of a carriage that is drawn straight along a level plane? (b) What motion has the carriage?

4. Compare the several velocities of the small front wheels of the carriage with those of the larger hind wheels.

5. (a) Can there be motion without direction? (b) How is the direction of the motion of a particle, when moving in a curve, represented?

6. How is the angular velocity of the earth's motion about its axis expressed?

7. (a) What two kinds of motions has the earth? (b) Are they rectilinear or curvilinear?

8. A railway car moves from west to east at the rate of 10 miles per hour, and a man walks through the car from the rear toward the front of the car at the rate of 4 miles per hour. At what rate is he moving east?

9. A ship is sailing due south at the rate of 14 miles an hour, and a man is running due north on its deck at the rate of 6 miles an hour. In what direction is the man moving (*i.e.* toward the north or the south), and at what rate?

10. A body moves in a certain direction. Has it a component at right angles to that direction?

11. If a body move in a circular path (*e.g.* a stone whirled in a sling) its motion every instant is compounded of motions in what two directions?

12. A body moves due northeast at the rate of 20 miles per hour. What is its northerly and what its easterly velocity?

CHAPTER II.

MOLAR DYNAMICS.

SECTION I.
FORCE.

21. Dynamics. *Dynamics* is the science which treats of the "circumstances under which particular motions take place." This science will be treated under three heads: (1) Molar Dynamics,—that is, the dynamics of solids and fluids, including the study of sound waves; (2) Molecular Dynamics, including heat; (3) Ether Dynamics,—that is, radiation, including light and electricity.

The term *Physics* is a generic term which includes all these branches; *i.e. Physics is the science which treats of the dynamics of masses, molecules, and the ether.*

22. Force Defined. When a body at rest is set in motion, or one which is in motion has that motion accelerated (positively or negatively), or when a moving body is deflected from a straight course, experience teaches us that there is always a cause, and we have also learned to apply to this cause the name *force*.

FORCE IS THAT WHICH CHANGES, OR TENDS TO CHANGE, A BODY FROM ITS STATE OF REST OR OF UNIFORM MOTION IN A STRAIGHT LINE.

The inference from this definition is that *no force is required to keep a free body[1] moving with uniform velocity in a straight line*, but that force is required to change its motion either in magnitude or in direction.

[1] By a *free body* is meant a body that for convenience is supposed to be free from the action of all resisting forces such as friction, resistance of the air, etc.

It cannot be directly shown that a moving body, if left to itself, would continue to move forever with uniform velocity in a straight line; for common experience affords us no examples of bodies moving in this manner. The reason is that it is practically impossible to isolate a body from the action of force. But we have abundant evidence that the more nearly a body is freed from the action of force, the more nearly will it continue to move uniformly in a straight line. Example: a stone projected along a sheet of smooth ice.

23. Strain and Stress. When force is applied to a body, in addition to producing motion of the body as a whole it has another effect: it changes, to a greater or less degree, the shape and possibly the size of a body, producing *relative motion of its parts*, even when no motion of the body as a whole ensues. Examples: the deformation caused by force in a bow when bent, in a cord when twisted, in a rubber band when stretched, in soft bodies like jelly when pressed, and in air when compressed.

When the form or volume of a solid body is temporarily changed, owing to the action of force, the body is said to be in a state of *strain*. Owing to the strain, forces tending to resist further strain are called into play within the body. The name *stress* is given to such forces. When, holding the ends of a rubber band in your two hands, you pull it, you observe an elongation or strain, and you feel not only a resistance to further stretching, but also a force drawing your hands toward each other. This is stress in the band.

24. Rigidity. The amount of deformation which a given force will produce depends largely upon the nature of the material. A great force is required to change the shape of a body of iron, while the force required to produce the same change in a body of jelly is very small.

Rigidity is the name given to the property of *solid bodies* which enables them to offer resistance to change of shape.

A *fluid* is a body of matter which possesses no rigidity.

SECTION II.

NEWTON'S THREE LAWS OF MOTION OR AXIOMS.
MOMENTUM.

The relations between matter and force are concisely expressed in what are known as *The Three Laws of Motion*, first enunciated by Sir Isaac Newton.

25. Newton's First Law of Motion. Inertia. *A body at rest remains at rest, and a body in motion moves with uniform velocity in a straight line, unless acted upon by some external force.*

A body is said to be acted upon by an "external force" *when the action is between that body and some other body* (in contradistinction to an action between parts of the same body). This law expressly declares that any change in the motion of a body, whether of velocity or of direction, indicates the presence of an external force.

The inability of matter to change the state that it is in, whether it be of motion or of rest, or, in other words, the property in virtue of which *external force* is required to change the velocity of a body, is called *inertia;*[1] hence, the First Law of Motion is often called the "Law of Inertia."

The backward motion of passengers when a car is suddenly started, and their forward motion when the car is suddenly stopped, the difficulty in starting a vehicle and the comparative ease of keeping it in motion after it is put in motion, and the ceaseless motion of the planets are due to inertia.

26. Momentum. We know that some moving bodies are stopped much more easily than others. An empty car is stopped much more easily than a loaded car. It is comparatively easy to set a small cricket ball rolling along the ground

---

[1] "Matter is, as it were, the plaything of force; submitting to any change of *state* that may be impressed upon it, but rigorously persevering in the state in which it is left until force again acts upon it." — TAIT.

with considerable velocity, but to set a large cannon ball rolling with the same speed requires much effort. And while we can easily stop the former when it is thrown to us, we should find it very difficult to stop the latter traveling at anything like the same speed.

This difference cannot be due to difference of *weight* of the balls, for, as we do not lift them off the ground, we are not obliged to overcome their weight. *The forces required to produce the same change of velocity in different bodies are proportional to their masses, provided the forces act for the same time.*

Mass may be said to measure inertia, or passivity, *i.e.* the property in virtue of which more or less force is required to change the velocity of a body. Conversely, inertia is the measure of mass. In fact, the terms *mass* and *inertia* are often used interchangeably.

Again, we have an instinctive dread of bodies of small mass when moving with great speed. A ball tossed is a different affair from a ball thrown.

Thus we are led to the consideration of a complex quantity called *momentum*. *The momentum of a body is a quantity measured by the product of its mass and its velocity.* A unit of momentum is the momentum of a unit mass moving with a unit velocity.

If the measures of the mass and the velocity of a body be m and v, respectively, the product $m\,v$, or momentum of the body, has $m\,v$ times the momentum of a unit mass moving with unit velocity. Thus, if the momentum of a mass of 1 k. having a velocity of 1 m. per second be taken as a unit, then a mass of 5 k., moving with the same velocity, would have 5 units of momentum ; and if the latter mass should have a velocity of 10 m. per second, its momentum would be $5 \times 10 = 50$ units.

27. Newton's Second Law. *Change of momentum is in the direction in which the force acts, and is proportional to its intensity and to the time during which it acts.*

This law (except as regards direction) is expressed in the following formula : $m\,v = f\,t$.

The product $f\,t$ is called the *impulse* of the force. In cer-

SIR ISAAC NEWTON.

[From the original painting by Sir Godfrey Knutter.]

tain cases, such as that of a blow from a hammer, it is quite impracticable to measure either the force or the (very short) time during which it acts, but its effect, *i.e.* the change of momentum, can be measured.

The product ft signifies that the momentum imparted to a body is proportional to the time (t) during which a force acts and to the intensity (f) of the force.[1] We infer from this equation that *a definite force acting upon any mass for a given time will generate in it a velocity which is inversely proportional to the mass.*

This law declares, by implication,[2] (1) that *an unbalanced* (§ 42) *force in a given time always produces exactly the same change of momentum, regardless of the mass of the body;* that *an unbalanced force never fails to produce a change of momentum;* hence any force, however small, can move any free body, of however great mass.

For example, a child can move a body having a mass equal to that of the earth, and the momentum that the child can generate in this immense body in a given time is precisely the same as that which he would generate by the exertion of the same force for the same length of time on a body having a mass of (say) 10 pounds. Momentum is the product of mass into velocity; so, of course, as the mass is large, the velocity acquired in a given time will be correspondingly small. The instant the child begins to act the immense body begins to move. Its velocity, infinitesimally small at the beginning, would increase at an almost infinitesimally slow rate, so that it might be years before its motion would become perceptible.[3]

---

[1] This formula virtually asserts that where there is no force there is no change of momentum (*i.e.* if $f=0, ft=0$). Hence, Newton's First Law is a corollary to the Second Law.

[2] No reference is made in the law to the *mass* of the body acted upon.

[3] It is easy to see how persons may get the impression that very large masses are immovable except by very great forces. The erroneous idea is acquired that bodies of matter are capable of resisting the tendency of forces to cause motion, and that the greater the mass, the greater the resistance ("quality of not yielding to force." — WEBSTER). The fact is, that *no body of whatever mass can resist motion;* in other words, "*a body free to move cannot remain at rest under the slightest unbalanced force.*" But as *time* is always required to generate change of momentum, there arises thence a deceptive appearance of resistance or holding back.

This law declares, by implication, (2) that *a force acting on a body in motion produces just the same effect as if it were acting on the same body at rest*, for no reference is made in the law to the *state* of the body acted upon.

Experiment. Draw back the rod d (Fig. 19) towards the left, and place the detent pin c in one of the slots. Place one of the brass balls on the projecting rod, and in contact with the end of the instrument, as at A. Place the other ball in the short tube B. Raise the apparatus to

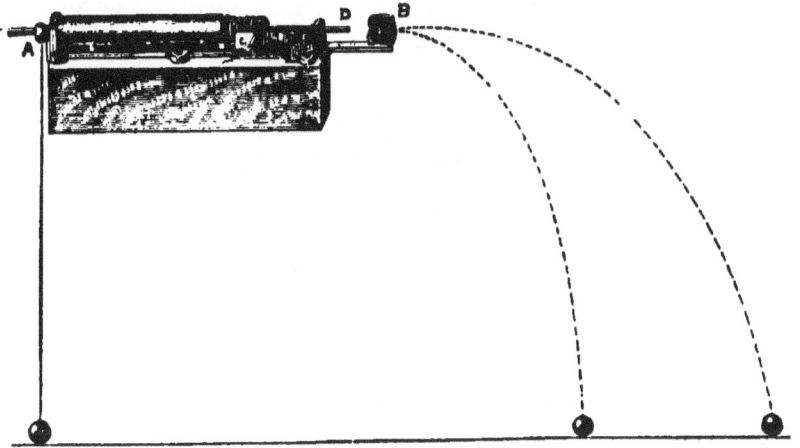

FIG. 19.

as great an elevation as practicable, and place it in a perfectly horizontal position. Release the detent, and the rod, propelled by the elastic force of the spring within, will strike the ball B, projecting it in a horizontal direction. At the same instant that B leaves the tube and is free to fall, the ball A is released from the rod and begins to fall. The sounds made on striking the floor reach the ears of the observer at the same instant; this shows that both balls reach the floor in sensibly the same time, and that the horizontal motion which one of the balls has does not affect the time of its fall, *i.e.* does not modify the effect of the force of gravity.[1]

---

[1] This principle of the independence of simultaneously acting forces was an experimental discovery made by Galileo. Before his time it was held that one cause must cease to act before another can commence to do so; and, accordingly, it was believed that when a projectile was shot into the air (instead of commencing its fall immediately), the force of projection must be expended before any tendency to fall could assert itself.

The law implies (3) that *if two or more forces act on a body, each produces its own change of momentum in its own direction, independently of the others.*

It declares, as we shall see later, that the operation of compounding forces is just the same as that of compounding the motions which the several forces acting simultaneously tend to produce.

28. Pulls and Pushes; Action and Reaction. Every force is either a *pull* or a *push*, and is accordingly called an *attractive* or a *repellent* force. It is evident that *there can be no pull or push except between at least two bodies or two parts of the same body;* i.e. there is no such thing as a one-sided pull. It is not possible for a person to pull without being himself pulled, or to push without being himself pushed.

Appearances sometimes seem to contradict the above statements. For example, a man standing on a wharf pulls a distant boat by means of a rope. The boat moves as the result of the pull, but, though he is bracing himself against the wharf, he is not willing, perhaps, to concede that he is likewise pulled. Let him stand in the boat and pull the rope which is attached at the other end to the wharf; both he and the boat move. What body, according to *appearances*, is pulled in this case? What bodies are *actually* pulled?

Force is always dual, inasmuch as it is always oppositely directed upon two bodies. By a conventionality of speech we say that one of the two bodies *acts* upon the other, and the latter *reacts* upon the former.

The wings of a bird act upon the air, giving a certain portion of it a rearward motion; the air reacts upon the wings, giving the bird a forward motion. The bat strikes the ball, imparting to it an acceleration; the ball reacts upon the bat, giving it a negative acceleration.

29. Newton's Third Law. *To every action there is an equal and opposite reaction.*

If action and reaction were not equal there might be a possibility that a person might raise himself by pulling on the soles of his feet or the hair of his head; that a vessel might be propelled in a calm by blowing against its sail with a powerful bellows (operated by steam) located on the deck of the same vessel; that a person sitting in a buggy might give himself a ride by pressing his feet against the dasher; that a person might advance, *i.e.* move his center of mass, without having the earth beneath him; or that a bird might fly without having the external air to act upon.

In case of an action between two bodies that are free from the action of resisting forces, the law implies that the *momenta* generated by the action and by the reaction are equal. The recoil of a rifle affords a good illustration of this. The explosion of the powder inside the barrel exerts equal and opposite impulses on the ball and on the gun, and causes them to move in opposite directions with equal momenta. Hence, if the speed of the recoil of the rifle be known, the speed of the ball can be computed, and *vice versa*.

For example, let the masses of the rifle and the ball be respectively 5 lb. and 1 oz., and the maximum velocity of the rifle be 10 feet per second. Then the momentum of the rifle at the instant is $(5 \times 10 =)$ 50 units. But the momentum of the ball at the same instant is also 50 units. Hence, $(50 \div \frac{1}{16} =)$ 800 feet per second is the maximum velocity of the ball.

EXERCISES.

1. How are we made aware of the existence of force?

2. A 10-lb. ball rests upon a table. (*a*) How great an upward pressure does the table exert upon the ball? (*b*) How do you explain this pressure?

3. Is perpetual motion possible?

4. A carriage is suddenly stopped, and the passengers are said to be "thrown out." Are they *thrown?*

5. What is a suitable name to apply to all interaction between bodies?

6. (*a*) Why may a man raise himself by pulling on a horizontal bar, but not by pulling on any part of his person? (*b*) In which case is he acted upon by an external force? (*c*) In the first case, which body receives the *action* and which the *reaction?* (*d*) State what receives the action in the second case, and what receives the reaction.

7. When do action and reaction neutralize each other and have no tendency to produce a change of motion?

8. What agent is the immediate cause of motion?

9. What distinction do you make between velocity and momentum?

10. Upon what does the momentum given to a ball fired from a gun by the expanding gases depend?

EXERCISES. 33

11. Inasmuch as equal forces are exerted for the same length of time by the gases on the ball and the gun, how will the momenta communicated to the two compare?

12. If there be 25 lbs. of matter in the gun and 1 oz. ($\frac{1}{16}$ lb.) in the ball, and the gun acquire a maximum velocity of 3 feet per second, what, at that instant, is the velocity of the ball?

13. Can any body be put in motion in no time? (Demonstrate from formula $Ft = MV$.)

14. Compare the momentum of a car weighing 50 tons, moving 10 feet per minute, with that of a lump of ice weighing 5 cwt., at the end of the third second of its fall.

15. With what velocity must a boy weighing 25 k. move to have the same momentum that a man weighing 80 k. has when running at the rate of 10 km. per hour?

16. Since $Ft = MV$, to what is change of momentum proportional?

17. If the same force act for the same length of time upon bodies having different masses, to what will the velocities produced be proportional?

18. Two boats of unequal masses are brought together by pulling on a rope. (a) Resistance being disregarded, how will their momenta at any given instant compare? (b) How will their velocities at the same instants compare?

19. If the motion of the moon in its orbit about the earth were to cease, these bodies would approach each other. The mass of the earth is about 80 times that of the moon. What part of the whole distance between them would the moon move before collision?

20. (a) Why does not a given force, acting the same length of time, give a loaded car as great a velocity as an empty car? (b) After equal forces have acted for the same length of time upon both cars, and have given them unequal velocities, which will be the more difficult to stop?

21. (a) The planets move unceasingly; is this evidence that there are forces pushing or pulling them along? (b) None of their motions are in straight lines; are they acted upon by external forces?

22. A certain body is in motion. Suppose that all hindrances to motion and all external forces be withdrawn from it, how long will it move? Why? In what direction? Why? With what kind of motion, i.e. accelerated, retarded, or uniform? Why?

23. If one body have four times the mass of another, how must the forces applied to them compare in order to give them equal momenta in equal times?

34 MOLAR DYNAMICS.

SECTION III.

MEASUREMENT OF FORCE.

30. Weight. Gravitation Units of Force. We have here to anticipate what will hereafter be more fully discussed, by defining the *weight* of any given mass as the measure of the force of *gravity* which exists between it and the earth. *Weight is a force.* Its magnitude is usually determined by measuring the strain (§ 23) which it produces in the supporting body, *e.g.* the elongation of the spring in a spring balance.

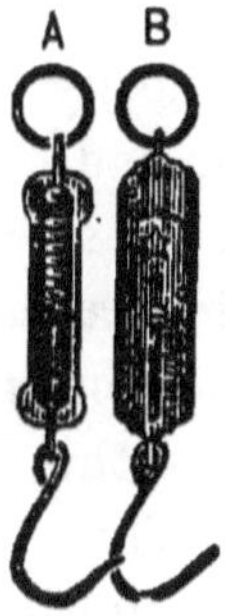

FIG. 20.

The household instrument called a spring balance is, strictly speaking, a *dynamometer*, *i.e.* a force-measurer. It contains a spiral spring, as seen in A (Fig. 20), carrying an index which moves over a scale, as shown in B. If a unit mass (*e.g.* 1 lb. or 1 k.) be hung upon the spring, the latter is lengthened by a certain definite quantity. If, grasping the ring in one hand and the hook in the other, you lengthen the spring by a muscular pull as much as it was lengthened by the force of gravity acting on the mass, the inference is that the muscular force which you exert is equal to the force of gravity exerted on the mass.

The units generally employed in measuring force are the *pound* and the *kilogram*, and are called the *gravitation units*.

All forces may be measured in the same units. To say that a man pulls a boat with a force of one hundred pounds is equivalent to saying that he pulls with a force that is equal to the force which acts between the earth and a body having a mass of one hundred pounds. A force of one pound, then, is an abbreviated expression for a force equal to the weight (at the locality in question) of a mass of one pound. The pound and the kilogram are primarily units of mass.

31. Force Tends to Produce Acceleration. *A constant force acting upon a free body always produces uniformly accelerated motion.* This is best illustrated by the fall or ascent of a body in a vacuum, the body being meantime acted on only by the constant force of gravity. In its ascent the force of gravity causes a uniform negative acceleration; in its fall it causes a uniform positive acceleration.

32. Measurement of Force by Direct Observation of Acceleration and Mass. If a force, f, be applied to a certain mass, m, for a unit of time, a certain momentum is generated in the mass. If the same force be applied to a greater mass for the same time, it will move with as many times less velocity as the mass is times greater, but the *product* of the mass and the velocity, *i.e.* the momentum, is the same. That is, the same force acting for the same length of time on free bodies having different masses may be measured by the change of momentum generated by it in a unit of time (*e.g.* a second), since this is constant and depends on nothing but the force. That is, $f = m\,a$, in which f represents any constant force acting on any mass m, a the acceleration, and $m\,a$ the rate of change of momentum.[1]

33. Fundamental Units. A system of units can be built on any selected units of *length, mass, and time*. All so-called dynamical units can be derived from these; hence, these are called *fundamental units*. Thus, a system of units derived from the *centimeter* (unit of length), the *gram* (unit of mass), and the *second* (unit of time), is called the *centimeter-gram-second system*, written the C. G. S. system.

34. C. G. S. Absolute Units of Force. The dynamical unit of force is that force which will impart to a unit mass a unit acceleration.

The unit of force in the C. G. S. system is called a *dyne*, and is *that force which is capable of giving to a gram mass an acceleration of one centimeter per second;* in other words, it is a constant force of the requisite intensity to impart in one

---

[1] Many physicists do not treat force as a cause (§ 22), but as a phenomenon, and define it as *the time rate of change of momentum.*

second to a gram-mass a velocity of one centimeter per second. Any force may be measured in dynes, and a spring balance might be graduated in dynes so as to measure force in *absolute units*.

It is important to observe that the dynamical units of force are *absolute*;[1] *i.e.* unlike the gravitation units of force, they are not affected by variations in the force of gravity, and are therefore *everywhere the same*.[2]

A dyne is a very small force. In expressing force of considerable magnitude the *megadyne* (a million dynes) is commonly used. A megadyne is rather more than the force of a kilogram.

35. British Absolute Units. Corresponding to the metric C. G. S. system is the British Foot Pound Second (F. P. S.) system. The British absolute unit of force is the *poundal*. A poundal is that force which is capable of giving to a pound-mass an acceleration of one foot per second.

36. Expression for Weight in Absolute Units. Any constant force which in one second produces in a mass of m grams an acceleration of a centimeters per second must be equal to $m \times a$ dynes (*i.e.* $f = m a$). The letter g is generally used instead of the letter a to denote the acceleration due to the force of gravity. By exact measurement the acceleration produced by the force of gravity on all free bodies (*i.e.* in a vacuum) is found to be, in the latitude of Boston at the level of the sea, 980.4 cm. (or nearly 32.2 feet) per second. Hence, the force of gravity acting on a mass of one gram must be (substituting, in the equation above, w (weight) for f, and g for a) $w = m g = 1 \times 980.4 = 980.4$ dynes. Consequently, it requires a force of 980.4 dynes to support (*i.e.* prevent from falling) a mass of one gram; or the weight of a gram-mass at sea level in latitude 42° is 980.4 dynes.[3] A dyne is therefore

---

[1] The dynamical system of units is frequently called the *absolute system*.

[2] The relation expressed in the formula $f = m a$ is based on the supposition that the unit of force is the dynamical unit. When measured in gravitation units, f is *not equal* to, but only *proportional* to, $m a$.

[3] The equation $w = m g$ expresses the fact that (using C. G. S. units) the number of dynes which a given mass weighs is g times the number of grams in that mass.

about $\frac{1}{980}$ of the weight of a gram-mass, or more exactly $\frac{1}{980.9}$ of the weight of a gram-mass at Paris. In the gravitation system the weight of a gram-mass is a gram-force; hence, 1 gram-force = 980.9 dynes at Paris. Gravitation units in grams-force at Paris are readily changed into dynes by multiplying by 980.9; at Boston, by multiplying by 980.4; and generally by multiplying by the value of g at any place.[1]

37. Expression for the Mass of a Body in Terms of its Weight. Since $w = mg$, $m = \dfrac{w}{g}$; that is, mass is measured by its weight in dynes or poundals, divided by the acceleration in centimeters or feet per second produced by gravity. Although w and g vary with location (§ 66), they vary proportionally; hence, the ratio $\dfrac{w}{g}$ (*i.e.* the mass) does not change, but is constant for the same body.

38. Problems and Solutions. A body suspended from a spring balance is found to weigh at Paris ½ k. Required its weight in dynes. Solution: ½ k. = 500 g.: 500 × 980.9 = 490,450 dynes.

Required to find the force which, acting for 10 seconds, gave to a mass of 10 g. a velocity of 1000 cm. per second. Solution:

$$F = \frac{mv}{t} = 10 \times \frac{1000}{10} = 1000 \text{ dynes.}$$

Required to find the mass in which a force of 1500 dynes produces an acceleration of 2 cm. per second. Solution: $m = \dfrac{F}{a} = \dfrac{1500}{2} = 750$ g.

Required to find the acceleration which a force of 2000 dynes can give a mass of 4 g. Solution: $a = \dfrac{F}{m} = \dfrac{2000}{4} = 500$ cm. per second.

39. Two Systems of Measurement of Force. We have found in the foregoing discussions that there are two methods of measuring force: one specially adapted to measuring balanced forces (see p. 41), called the *statical* or *gravitation* system; the other specially adapted to measuring unbalanced forces (see

---

[1] The weight of a pound-mass is equal to 32.2 poundals; or the poundal as a unit of force is $\frac{1}{32.2}$ of a pound-force or nearly equal to the weight of half an ounce.

p. 42), called the *dynamical* or *absolute* system; though a force, whether balanced or unbalanced, may always be measured by either system. The gravitation system is so called because by it forces are compared with the force of gravity as a standard. The two methods of measuring force give rise to *two systems of units*, called respectively the *gravitation* and the *absolute* systems, either one of which is easily convertible into the other.

40. Galileo's Experiment. Galileo let fall from the leaning tower at Pisa[1] iron balls of different masses, and found that they fell with equal acceleration and reached the ground at the same instant. This celebrated experiment established two important facts:

Fig. 21.

(1) *At any given place the acceleration due to gravitation is independent of the mass of the falling body.*

This fact may also be deduced from the formula $f = ma$. For, if f be proportional to m, it follows that a must be the same for all bodies. That the force of gravity f is proportional to mass m at the same place was demonstrated by Galileo's experiment. For, let f and f' be the intensities of two forces drawing two bodies, whose masses are respectively m and m', to the earth; then

$$f = ma, \text{ and } f' = m'a';$$

but, as proved by Galileo's experiment,

$$a = a';$$

---

[1] This building (Fig. 21), consisting of a series of open galleries one above another reaching to a total hight of 179 feet, is admirably adapted to the purpose here stated.

hence (dividing), $\dfrac{f}{f'} = \dfrac{m}{m'}$,

and, in general, $f \infty\, m$;

i.e. (2) *the intensity of the earth's attraction at the same place varies as the mass.*

In other words, the deductions from this experiment are: (1) that all free bodies, whatever their mass, fall toward the earth with equal accelerations; and (2) that if one body possess twice the mass of another, twice the force is required to give it the same acceleration.

Proposition (1) is seemingly contradicted by everyday experience, for if a coin and a piece of tissue paper be dropped from a hight they fall with very different velocities. But if a coin and a feather be placed in a long glass tube (Fig. 22), the air exhausted, and the tube turned end for end, it will be found that the coin and the feather fall in the vacuum with equal velocities. It is evident, then, that when there is a difference in the acceleration of falling bodies at the same place it is not due to the force of gravitation, but to some other force: *e.g.* the resistance of the air.

FIG. 22.

EXERCISES.

1. Define a gravitation unit of force and an absolute unit of force, and state wherein the latter for scientific purposes is preferable to the former.

2. A constant force acting on a free body produces what kind of motion?

3. Explain the meaning of the equation $w = mg$.

4. To what is the acceleration produced in equal masses proportional, — *i.e.* if m is constant, a will vary as what?

5. On what condition will equal forces produce equal accelerations?

6. Suppose that you fill a box with sand, place it on a toy cart, pull the cart by a string with a constant force along a smooth floor for a certain number of seconds, and observe the acceleration given the load (cart, box, and sand), then remove the sand and replace it with lead shot.

How can you tell, by pulling the load with the same force as before, when it has the same mass as the former load?

7. (*a*) Has the same mass equal weights in Paris and Boston? (*b*) How sensitive must a spring balance be to discover any difference?

8. (*a*) When we speak of a force of one pound, what do we mean? (*b*) When we speak of a force of one dyne, what do we mean? (*c*) When we speak of a mass of one pound, what do we mean?

9. (*a*) If one mass be four times another, how many times as much force is necessary to produce the same acceleration in the former as in the latter? (*b*) How many times greater is the force of gravity acting on a mass of one hundred pounds than on a mass of one pound? (*c*) If a hundred-pound iron ball and a one-pound iron ball be let drop from the same hight at the same instant, which ought to reach the ground first?

10. A body weighing 4 g. is moving with an acceleration of 12 cm. per second. What is the force acting?

11. A body acted on by a force of 100 dynes receives an acceleration of 20 cm. per second. What is its mass?

12. A body of mass 30 g. is moved by a constant force of 50 dynes. What is its acceleration?

13. What acceleration will a force of 20 dynes produce on a mass of 10 g.?

14. What velocity will a force of 20 dynes acting on 1 k. impart to it in 5 minutes?

15. (*a*) What is the weight in dynes of a mass of 1 k. in Boston? (*b*) How many more dynes does it weigh in Paris?

16. A constant force of 20 dynes acts on a mass of 5 g. and gives it a velocity of 500 cm. per second. How many seconds does it act?

17. How is the value in dynes of a gram weight at any locality determined?

18. How many times greater is the static unit of force (one gram) than the dynamic unit?

19. A man jumps from an elevation with a 50-pound weight in his hand. What is the pressure of the weight on his hand during his descent?

20. How far can a force of 10 dynes move a kilogram mass in a minute?

Solution: $f = ma$, or $10 = 1000\, a$; whence $a = .01$ cm. per second. Distance traversed from rest in one minute $= \frac{1}{2} a\, t^2 = \frac{.01}{2} \times 60^2 = 18$ cm.

SECTION IV.

COMPOSITION AND RESOLUTION OF FORCES.

41. Graphical Representation of Force. A force is defined when its *magnitude, direction,* and *point of application* are given. We may represent forces graphically by straight lines whose lengths bear to one another the same relation as the numerics of the forces, while the directions of these lines indicate the directions of the forces, and the points from which the lines are drawn indicate the points of application. Thus, on a scale of 1 cm. = 1 k. the line A B (Fig. 23) represents a force of 3.2 k. acting toward the right with its point of application at A; and the line D E represents a force of 2 k. acting parallel to the first, with its point of application at D.

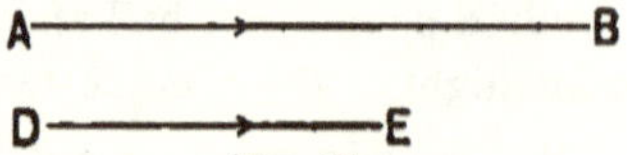

FIG. 23.

42. Composition of Forces Acting in the Same Line; Equilibrium of Forces; Balanced Forces.

Experiment. Insert two stout screw-eyes into the opposite extremities of a block of wood. Attach a spring balance to each eye. Let two persons pull on the spring balances at the same time, and with equal force, as shown by the indexes, but in opposite directions. The block does not move. One force just neutralizes the other, and the result, so far as any movement of the block is concerned, is the same as if no force acted on it.

When one force opposes in any degree another force, each is spoken of as a *resistance* to the other. Let f represent the number of pounds of any given force, and let a force acting in any given direction be called *positive*, and indicated by the plus (+) sign, and a force acting in an opposite direction to the force which we have denominated positive be called *negative*, and indicated by the minus (−) sign. Then if two forces, $+f$ and $-f$, acting on a body at the same point or along the same line be equal, they are said to be *balanced*, and the result is that no change of motion is produced.

Viewed algebraically, $+f-f=0$; or, correctly interpreted, $+f-f=$ (is equivalent to) 0, *i.e.* no force. In all such cases there is said to be an *equilibrium of forces*, and the body is said to be in *a state of equilibrium*. Equilibrium is the condition of two or more forces which are so opposed that their combined action on a body produces no change in its rest or motion.

A force that produces equilibrium with one or more forces is called an *equilibrant*. That branch of dynamics which treats of the relation of force to the motion which it produces is called *kinetics*, and that branch which treats of equilibrium of forces is called *statics*.

43. Unbalanced Forces. If one of the forces be greater than the other, the excess is spoken of as an *unbalanced force*, and its direction is indicated by one or the other sign, as the case may be. Thus, if a force of $+8$ lbs. act on a body toward the east, and a force of -10 lbs. act on the same body along the same line, then the unbalanced force is -2 lbs.; *i.e.* the result is the same as if a single force of 2 lbs. acted on the body toward the west. Such an equivalent force is called a *resultant*. *A resultant force is a single force that may be substituted for two or more forces and produce the same result that the simultaneous action of the several forces would produce.*

The resultant of any number of forces acting in the same straight line is equal to the algebraic sum of the forces. An equilibrant of several forces is equal in magnitude to their resultant, but opposite in direction. The process of combining several forces so as to find their resultant is called *composition of forces*. The forces combined are called *components*. The converse operation, of finding component forces which shall have the same effect as a given force, is called *resolution of forces*.

An unbalanced force always produces acceleration. Hence, a body acted on by an unbalanced force cannot be at rest.

PRESSURE, TENSION.

44. Pressure; Tension. A balanced force does not produce acceleration, but causes either a *pressure* or a *tension*. A force exerts pressure when it tends to *compress* or shorten in the direction of its action the body on which it acts. Examples : pressure exerted on the springs of a carriage, on air when it is compressed in an air gun, etc. A force causes tension when it tends to *lengthen* in the direction of its action a body on which it acts. A body thus subjected to a force tending to elongate it is said to be in a *state of tension*, the stress to which it is subjected is called its *tension*, and its strength to resist being pulled apart is called its *tensile strength*.

Equilibrium is often maintained by the reaction of a surface with which the body acted on is in contact. A simple illustration is that of a body supported on a horizontal surface, as that of a table. Here the reaction caused by the compression of the material of which the table is composed is equal to the weight of the body.

EXERCISES.

1. Explain the use of a line to represent a force.

2. (*a*) When a force of 100 lbs. is represented by a line 5 in. long, what is the scale ? (*b*) What force will a line ¼ in. long represent on the same scale ?

3. (*a*) Represent on a scale of ¼ in. = 1 lb. the resultant of forces of 5 lbs. and 7 lbs. acting in the same direction. (Always place arrowheads in lines representing forces to indicate the direction of the forces.) (*b*) Show, by points *A*, *B*, and *C* placed in the line, the components of this resultant. (*c*) Represent the same two forces acting in opposite directions upon the same point *A*. (*d*) How will you represent the resultant of these two opposing forces ?

4. Three men, *A*, *B*, and *C*, pull on a rope in the same direction with forces, respectively, of 50 lbs., 60 lbs., and 70 lbs. *A* is nearest the end of the rope, *B* next, and *C* next. (*a*) What is the tension of the rope between *A* and *B* ? (*b*) What between *B* and *C* ? (*c*) A man, *D*, just beyond *C* pulls with a force of 75 lbs. in the opposite direction. With what force must a man, *E*, pull, that there may be equilibrium ? (*d*) When there is equilibrium, what is the tension of the rope between

C and D? (e) How great must be the tensile strength of the rope between C and D? (f) Write the equation showing the algebraic addition of the forces in case of equilibrium.

5. The hooks of two spring balances are connected by a string and the balances are pulled. (a) If one registers 5 lbs., what does the other register? (b) What is the tension in the string?

6. How is change of motion produced?

7. What other effects besides change of motion may a force produce?

SECTION V.

COMPOSITION OF PARALLEL FORCES. MOMENTS OF FORCES.

45. Composition of Parallel Forces Acting in the Same Direction and in the Same Plane.

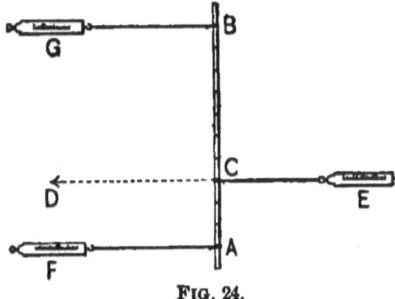

Fig. 24.

Experiment. A B (Fig. 24) represents a rod in a horizontal position with three strings loosely looped around it so that they may be slid along the rod. Dynamometers are attached to the free ends of the strings. The strings are all stretched in parallel directions in a plane parallel to the top of the table. (Great care must be taken in the manipulation to keep the three strings exactly parallel.) The dynamometers register the tensions in the several strings, i.e. the forces applied through them to the rod.

Observe: (1) When there is equilibrium the dynamometer E registers as much as do F and G added together. But the force applied at C is the equilibrant of the other forces, and this is equal to their resultant acting in the direction C D. (II) The point of application of the resultant (or equilibrant) is between the points of application of the components. (III) This point is nearer the greater force. (IV) The distance of this point from the smaller force is as

DYNAMICAL COUPLE. 45

many times greater than its distance from the larger force as the larger force is times the smaller force. For example, if A F be 14 lbs. and B G be 6 lbs. (14 : 6 = 7 : 3), then distances C A and C B will be as 3 : 7. In other words, the component forces are said to *vary inversely as*, or to be *inversely proportional to*,[1] their distances from their resultant. These observations are summarized as follows : *The resultant of two parallel forces in the same direction is equal to their sum, and the distances of their points of application from the point of application of the resultant vary inversely as the intensities of the components.*

Corollary : *The condition of equilibrium is that the algebraic sum of the forces* (positive and negative) *must be zero.*

When more than two forces act on a body in the same plane and in the same direction, the resultant of any two of them (and its point of application) is found, then the resultant of this resultant and a third force, and so on until all have been used.

46. Dynamical Couple. *Two equal forces applied to the same body in parallel and opposite directions not in the same line constitute* what is called "*a couple.*" The effect of a couple is to produce *rotation*, but no motion of translation.

Since the two forces which constitute a couple are equal and opposite, their resultant is zero, and therefore no single force can equilibrate a couple.

47. Moment of a Force. The value of a force for producing rotation about a given axis is called its *moment* with reference to that axis. Point C (Fig. 25) may represent

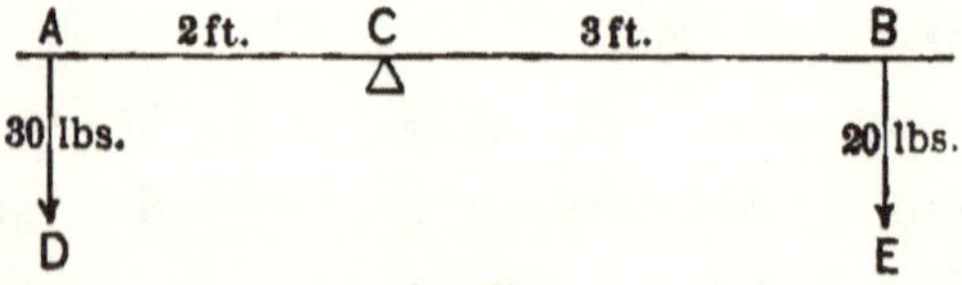

FIG. 25.

---

[1] The pupil should acquire immediate familiarity with these expressions, which occur so frequently in Physics, and in this connection should practice writing inverse proportions. Thus, for the quantities here given, 14 : 6 = $\frac{1}{7}$: $\frac{1}{3}$, *i.e.* the forces are proportional to the reciprocals of their respective distances from the resultant.

the extremity of the axis about which A B is supposed to rotate. The perpendicular distance (C A or C B) from the axis of rotation to the line of direction in which a force acts (A D or B E) is called the *leverage* of the force.

The moment of a force is measured by the product of the intensity of the force into its leverage.

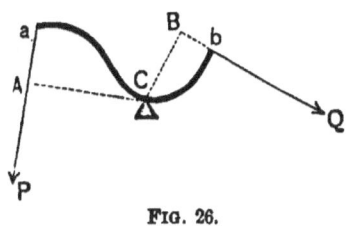

FIG. 26.

For example, the moment of the force A D (Fig. 25) is expressed numerically by the number $(30 \times 2 =) 60$, and the moment of B E is $(20 \times 3 =) 60$. By definition the line A C (Fig. 26) is the leverage of force a P, and B C of the force b Q.

48. Equilibrium of Moments. The moment of a force is said to be *positive* when it tends to produce right-hand rotation, *i.e.* rotation in the direction in which the hands of a clock move, and *negative* when its tendency is in the reverse direction. *If two forces act at different points of a body which is free to rotate about a fixed point, they will produce equilibrium when the algebraic sum of their moments is zero.* Thus, the moment of the force applied at A (Fig. 25) is $-(30 \times 2) = -60$. The moment of the force applied at B in an opposite direction is accordingly $+(20 \times 3) = +60$. Their algebraic sum is zero, and consequently there is equilibrium between the moments, and no tendency to rotation.

When more than two forces act in this manner there will be equilibrium if the algebraic sum of all the moments (positive and negative) be zero. Thus, the equation of moments acting about the axis D (Fig. 27) is $([f]\ 45 + [e]\ 25 + [a]\ 30) + ([c] - 30$

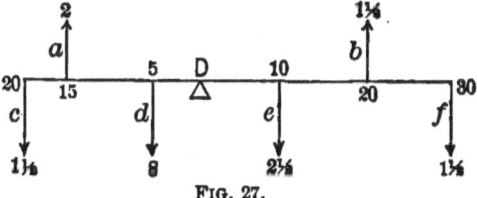

FIG. 27.

$[d] - 40 [b] - 30) = 0$; the sum of all the moments being zero, there is equilibrium of moments.

49. Moment of a Couple. The moment of a couple, or its value in producing rotation, is the sum of the moments of its two components about the axis of rotation. Let F and F_1 (Fig. 28) constitute a couple whose points of application are A and B. To find the rotating value of the couple, let P be the axis of rotation; then the moments of F and F_1 relatively to P are F × A P, and F_1 × B P. The total resultant moment of the two forces is (F × A P) + (F_1 × B P), or (since F = F_1) F × A B.

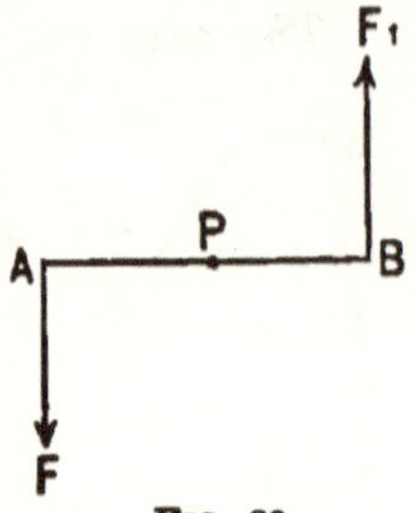

FIG. 28.

EXERCISES.

1. Two parallel forces of 8 lbs. and 12 lbs. act in the same direction, respectively at points A and B, 12 inches apart. Find the magnitude and position of their resultant.

2. The smaller of two parallel forces having the same direction is 5 inches from the resultant. What is the distance of the resultant from the other force?

3. Two men carry a weight of 100 lbs. suspended from a pole 15 feet long; each man is 18 inches from his end of the pole. Where must the weight be attached in order that one man may bear ¾ of it?

4. Take from the last problem the number of pounds supported by each man and the respective distances of each from the weight, and make an inverse proportion which shows the relation that must exist between these quantities.

5. How can a force of 4 lbs. be made to produce equilibrium with a force of 12 lbs.?

6. Draw a line 2 inches long. Represent on a scale of ¼ inch = 1 lb. a force of 8 lbs. applied at a point A, ¼ of 1 inch from one end of the line and at right angles to it. Take for the axis of rotation a point B, ¾ inch from the same end of the line. From point C, ¼ inch from the other end

of the line draw a line which will represent a force that will produce equilibrium with the first force, and thereby prevent rotation.

7. Repeat the work of the last problem, but assume that the force applied at A acts obliquely on the line.

8. Can a single force produce equilibrium with a couple?

9. (a) A plank weighing 40 lbs. is placed across a log so as to be balanced. A boy weighing 60 lbs. sits on one end of the plank. Where shall another boy weighing 90 lbs. sit that he may balance the first? (b) What pressure will be exerted upon the log?

10. Two horses harnessed abreast are ploughing. How can you arrange that one horse shall pull only two thirds as much as the other?

11. The maximum muscular force which a certain man can exert is 200 lbs. With what leverages can he raise a stone weighing a ton?

12. How can pressure be multiplied indefinitely?

13. Three forces of 2, 10, and 12 units act on a body along parallel lines. Show how they may be adjusted so as to be in equilibrium?

14. A force of 10 units has a moment about a certain axis of 75 units. How many units of distance is the axis from the line of action of the force?

SECTION VI.

CENTER OF MASS OR CENTER OF INERTIA.

50. Center of Mass Defined. Let Fig. 29 represent any body of matter, *e.g.* a stone. Every particle of the body is acted upon by the force of gravitation. The gravitation forces acting on the particles form a set of parallel forces, the resultant of which equals their sum (§ 45), and has the same downward direction as its components. In whatever position the body may be, the resultant passes through a definite point in it; this point is called the *center of mass* or *center of inertia* of the body. The center of mass (c.m.) of a body is, therefore, the point of appli-

FIG. 29.

CENTER OF MASS DEFINED.

cation of the resultant of all the gravitation forces; and for many practical purposes *the whole mass, weight, or inertia of the body may be supposed to be concentrated at this point.*[1]

Let G (Fig. 29) represent the c.m. of the stone. For practical purposes, then, we may consider that the force of gravitation acts only at this point, and in the direction G F. If the stone fall freely, this point cannot deviate from a vertical path, however much other points of the body may rotate about this point during its fall. Inasmuch, then, as the c.m. of a falling body always describes a definite path, a line, G F, that represents this path, or the path in which a body supported tends to move, is called the *line of direction*. It may be defined as the straight line in which lie the center of mass of the body and the center of mass of the earth; its direction is always vertical.

To support any body, then, *it is only necessary to provide a support for its center of mass. The supporting force must be applied somewhere in the line of direction.* The difficulty of poising a book, or any other object, on the end of a finger consists in keeping the support under its center of mass, *i.e.* in the line of direction.

Fig. 30 represents a toy called a " witch," consisting of a cylinder of pith terminating in a hemisphere of lead. The toy will not lie in a horizontal position, as shown in the figure, because the support is not applied immediately under its c.m. at G; but when placed horizontally it immediately assumes a vertical position. It appears to the observer to rise; but, regarded in a technical sense, it really falls, because its c.m. takes a lower position.

Fig. 30.

---

[1] The expression *center of mass* does not necessarily signify that point occupying a central position among the particles of a body, but a point where, for convenience in some dynamical problems, we may consider all the mass (or inertia) to be concentrated. The center of mass is often called the *center of gravity*. By the *place* or the *location* of a body mathematicians mean the point where its center of mass is situated. Thus, in dynamical problems the distances between celestial bodies, as the sun, moon, and earth, are the distances between their centers of mass.

50 MOLAR DYNAMICS.

51. How to Find the Center of Mass of a Body. Imagine a string to be attached to a potato, as in Fig. 31, and to be suspended from

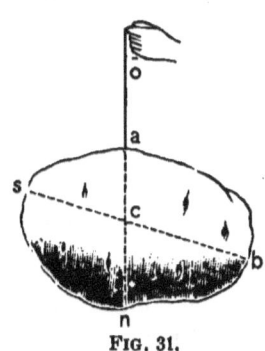

FIG. 31.

the hand. When the potato is at rest, there is an equilibrium of forces, and the c.m. must be somewhere in the line of direction o n. Suspend the potato from some other point, as b, and the c.m. must be somewhere in the new line of direction, b s. Since the c.m. lies in both the lines a n and b s, it must be at c, their point of intersection. It will be found that, from whatever point the potato is supported, the point c will always be vertically under the point of support. In a similar manner the c.m. of any body may be found. But the c.m. of a body may not be coincident with any particle of the body; for example, the c.m. of a ring, a hollow sphere, etc.

The center of mass of any symmetrical body of uniform density coincides with its geometrical center. Examples: the middle point of a material straight line; that point on a straight line joining the vertex to the middle of the base of a triangle which is situated at a distance from the vertex equal to two thirds the length of the line; the geometrical center of any polygon, of a sphere, of a circular cylinder.

52. Equilibrium of Bodies. A body will rest in equilibrium when its line of direction passes through its point of support. A body will be supported at its base when its line of direction falls within its base or lowest side. [The base of any body, *e.g.* a chair, is the polygon formed by joining by straight lines the points of support.] There are three kinds of equilibrium :

(1) A body so supported that when slightly disturbed it tends to return to its original position is said to be in *stable equilibrium*. This will be the case whenever such a disturbance raises its c.m.; for the weight of the body acting at its c.m. tends to bring this point as low as possible, and thus causes it to return to its former position. Evidently a body is in stable equilibrium when the supporting force is applied in the line of direction *above* its c.m.

(2) A body so supported that a slight disturbance tends to cause it to take a new position with its c.m. lower than before is in *unstable equilibrium*.

STABILITY OF BODIES. 51

(3) A supported body whose c.m. is neither raised nor lowered by a disturbance is in *neutral equilibrium*.

For example, a cylinder, if it be uniformly dense, is in *neutral* equilibrium when placed on its side upon a horizontal plane, and it rests equally well in all positions. But if, on account of unequal density, its c.m. be not in its axis, then its equilibrium is *stable* when its c.m. is below its axis, and *unstable* when it is above it.

53. Stability of Bodies. The ease or difficulty with which bodies supported at their bases are overturned varies with the hight to which their c.m. must be raised to overturn them. The letter c (Fig. 32) marks the position of the c.m. of each of the four bodies A, B, C, and D. If any one of these bodies be overturned, its c.m. must pass through the arc c i, and be raised through the hight a i. By comparing A with B, and supposing them to be of equal weight, we learn that *in overturning two bodies of equal weight and hight of c.m., the c.m. of that body which has the larger base must be raised higher,*

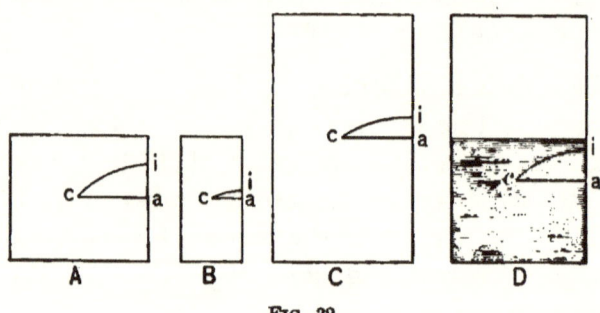

FIG. 32.

and *that body is, therefore, overturned with greater difficulty.* A comparison of A and C, supposing them to be of equal weight, shows that *when two bodies have equal bases and weights, the body having its c.m. higher is more easily overturned.* D and C have equal masses, bases, and hights, but D is made heavy at the bottom, and this *lowers its c.m. and gives it greater stability.*

EXERCISES.

1. Where is the c.m. of a box?
2. Why is a pyramid a very stable structure?
3. What is the object of ballast in a vessel?

FIG. 33.

4. State several ways of giving stability to an inkstand.
5. (a) In what position would you place a cone on a horizontal plane that it may be in stable equilibrium? (b) That it may be in neutral equilibrium? (c) That it may be in unstable equilibrium?
6. In loading a wagon, where should the heavy luggage be placed? Why?
7. Why are bipeds slower in learning to walk than quadrupeds?
8. Why is mercury placed in the bulb of a hydrometer?
9. How will a man by rising in a boat affect its stability?
10. Which is more liable to be overturned, a load of hay or a load of stone of equal weight?
11. What attitude does a man assume when carrying a heavy load on his back? Why?
12. What position do bodies floating in air or in water take?

FIG. 34.

13. (a) Explain how the toy horse (Fig. 33) stands upon the platform without falling off. (b) Explain how the toy may rock upon its support without falling off.

14. It is difficult to balance a lead pencil on the end of a finger; but by attaching two knives to it, as in Fig. 34, it may be rocked to and fro without falling. Explain.

FIG. 35.

15. If the end C of the triangular frame A B (Fig. 35) be raised and allowed to fall, the frame will rock to and fro on its support, and finally come to rest in its original position. (a) What kind of equilibrium has it? (b) If the weight at the end B be removed, will the frame be supported by the table? (c) Why?

SECTION VII.

COMPOSITION OF FORCES ACTING AT ANGLES WITH ONE ANOTHER.

54. Parallelogram of Forces. If two forces having a common point of application act at an angle with each other, their resultant and equilibrant may be ascertained by means of the "parallelogram of forces," as the following experiment will illustrate:

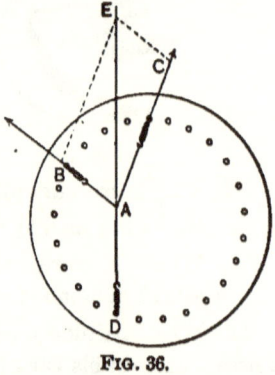

FIG. 36.

Experiment. Insert pegs in any three holes of the circle in the top of the circular table, Fig 36. Join these by threads attached to spring balances as shown in the figure. Stretch the balances so as to indicate any desired pull in each of the threads. Place under the threads a sheet of white paper. Locate on the paper the common point of application A of the three forces. Draw lines A B, A C, and A D to represent the directions in which the forces act. Since the point A does not move, it is evident that the three forces are in equilibrium, and that any one of the three forces is the equilibrant of the other two. Select any one for an equilibrant (*e.g.* A D) and extend it in an opposite direction from A, representing (on some suitable scale) a force A E equal to and opposite to the force A D as indicated by the dynamometer D. On the same scale lay off distances A B and A C representing the magnitudes of the forces acting in the directions of these lines. The line A E is by definition the resultant of A B and A C. Connect E with C and B. The figure, if the work be done with care, will be found to be a *parallelogram*. The diagonal E A represents the magnitude of the resultant of the forces A B and A C, and the same line with the direction reversed (*i.e.* A E) represents the equilibrant.

If two forces applied at a point be represented in magnitude and direction by the adjacent sides of a parallelogram drawn from the common point of application, their resultant will be represented in magnitude and direction by the diagonal which passes through that point.

54 MOLAR DYNAMICS.

This proposition is applicable whether the forces act on a particle or on a rigid body of any size provided they lie in the same plane. Thus,

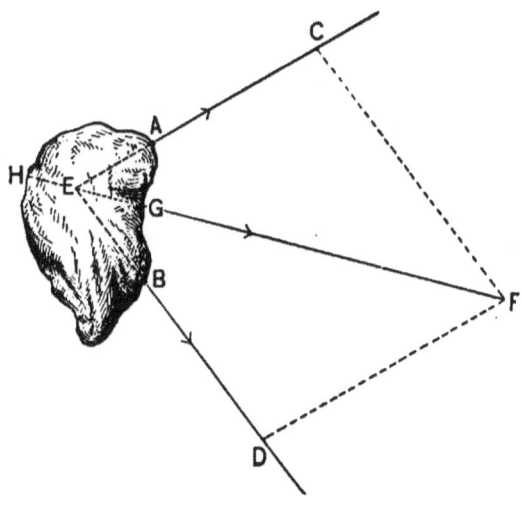

FIG. 37.

let two forces applied at points A and B of a stone (Fig. 37) act in the directions A C and B D, respectively. The direction of the resultant must pass through E, the point where the lines of direction of the given forces when produced backwards intersect. If, now, the lines E C and E D be laid off to represent the relative intensities of the forces, the diagonal E F of the parallelogram constructed thereon will represent their resultant, and its point of application may be G or any other point in the line G H.

55. Composition of More than Two Forces in the Same Plane. *When more than two components are given, find the resultant of any two of them, then that of this resultant and a third, and so on till every component has been used.* Thus, in Fig. 38, A C is the resultant of A B and A D, and A F is the resultant of A C and A E, *i.e.* of the three forces A B, A D, and A E.

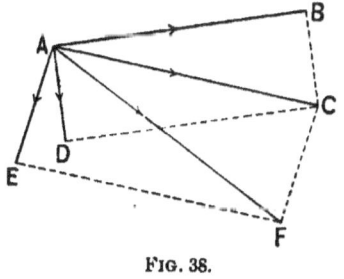

FIG. 38.

RESOLUTION OF FORCES. 55

Generally speaking, *a motion may be the result of any number of forces.* When we see a body in motion, we cannot determine by its behavior how many forces have concurred to produce its motion.

56. Resolution of Forces. Assume that a ball has an acceleration in a certain direction A C (Fig. 39), and that one of the forces that produces this acceleration is represented in intensity and direction by the line A B; what must be the intensity and direction of the other force? Since A C is the resultant of two forces acting at an angle to each other, it is the diagonal of a parallelogram of which A B is one of the sides. From C, draw C D parallel and equal to B A, and complete the parallelogram

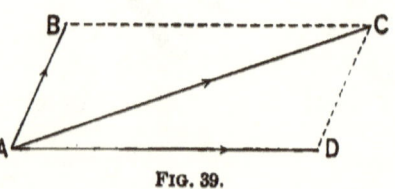

FIG. 39.

by connecting the points B and C, and A and D. Then, according to the principle of composition of forces, A D represents the intensity and direction of the force which, combined with the force A B, would give the ball an acceleration A C. The component A B being given, no other single force than A D will satisfy the question. Had the question been, "What forces can produce the motion A C?" an infinite number of answers might have been given.

It is often necessary to resolve a force in order to ascertain the effective force in a certain direction. Thus, when boat sails are exposed obliquely to the wind, the pressure effectual in moving the boat is only a component of the whole force of the wind. The line a f (Fig. 40) represents the force of the wind acting on the sail c d at the point a. Resolving this

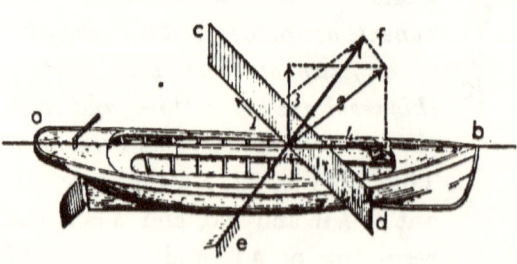

FIG. 40.

force we obtain the components 2 (normal to the sail) and 1 (a useless component called a tail wind). The boat does not move in the direction of the pressure on its sail, because it is more easily moved lengthwise than breadthwise. Hence the normal pressure must be resolved into two components, one 4 along the direction of least resistance, *i.e.* the direction of easy motion, the other 3 at right angles to it. The component 4 drives the boat forward. The component 3 tends to cause a slow broadside motion called leeway, but this may be partly counteracted by a deep keel or a center-board, so that the boat will sail approximately along the line a b.

EXERCISES.

1. What is the greatest and what the least resultant of two forces of 15 lbs. and 17 lbs.?

2. Draw upon paper pairs of lines making about the same angles with each other as A B and A C in the four diagrams, Fig. 41, and having

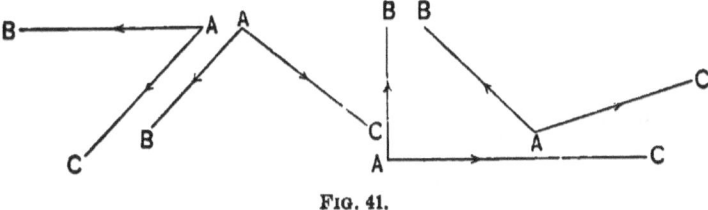

FIG. 41.

about the same directions ; assign numerical values arbitrarily to each component, drawing to scale, and find the direction and the numerical value of the resultant of each pair of components.

3. Two forces of 20 lbs. and 30 lbs. act at an angle of 90°. Find the intensity of their resultant without constructing a parallelogram.

4. Resolve a force of 40 lbs. into two components at right angles to each other, one of the forces to be 15 lbs.

5. A weight of 50 k. is supported by two strings inclined to the vertical at 30° and 60°. Find the tension of each string.

6. What three conditions are requisite that a force may be in equilibrium with two parallel forces?

7. Draw a line to represent a force of 20 lbs. acting at an angle of 30° with a horizontal line, and find its efficiency in a vertical direction.

8. The resultant of two equal forces acting upon a point at an angle of 90° is 10 lbs. Find the value of each component.

SECTION VIII.

CURVILINEAR MOTION.

57. How Curvilinear Motion is Produced. Motion is *curvilinear* when its direction changes at every point. But according to the First Law of Motion, every moving body proceeds in a straight line unless compelled to depart from it by some external force. Hence, curvilinear motion can be produced only by an external force acting continuously upon the body at an angle to the straight path in which the body tends to move, so as constantly to change its direction. In case the body moves in a circle, this force acts at right angles to the path of the body or towards the center of motion; hence, this deflecting force has received the name of *central force*.

Thus, suppose a ball at A (Fig. 42), suspended by a string from a point d, to be struck by a bat in such a manner that it tends to move in the direction A o. As it is restrained from taking that path by the tension of the string, which operates like a force drawing it toward d, it takes, in obedience to the two forces, an intermediate course. At c its motion is in the direction c n, in which path it would move but for the string, in accordance with the First Law of Motion. Here, again, it is compelled to take an intermediate path. Thus, at every point the tendency of the moving body is to preserve the direction it has at that point and consequently to move in a straight line. The only reason it does not so move is that it is at every point forced from its natural path by the pull of the string. But if the string be cut when the ball reaches the point i, the ball, having no force operating to change its motion, continues in the direction in which it is moving at that point, *i.e.* in the direction i h, which is tangent to its former circular path.

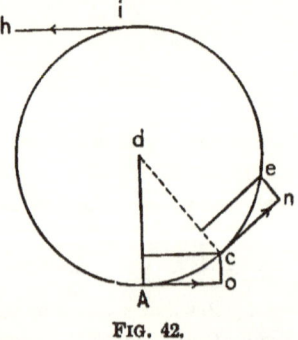

FIG. 42.

If the free end of the string be held in the hand, the ball while revolving about the hand appears to pull the hand. But it is evident that no force acts upon the ball except the pulling force exerted by the hand through the string, and that this apparent pull on the part of the ball is only the effect of the *reaction* of the force exerted by the hand upon the ball. This reaction is erroneously called "centrifugal force."

There is no centrifugal force; there is no "tendency to fly off from the center"; and there is no tendency of any kind that is not fully explained by the First Law of Motion.

58. Law of Central Force. *For a body moving in a circular orbit the central force is proportional to the mass of the body and to the square of its velocity, and inversely proportional to its distance from the center of motion.*

Let f represent the central force, m the mass of the revolving body, v its velocity, and r the radius of the circle, and the law may be expressed in the following formula:

$$f = \frac{mv^2}{r}.$$

The farther a point is from the axis of motion, the farther it has to move during a rotation; consequently the greater must be its velocity to complete a revolution in a given time. Hence, of bodies upon the earth's surface, those situated at the equator have the greatest velocity due to the earth's rotation, and consequently the greatest tendency to fly off from its surface. The effect of this is to neutralize, in some measure, the force of gravity. It is calculated that a body weighs about $\frac{1}{289}$ less at the equator than at either pole, in consequence of the greater tangential tendency at the former place. But 289 is the square of 17; hence, if the earth's velocity were increased seventeenfold, objects at the equator would weigh nothing, *i.e.* the tangential tendency would be equal to their weight.

59. Tendency to Rotate Around the Shortest Axis. It can be demonstrated mathematically, as well as experimentally, that *a freely rotating body is in stable equilibrium only when rotating about its shortest diameter;* hence the tendency of a rotating body to take this position.

ROTATION AROUND THE SHORTEST AXIS.

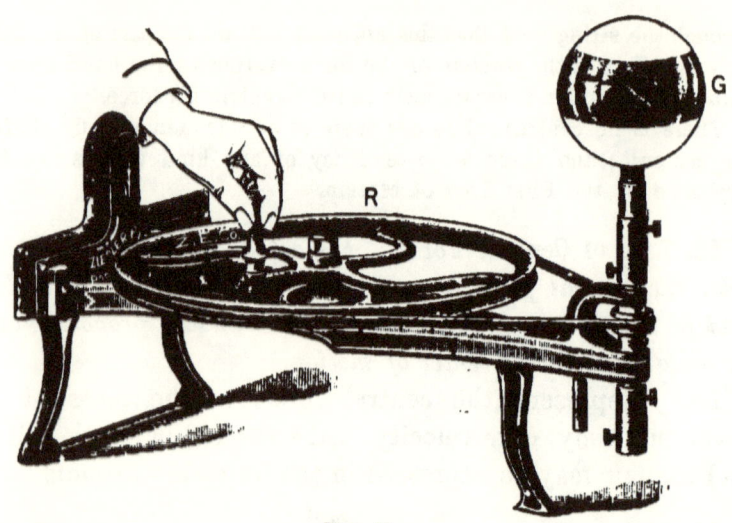

FIG. 43.

Experiment. Arrange some kind of rotating apparatus, *e.g.* R (Fig. 43). Suspend a skein of thread, a (Fig. 44), by a string, and cause it to rotate; it assumes the shape of the oblate spheroid a'. A chain, b, assumes a similar form. Pass a string through the longest diameter of an onion, c, and cause it to rotate; the onion gradually changes its position so as to rotate on its shortest axis.

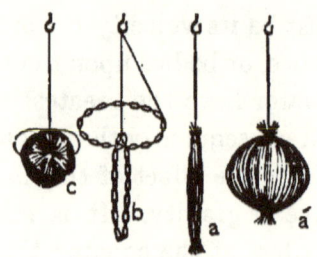

FIG. 44.

Mount a glass globe, G (Fig. 43), about one tenth full of colored water, and cause it to rotate. The liquid gradually leaves the bottom, rises, and forms an equatorial ring within the glass. In a similar way the water of the earth's great ocean is "heaped up" at the earth's equator.

EXERCISES.

1. (a) What is the cause of the stretching force exerted on the rubber cord when you swing a return ball about your hand? (b) Suppose that you double the velocity of the ball, how many times shall you increase this stretching force?

2. Why do wheels and grindstones, when rapidly rotating, tend to break, and the pieces to fly off?

MOLAR DYNAMICS.

3. On what does the magnitude of the pull between a rotating body and its center of motion depend?

4. (*a*) Explain the danger that a carriage will be overturned in turning a corner. (*b*) How many fold is the tendency to overturn increased by doubling the velocity of the carriage?

5. Account for the *curvilinear* orbits of the planets.

6. How are their motions in their orbits and around their axes *maintained*?

7. In what way should the rails be laid in order to neutralize the tangential tendency of a railroad train when going around a curve?

8. State and explain the posture of a bicycle rider in turning a curve.

9. In what way is the weight of terrestrial bodies nullified in some degree by the earth's motion?

10. A circus rider going around a ring inclines inward so that the line of direction of his body falls without his base. How is he supported?

SECTION IX.

THE PENDULUM.

60. Dynamics of the Pendulum. When a pendulum bob, B (Fig. 45), is raised, the force of gravity acting on it, represented by the line B G, may be resolved into two components, one of which, B C, acts upon the point of support S, while the other, B D, acts at right angles to it, producing acceleration toward O. Its inertia carries it beyond O against the action of gravity, which gives it a negative acceleration and brings it to rest at M. The backward swing from M to B is explained in the same way. Thus the pendulum oscillates under the action of gravity, which reverses its acceleration at point O during each swing.

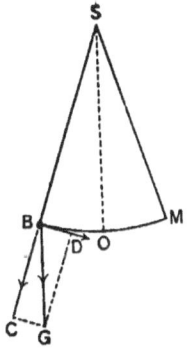

FIG. 45.

The motion from one extremity of the arc through which a pendulum swings to the other is called a *vibration* or an *oscillation*. The time occupied by the pendulum in moving once over this arc is called the *time* or *period of vibration*, and the angle B S O is called the *amplitude of vibration*.

61. Laws of the Pendulum.

Experiment 1. From a bracket suspend by strings leaden balls, as in Fig. 46. Draw B and C to one side, and to different hights, so that B may swing through a short arc and C through a longer arc, and let both drop at the same instant. C moves faster than B, and completes a longer journey at each swing, but both complete their swing, or vibration, in the same time.

Hence, (1) *the time occupied by the vibration of a pendulum is independent of the length of the arc.*

Of only very small arcs may this law be regarded as practically true. The pendulum requires a somewhat longer time for a long arc of vibration than for a short one, but the difference becomes perceptible only when the difference between the arcs is great, and then only after many vibrations.

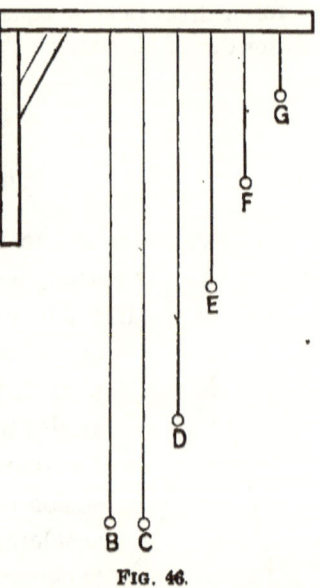

FIG. 46.

Experiment 2. Set all the balls swinging; only B and C swing together; the shorter the pendulum, the faster it swings. Make B 1 m. long and F ¼ m. long. With watch in hand, count the vibrations made by B. It completes 60 vibrations in a minute; in other words, it "beats seconds." A pendulum, therefore, to beat seconds must be 1 m. long (more accurately in the latitude of Boston at sea level .9935 m., or 39.117 in.). Count the vibrations of F; it makes 120 vibrations in the same time that B makes 60 vibrations. Make G one ninth the length of B; the former makes three vibrations while the latter makes one, consequently the time of vibration of the former is one third that of the latter.

Hence, (2) *the time of one vibration of a pendulum varies as the square root of its length.*

The length l of a simple pendulum which shall swing in a time t, or the time of swing for a length l, can be found from the formulæ:

62   MOLAR DYNAMICS.

$$l = .9935 \times t^2, \text{ whence } t = \sqrt{\frac{l}{.9935}} \text{ for } l \text{ meters};$$

or $$l = 39.117 \times t^2, \text{ whence } t = \sqrt{\frac{l}{39.117}} \text{ for } l \text{ inches.}$$

By experiments too difficult for ordinary school work, it has been ascertained that (3) *the time of vibration of a pendulum varies inversely as the square root of the force of gravitation* (upon which the value of g depends).

To sum up the above three laws of the pendulum, we have the formula:[1]

$$t = \pi \sqrt{\frac{l}{g}}, \text{ whence } g = \frac{\pi^2 l}{t^2},$$

in which l = length of pendulum; t = time of one vibration in seconds.

62. Simple Pendulum; Center of Oscillation. A *simple pendulum* is a material particle supported by a weightless thread. Such a pendulum can exist only in the imagination, but the conception is useful. Every real pendulum is a *compound pendulum*, which may be supposed to be composed of as many simple pendulums bound together as there are particles in the pendulum. Those particles nearest the point of suspension tend to quicken, and those farthest away tend to check, the motion of the combination. It is apparent that there must be in every compound pendulum a particle so situated that its motion is neither quickened nor checked by the combined action of the particles above and below it. The location of this particle is called the *center of oscillation*. The *real length* of a compound pendulum is the distance of this point from the point of suspension, and it is this length that is referred to in the laws of the pendulum.

[1] The student may find the development of this formula in Chapter VII of Maxwell's " Matter and Motion."

The center of oscillation of a pendulum may be found approximately as follows : A small lead ball suspended by a thread is a near approximation to a simple pendulum, and the distance from the center of the ball to the point of suspension may be taken as the length of this pendulum. Suspend from the same support this pendulum and the pendulum whose center of oscillation is to be found. For example, let the pendulum be a lath (Fig. 47) suspended at its upper extremity, A. Lengthen or shorten the ball pendulum till it swings in the same time as the lath. Then the true lengths of the two pendulums must be the same. Lay off on the lath from its point of suspension a distance equal to the distance from the point of suspension to the center of the ball, and this will give the center of oscillation of the lath pendulum. This point, in case the lath be of uniform dimensions and density throughout, will be at C, or at a distance of two thirds the length of the lath from its point of suspension, A.

Fig. 47.

If a weight (or bob) be attached to the lower end of A B, its center of oscillation is moved lower and the period of vibration is lengthened. If the bob of a pendulum be raised (which usually may be done by turning a thumb-screw just beneath it), the pendulum is shortened and its time of vibration is decreased.

63. Uses of the Pendulum. The isochronism of the pendulum is utilized in the measurement of time, *i.e.* in subdividing the solar day into hours, minutes, and seconds. The office of the pendulum in clocks is to regulate the rate of motion of the works. The balance wheel replaces the pendulum in watches and some clocks.

One of the most important uses of the pendulum from a scientific standpoint is that in determining the acceleration due to the force of gravitation at any place.

The time of vibration is less at a place where the force of gravitation is greater because the accelerating force for the same mass is greater, and hence the pendulum will move faster.

Hence, it is apparent that by determining the time of vibration of a pendulum of the same length at different distances from the center of mass of the earth (*e.g.* at the top and bottom of a mountain, or at sea level at different latitudes), the relative value of g at these places, *i.e.* the acceleration produced by gravitation, may be ascertained.

At the poles of the earth the length of a seconds pendulum is 99.62 cm. and $g = 983.2$ cm. per second. At the equator, $l = 99.10$ cm.; $g = 978.1$ cm. per second.

EXERCISES.

1. (*a*) What is the length of a pendulum that beats half-seconds? (*b*) Quarter-seconds? (*c*) That makes one vibration in two seconds? (*d*) That makes two vibrations per minute?

2. State the proportion that will give the number of vibrations per minute made by a pendulum 40 cm. long.

3. How will the periods of vibration compare in the case of two pendulums whose lengths are, respectively, 4 feet and 49 feet?

4. Two pendulums make, respectively, 50 and 70 vibrations per minute. Compare their lengths.

5. How long must a pendulum be to make one vibration in 5 seconds in Boston?

6. One pendulum is 20 inches long, and vibrates four times as fast as another. How long is the other?

7. (*a*) What effect on the time of vibration of a pendulum has the weight of its bob? (*b*) What effect has the length of the arc? (*c*) What affects the time of vibration of a pendulum?

8. How can you quicken the vibration of a pendulum threefold?

9. A clock loses time. (*a*) What change in the pendulum ought to be made? (*b*) How would you make the correction?

10. Two pendulums are 4 and 9 feet long, respectively. While the short one makes one vibration, how many will the long one make?

11. What is the time of vibration of a pendulum (39.09 ÷ 4 =) 9.77 inches long?

12. The number of vibrations made by a given pendulum in a given time varies as the square root of the force of gravity. Force of gravity at any place is expressed by the value of g (*i.e.* by the acceleration which it produces). If at a certain place a pendulum 39.09 inches long make 3600 vibrations in an hour, and the value of g be 32.16 feet, what is the acceleration at a place where the same pendulum makes 3590 vibrations in the same time?

13. A pebble is suspended by a thread 2 feet long; required the number of vibrations it will make in a minute.

SECTION X.

GRAVITATION.

64. Gravitation is Universal. We know that there is a stress between the earth and all bodies on or near it, and we have learned to call this the *weight* of bodies. Sir Isaac Newton was the first to show that this stress is not limited to the earth and terrestrial bodies, but exists between particles separated by any distance, however great. So that *there is a stress between every particle of matter in the universe and every other particle.* This mutual action is called *Universal Gravitation.*

That there is a stress between the sun and the earth, and the earth and the moon, is shown by their curvilinear motions in their orbits. Tides and tidal currents on the earth are due to the stress between the sun and the moon and the masses of water on the earth's surface.

65. Law of Gravitation. The Law of Universal Gravitation is as follows:

The gravitation stress between every two particles of matter in the universe varies directly as the product of their masses, and inversely as the square of the distance between them.

If the masses of two bodies be represented by m and m', the distance between their centers of mass by d, and the gravitation stress by g, this relation is expressed mathematically thus: $g \propto$ (varies as) $\frac{mm'}{d^2}$. For

66   MOLAR DYNAMICS.

example, if the mass of either body be doubled, the product (mm') of the masses is doubled, and consequently the stress is doubled. If the distance between their centers of mass be doubled, then $\left(\dfrac{1}{2^2}=\dfrac{1}{4}\right)$ the stress becomes one fourth as great.

66. Law of Weight. *The weight of a body at or above the surface of the earth varies inversely as the square of the distance from the center of mass of the earth.*

Since the earth is not a perfect sphere,[1] it follows from the law that the weight of the same body differs at different places on the earth's surface. Its loss of weight in being transported from the poles to the equator, due to this increase of distance from the center of mass of the earth, is estimated to be $\frac{1}{588}$ of its weight at the poles. But we have previously seen (p. 58) that the tangential tendency at the equator diminishes the weight of a body $\frac{1}{289}$. Now in consequence of difference in distance from the center of mass of the earth and difference in velocity due to the earth's rotation, a body weighs at the equator $\frac{1}{588} + \frac{1}{289} = \frac{1}{192}$ less than at the poles.

We infer from the law of gravitation that a body weighs more at the earth's surface than above it; in other words, bodies become lighter as they are raised above the earth's surface. But since the force diminishes as the square of the distance from the center (not from the surface) of the earth, and as the surface is about 4000 miles from the center of mass, the diminution for a few miles or for any distance which we are able to raise bodies is scarcely perceptible; hence, in all commercial transactions we may, without important error, buy and sell as if the weighing always took place at the same distance from the center of mass of the earth, in which case mass is strictly proportional to weight.

EXERCISES.

1. (*a*) Which is independent of mass, weight or acceleration? (*b*) Which varies as the mass?

2. Why does a 100-lb. iron ball fall with no greater acceleration than a 1-lb. ball of the same material?

3. (*a*) Which falls with greater acceleration in the air, an iron ball or a wax ball? Why? (*b*) How would their accelerations compare in a vacuum? (*c*) Is acceleration independent of kind of matter?

[1] The earth is a spheroid, its polar diameter being about 43 kilometers (nearly 27 miles) shorter than its equatorial diameter.

4. If the earth's mass were doubled without any change of volume, how would the change affect your weight?

5. On what principle may you determine that the mass of one body is ten times the mass of another body?

6. How many times must you increase the distance between the centers of mass of two bodies in order that the gravitation stress between them may become one fourth as great?

7. (*a*) If a body on the surface of the earth be 4000 miles from the center of mass of the earth, and weigh at this place 100 lbs., what would the same body weigh if it were taken 4000 miles above the earth's surface? (*b*) What 2000 miles above the earth? (*c*) What 100 miles above the earth?

8. If at sea level in Boston $g = 980.4$ cm., what is the value of g at a point 5 miles above sea level?

9. What retains the planets in their orbits?

10. If there were but one body of matter in existence, (*a*) would it have weight? .(*b*) Would it have mass?

11. What is the character of the motion produced by a constant force acting on a free body?

SECTION XI.

WORK, ENERGY, AND POWER.

67. Work. Whenever a force causes a change of motion or maintains motion against resistance, it is said *to do work*. A force to do work *must effect a change of position*. *Force* and *space* are essential conditions of work. *An unbalanced force always does work*, inasmuch as it always causes a change of motion.

The body that moves another body is said to do work upon it; and the body moved is said to have work done upon it.

When the heavy weight of a pile driver is raised, work is done upon it; when it descends and drives the pile into the earth, work is done upon the pile, and the pile in turn does work upon the matter in its path.

68. Energy. *The energy of a body is its capacity for doing work.* It is measured by the quantity of work which the body is capable of doing. The work done by a body, or

68 MOLAR DYNAMICS.

done on a body, is a measure of its loss or gain of energy; hence, the unit of work is also the unit of energy.

The act of doing work either consists in a transfer of energy from the body doing work to the body on which work is done, as when the wind propels a vessel, *or it consists in a transformation of one kind of energy into another kind.* When the pile driver strikes the pile and the pile is forced into the earth, a part of the energy in each act is transformed into *heat,* which we shall learn, farther on, is *molecular energy.* Work, therefore, may be defined as *the act of transmitting or transforming energy.*

69. Kinetic and Potential Energy. Every moving body can impart motion, therefore it can do work upon another body; hence, *every moving body possesses energy.* The energy which a body possesses in consequence of its motion is called *kinetic energy.*

When a body is projected upward its kinetic energy diminishes as it rises and finally becomes *nil,* but it is not lost, for it reappears as the body falls. Its energy becomes, while rising, *stored up* in virtue of its higher *position.* Energy in store, *i.e.* not in an active state, is called *potential energy.* It is the capacity for doing work possessed by a mass *in virtue of its position being such that it is possible for it to move, and in virtue of the existence of a stress which tends to move it.* Hence, it is convertible into kinetic energy without the agency of any additional work except that of removing obstacles to the conversion. Potential energy implies a tendency to motion, as truly as kinetic energy implies motion.

Illustrations of energy in the potential state:

A stone lying on the ground is devoid of energy. Raise it and place it on a shelf; in so doing you perform work upon it. As you look at it lying motionless upon the shelf, it *appears* as devoid of energy as when lying on the earth. Attach one end of a cord to it and pass it over a pulley, and wind a portion of the cord around the shaft connected with a sewing machine, lathe, or other convenient machine. Suddenly withdraw the shelf from beneath the stone. The stone moves; it communicates motion to the machinery, and you may sew, turn wood, etc., with the energy given to the machine by the stone.

The work done on the stone or the energy transmitted to the stone in raising it was not lost; it reappeared while the stone was descending. There is a very important difference between the stone when lying on the ground and the stone when lying on the shelf; the former is powerless to do work; the latter can do work. Both are alike motionless, and you can see no difference, except an *advantage* that the latter has over the former in having a *position* such that it *can move*. What gave it this advantage? Work. *A body*, then, *may possess energy due merely to* ADVANTAGE OF POSITION, *derived from work performed upon it.*

We are as much accustomed to store up energy for future use as to store up provisions for the winter's consumption. We store it when we wind up the spring or weight of a clock, to be doled out gradually in the movements of the machinery. We store it when we bend the bow, condense air, or raise any body above the earth's surface.

We see, then, that energy may exist in bodies by virtue of their *actual motion*, or it may exist in bodies by virtue of their having an *opportunity and a tendency to move*, as in the stone lying on the shelf. But it should be remembered that a body does work only when moving; hence, the potential energy of a body must become kinetic before the body can do work.

70. Energy of Chemical Separation. Matter may possess potential energy in virtue of chemical separation and chemical affinity, and the potential energy is a measure of the work done in effecting the separation. For example, the entire value of coal consists in its potential energy, which was stored up by the work performed through the agency of the sun's energy in separating the carbon of carbon dioxide from the oxygen. Gunpowder possesses potential energy sufficient to do a quantity of work, *e.g.* in blasting, which would require many laborers a long time to do.

A body possesses potential energy when, in virtue of work done upon it, it occupies a position of advantage, or its con-

stituent particles occupy positions of advantage, so that the energy expended can be restored at any time by the return of the body to its original position, or by the return of its particles to their original positions.

71. Practical Units of Work and Energy. The practical unit adopted is *the work done or energy imparted in raising 1 pound through a vertical hight of 1 foot.* It is called a *foot-pound*. The metric unit is the work done or energy imparted in raising 1 k. a vertical hight of 1 m., and is called a *kilogrammeter*. The kilogrammeter is equivalent to 7.2331 foot-pounds. Since the work done in raising 1 pound 1 foot high is 1 foot-pound, the work of raising 1 pound 10 feet high is 10 foot-pounds, which is the same as the work done in raising 10 pounds 1 foot high; and the same, again, as raising 2 pounds 5 feet high.

There are many kinds of work besides that of raising weights. But since, with the same resistance, the work of producing motion against any given resistance is independent of direction, it is easy, in all cases in which the resistance and the space through which the resistance is overcome are known, to find the equivalent in work done in raising a weight vertically. By thus securing a common standard for measurement of work, we are able to compare any species of work with any other. For instance, let us compare the work done in sawing through a stick of wood by a man whose saw must move 10 m. against an average resistance of 12 k., with that done by a bullet in penetrating a plank to a depth of 2 cm. against an average resistance of 200 k. Moving a saw 10 m. against 12 k. resistance is equivalent to raising 12 k. mass 10 m. high, or doing 120 kgm. of work ; a bullet moving 2 cm. against 200 k. resistance does as much work as is required to raise 200 k. mass 2 cm. high, or 200 × .02 = 4 kgm. of work. 120 ÷ 4 = 30 times as much work done by the sawyer as by the bullet.

72. Absolute Units of Work.[1] If force be measured in dynes, and distance in centimeters, the work done is ex-

[1] The pupil will, perhaps, be assisted by the accompanying diagram (Fig. 48) in his first attempts to acquire and classify the units of force, energy, and work in the several systems. In this connection he should consult the "Reduction of measures to and from the C. G. S. system" in the Appendix.

pressed in a C. G. S. unit called an *erg*. *An erg is the work done or energy imparted by a force of one dyne acting through a distance of one centimeter.*

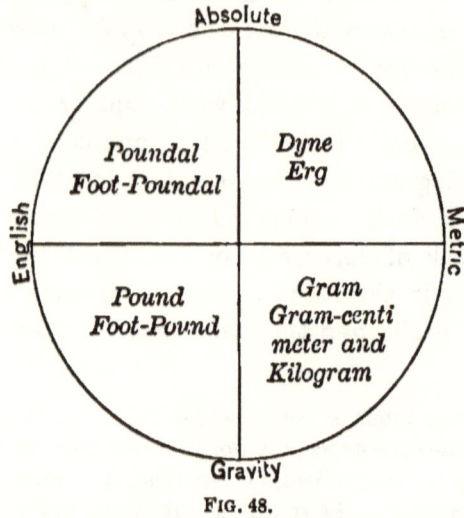

FIG. 48.

The F. P. S. unit of work or energy is the *foot-poundal*, and is the work done or energy imparted by a force of one poundal acting through a distance of 1 foot.

73. Formulas for Calculating Work or Energy Imparted. *Force* and *space* (or distance), being essential conditions of work, are necessarily the quantities employed in calculating work. A given force acting through a space of 1 foot does a certain quantity of work; it is evident that the same force acting through a space of 2 feet would do twice as much work. Hence, the general formula:

$$W = fs, \qquad (1)$$

in which f represents the force employed, s the space through which the force acts, and W the work done.

In case a force encounters resistance, the magnitude of the force necessary to produce motion varies with the resistance. Often the work done upon a body is more conveniently determined by *multiplying the resistance by the space through which it is overcome*, and our formula becomes by substitution of r (resistance) for f (the force which overcomes it)

$$rs = W. \qquad (2)$$

For example, a ball is shot vertically upward from a rifle in a vacuum; the work done upon the ball (by the explosive force of the gunpowder) may be calculated by multiplying the average force (difficult to ascertain) exerted upon it by the space through which the force acts (a little greater than the length of the barrel); or by multiplying the resistance to motion offered by gravity, *i.e.* its weight (easily ascertained), by the distance the ball ascends.

Let us calculate the energy stored in a bow by an archer whose hand, in bending the bow by pulling on the string, moves 6 inches ($\frac{1}{2}$ foot) against an average resistance of 20 pounds. Here $rs = 20 \times \frac{1}{2} = 10$ foot-pounds of work done upon the bow, or 10 foot-pounds of energy stored in the bow.

74. Formula for Calculating the Kinetic Energy of a Body when its Mass and Velocity are Known.

Suppose a body to have a mass m and a velocity v; it can do a definite quantity of work before it is thereby brought to rest. If it be moving upward, a mutual work is performed by it and by the earth, consisting in each destroying the other's momentum. If its velocity be such that it will rise to a hight s, then its kinetic energy is such that it will do $(f = mg)$ $m\,g\,s$ absolute units of work, or

$$E_k \text{ (kinetic energy)} = m\,g\,s. \qquad (1)$$

We may find, then, to what vertical hight a body having a given velocity would rise if directed upward, and from formula (1) determine its kinetic energy; but a formula may be obtained which will give the same result with less trouble; thus, substituting g for a in the formula $v = at$ (p. 11), we have $v = gt$; whence

$$t = \frac{v}{g}, \text{ or } t^2 = \frac{v^2}{g^2}.$$

Again, $s = \frac{1}{2} gt^2$; substituting the value of t^2 in this equation we have

$$s = \tfrac{1}{2} g \times \frac{v^2}{g^2} = \frac{v^2}{2g}.$$

Substituting for s in equation (1) its value we have

$$E_k = \frac{m v^2}{2}, \qquad (2)$$

a formula which will determine the kinetic energy of a body in absolute units when its mass and velocity are known, since the energy is the same whatever be the direction of the motion.

Hence, the kinetic energy of a body is half the product of its mass by the square of its velocity.

If the result be desired in gravitation units, *i.e.* in gram-centimeters or foot-pounds, the number of absolute units must be divided by g, since g ergs (980) are equivalent to one gram-centimeter, or g foot-poundals (32.2) are equivalent to one foot-pound.

75. Energy Contrasted with Momentum. It is evident from formula (2) that *when the mass of a body remains the same, its energy is proportional to the square of its velocity; while its momentum, as we have learned, is proportional to its velocity.* In other words, the effect of increasing the velocity of a moving body would seem to be to increase its working power much more rapidly than its momentum.

Furthermore, we have seen (p. 28) that momentum $= M V = ft$; and again (p. 71), W (work done) or E_k (kinetic energy imparted) $= fs$.

The momentum, then, imparted to a body is the product of the force into the *time* it acts; energy imparted is the product of force into the *space* through which it acts. It is evident, therefore, that *when time is considered, force may be measured by the momentum, and when the space is considered, by the energy which it imparts.* If we know only the momentum of a body, we can tell *how long* it would move against a given resistance, but not how far. If we know only the kinetic energy of a body, we can tell *how far* it would move against a given resistance, but not how long.

The following illustration will help to make the distinction plain: We realize that there is an important difference between the slow motion of a gun's recoil and the rapid motion of the bullet projected from it, though, as we have seen, the momenta are the same. By a suitable force the gun may be brought to rest in one second, and in that time will move (say) 6 inches. The same force will in one second bring the bullet to rest, but, as may be easily shown (assuming the mass of the gun and the bullet to be, respectively, 10 pounds and $\frac{1}{2}$ ounce), the bullet will have moved about 250 feet. Hence, the bullet has a much greater penetrating power than the gun.

A bullet moving with a velocity of 400 feet per second will penetrate, not twice, but four times as far into a plank as one having a velocity of 200 feet per second. A railway train having a velocity of 20 miles an hour will, if the steam be shut off, continue to run four times as far as it would if its velocity were 10 miles an hour. The reason is now apparent why light substances, even so light as air, exhibit great energy when their velocity is great.

EXERCISES.

1. Does the energy expended in raising the stones to their places in the Egyptian pyramids still reside in the stones?
2. What kind of energy is that contained in gunpowder?
3. Can a person lift himself, or put himself in motion, without exerting force upon some other body?
4. (*a*) Can a body do work upon itself? (*b*) Can a body generate energy in itself, *i.e.* increase its own energy?
5. (*a*) Suppose that an average force of 25 pounds is exerted through a space of 10 inches in bending a bow, what amount of energy will it give the bow? (*b*) What kind of energy will the bow, when bent, possess?
6. (*a*) What amount of kinetic energy does a mass of 20 pounds, moving with a velocity of 300 feet per second, possess? (*b*) What amount of work can the body do?

7. How many fold is the kinetic energy of a body increased by doubling its velocity?

8. How high will 1200 foot-pounds of energy raise 100 pounds?

9. A force of 500 pounds acts upon a body through a space of 20 feet. One fourth of the work is wasted in consequence of resistances. How much available energy is imparted to the body?

10. How much energy is stored in a body weighing 1000 pounds, at a hight of 200 feet above the earth?

11. A horse draws a carriage on a level road at the uniform rate of 5 miles an hour. (a) Does energy accumulate? (b) What kind of energy does the carriage possess? (c) Suppose that the carriage were drawn up a hill, would energy accumulate? (d) What kind of energy would it possess when at rest on the top of the hill? (e) How would you calculate the quantity of energy it possesses when at rest on top of the hill? (f) Suppose that the carriage is in motion on top of the hill, what two formulas would you employ in calculating the total energy which it possesses?

12. How much work is done per hour if 80 k. be raised 4 m. per minute?

13. (a) What energy must be imparted to a body weighing 50 g. that it may ascend 4 seconds? (b) How many times as much energy must be imparted to the same body that it may ascend 5 seconds? (c) Why?

14. Compare the momenta, in the two cases given in the last question, at the instants the body is thrown.

15. How much energy is stored in a body which weighs 50 k. at a hight of 80 m. above the earth's surface?

16. How much kinetic energy would the same body have if it had a velocity of 100 m. per second?

17. Suppose it to fall in a vacuum, how much kinetic energy would it have at the end of the fourth second?

18. If it should fall through the air, what would become of the part of the energy lost in consequence of the resistance offered by the air?

19. A projectile of mass 25 k. is thrown vertically upward with an initial velocity of 29.4 m. per second. How much energy has it?

20. What becomes of its energy during its ascent?

21. (a) Compare the momentum of a mass 50 k. having a velocity of 2 m. per second, with the momentum of a body of a mass 50 g. having a velocity of 100 m. per second. (b) Compare their energies.

22. Which, momentum or energy, will enable one to determine the amount of resistance that a moving body can overcome?

23. Explain how a child who cannot lift 30 k. can draw a carriage weighing 150 k.

24. How many and what transformations take place during a single swing of a pendulum?

25. What quantity of energy will be expended if a force of 60 pounds move a body a distance of 20 feet?

26. Describe the transformations of energy that take place during a single swing of the pendulum.

27. A blacksmith raises a hammer and strikes an anvil. State all the transformations of energy that occur.

28. How much work is done against gravity by a man weighing 150 pounds in climbing a mountain a mile high?

29. In winding a clock a weight of 8 pounds is raised a yard from the bottom of the clock. (*a*) How much energy is stored up? (*b*) When the weight has fallen 1 foot, how much potential energy has it left? (*c*) How much potential energy will it have after it has fallen 2 feet? (*d*) How has its lost energy been expended? (*e*) Had the weight fallen freely, what amount of kinetic energy and what amount of potential energy would it have after falling 2 feet?

30. Does the motion that one body can communicate to another depend upon the momentum of the former, or upon its energy?

76. Power. The *power* (sometimes called *activity*) of any agent, *e.g.* a steam engine, an animal, etc., is the *rate* at which it does or can do work, and is measured by the quantity of work it either does or can do *per unit of time*, and is determined by the formula

$$P \text{ (power)} = \frac{w \text{ (work)}}{t \text{ (time)}}.$$

In estimating the total quantity of work done, the time consumed is not taken into consideration. The work done by a hod-carrier in carrying 1000 bricks to the top of a building is the same whether he does it in a day or a week. But in estimating power or *the rate* at which the agent is capable of doing work, it is evident that time is an important element. The work done by a horse in raising a barrel of flour 20 feet is about 4000 foot-pounds; even a mouse could do the same quantity of work in time, but he has not the power of a horse.

EXERCISES.

The unit in which activity or rate of doing work is estimated is called (inappropriately) a horse-power. A horse-power is 550 foot-pounds per second, 33,000 foot-pounds per minute, or 1,980,000 foot-pounds per hour.

The logical unit of power is a unit of work in a unit of time, as one erg per second. The absolute unit of power, chiefly used in measuring electrical power, is the *watt*, or 10,000,000 ergs per second. A moderate estimate of man's power is 100 watts.

1 erg per second = .0000001 watts,
1 horse-power = 746 watts,
1 foot-pound per minute = 226,043 ergs per second.

EXERCISES.

1. For which is a truck horse valued, his energy or his power?
2. Do we speak of the power or the energy of the steam engine?
3. Which do we apply to levers and machines in general, power or force?
4. "Energy is the power of doing work." Is this true?
5. Shall we say that the power, or the energy, of the horse is greater than that of man?
6. How much work can a 2 horse-power engine do in an hour?
7. (a) What quantity of work is required to raise 50 tons of coal from a mine 200 feet deep? (b) An engine of how many horse-power would be required in order to do it in 2 hours?
8. A car of mass 3 tons is drawn by a horse at a speed of 180 feet per minute. The index of the dynamometer to which the horse is attached stands at 800 lbs. (a) At what rate is the horse working? (b) Express the rate in horse-power.
9. A dynamometer shows that a span of horses pull a plough with a constant force of 1500 lbs. What power is required to work the plough if they travel at the rate of 2 miles per hour?
10. What horse-power in an engine will raise 1,350,000 lbs. 50 feet in an hour?
11. How long will it take a 3 horse-power engine to raise 10 tons 50 feet?
12. How far will a 2 horse-power engine raise 3000 lbs. in 10 seconds?

MOLAR DYNAMICS.

13. A force of 10,000 dynes acting through a space of 100 meters per second furnishes a power of how many watts?

14. The wind moves a vessel with a uniform velocity of 5 miles an hour against a constant resistance of 2000 lbs. What power is furnished by the wind?

15. If a 2 horse-power engine can just throw 1050 lbs. of water to the top of a steeple in 2 minutes, what is the hight of the steeple?

16. Supply the following ellipses by selecting appropriate words from the following: *viz.* force, work, energy, power. When — acts through space — is performed, and — is imparted. The rate at which — is performed determines the — of the agent. The — of a bullet flying through vacant space. The — of a horse. The — of wind. The — of a bent bow. What — must a bullet of mass 1 ounce have that it may rise 4 seconds? What — is consumed by a steamer in crossing the ocean? What — is necessary that it may traverse 300 knots per day, and what must be the average — exerted to overcome the resistances at the required rate?

17. It is estimated that 300,000 cubic feet of water plunge over the Niagara escarpment 150 feet downward every second. What power can these falls furnish?

18. A rifle bullet whose mass is 1 ounce is projected vertically upward with a velocity of 161 feet per second. What quantity of work does it do in rising against the force of gravity?

SECTION XII.

MACHINES.

77. Uses of Machines.

Experiment 1. Suspend, as in Fig. 49, a fixed pulley, A, and a movable pulley, B. Let the scale-pan C counterbalance the pulley B, so that there shall be equilibrium. Suspend from B two balls, L L, of equal weight, and suspend on the side where the pan is a single ball, K, equal to one of the former. The single ball supports the two balls; *i.e.* by the use of the machine, a force[1] of 1 is enabled to balance a force of 2. So far no work is done. Place a very small weight in the pan; this additional weight destroys the balance, the balls L L rise, and work is done upon the balls.

[1] A perpetuation of the "time-honored" custom of calling the force applied to a machine *power*, and the force exerted by the machine to overcome some external resistance *weight* — all of which is exceedingly confusing to the pupil — is indefensible.

USES OF MACHINES. 79

As the weight K plus a very small weight causes the motion, we shall regard this as the force (*f*); and as the weights L L are the bodies moved (the pulleys and pan, being parts of the machine, may be disregarded), they may be regarded as the resistance (*r*) overcome, or the body on which work is done. Measure the respective distances through which *f* acts and *r* moves during the same time: *r* moves only one half as great a distance as that through which *f* acts; *i.e.* if *r* rise 2 feet, *f* must act through 4 feet. Suppose that *r* is 2 pounds, then *f* is 1 + pounds. Now 2 (pounds) × 2 (feet) = 4 foot-pounds of work done on *r*. Again, 1 + (pounds) × 4 (feet) = a little more than 4 foot-pounds of work (or energy) expended.

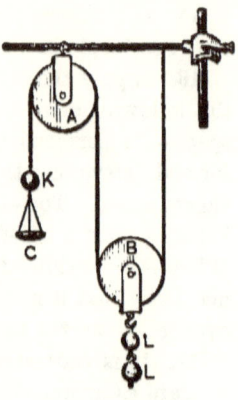

FIG. 49.

It thus seems that, although a machine will enable a small force to balance a large force, when work is performed the work applied to the machine is greater, rather than less, than the work which the machine transmits to the resistance. The work applied is greater than the work transmitted by the amount of work wasted in consequence of friction and other resistances. So that *by the employment of a machine there is never a gain, but always a loss of work.*

What, then, is the advantage gained in using this machine? Suppose that *r* is 400 pounds, and that the utmost force that a man can exert is a little more than 200 pounds. Then without the machine the services of two men would be required to move the resistance; whereas one man can move it with a machine, although he will be obliged to move twice as far as the resistance moves, a matter of little consequence in comparison with the advantage of being able to do the work alone. The advantage gained in this instance seems to be one of *convenience.* Men, however, are accustomed to speak of it as "*a gain of force*" (or more commonly and inaccurately, "*of power*"), inasmuch as a small force overcomes a large resistance.

Experiment 2. If, instead of applying the small additional weight to the pan, it be suspended from one of the balls L L, the weight of these balls, together with the additional weight, becomes the cause of motion, and K is the resistance. In this case there is a loss of force, because the force employed is greater than the force overcome. Measure the distances traversed, respectively, by K and L L in the same time. K moves twice as far as L L, and of course with twice the speed. There is *a gain of speed* at the expense of *force*.

It thus appears that, if it should be desirable to move a resistance with greater speed than it is possible or convenient for the force to act, it may be accomplished through the mediation of a machine, by applying to it a force proportionately greater than the resistance. This apparatus is one of many *contrivances called machines, through the mediation of which force can be applied to resistance more advantageously than when it is applied directly to the resistance.*

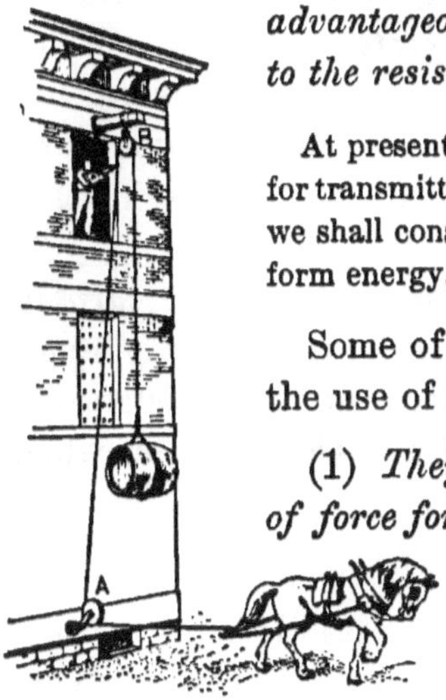

At present we deal with machines employed as means for transmitting and modifying motion and force. Later we shall consider machines whose function is to transform energy, such as the steam engine, dynamo, etc.

Some of the many advantages derived from the use of machines are:

(1) *They may enable us to exchange intensity of force for speed, or speed for intensity of force.* A gain of intensity of force or a gain of speed is called a *mechanical advantage.*

FIG. 50.

(2) *They may enable us to employ a force in a direction that is more convenient than the direction in which the resistance is to be moved.*

(3) *They may enable us to employ other forces than our own muscular force in doing work;* e.g. the muscular force of animals; the forces of wind, water, steam, etc.

GENERAL LAW OF MACHINES.   81

How are the last two uses illustrated in Fig. 50 ? The pulleys employed are called fixed pulleys, *i.e.* they have no motion except that of rotation. Is any mechanical advantage gained by fixed pulleys ? What is the use of a fixed pulley ? Pulley B (Fig. 49) is a movable pulley. What advantage is gained by means of a movable pulley ?

78. General Law of Machines. From the experiments and discussion above we derive the following formula for machines :

$$fs = rs' + w, \qquad (1)$$

in which f represents the force applied, and s the distance through which f acts ; r represents the resistance overcome, and s' the distance through which its point of application is moved; w represents the wasted work. A machine in which there is no wasted work is a *perfect machine.* Such a machine is purely ideal, as none exists. If in our calculations we regard a machine as perfect (though subsequently suitable allowance must be made for the wasted work), then our formula becomes $\qquad fs = rs',$ $\qquad\qquad$ (2) whence $r : f :: s : s'$; *i.e. the force and the resistance vary inversely as the distances which their respective points of application move.* In other words, the ratio of the resistance to the force is the reciprocal of the ratio of the distances which these points move ; thus, if

$$r : f = 4, \text{ then } s' : s = \tfrac{1}{4}.$$

This law applies to machines of every description ; hence it is called the *General* or *Universal Law of Machines.* When r is greater than f, there is a gain of intensity of force, and $\dfrac{r}{f} =$ *the ratio of gain of intensity of force.* When s' is greater than s, there is a gain of speed, and $\dfrac{s'}{s} =$ *the ratio of gain of speed.*

Since fs, the work done upon a machine, is always greater than rs', the work transmitted by the machine, we infer that *no machine creates or increases energy.* No machine transmits more energy than it receives. A machine may enable us to gain intensity of force, but not energy. By taking s great enough, f can be made as small as we please; in this case *in proportion as force is gained, time, distance, or speed is lost.*

79. Efficiency of Machines. The *efficiency* of a machine is a fraction, usually a *per cent*, expressing the ratio of the energy given out by the machine and utilized to the total energy expended upon the machine. The limit of the efficiency of a machine is *unity*, which is the efficiency of an "ideal," or perfect, machine, in which no energy is lost. The object of improvements in machines is to bring their efficiency as near to unity as possible.

For instance, if 50 foot-pounds of energy be expended on a machine, and friction convert 8 foot-pounds into heat, and 5 foot-pounds be lost in consequence of the utilization of only a component of the working force, so that the machine is able to give out only 37 foot-pounds, its efficiency is $\frac{37}{50} = 74$ per cent. If the friction can be reduced one half, and an improvement can be made in the machine which will render the entire working force effective, then there will be wasted only 4 foot-pounds of energy, and its efficiency will be raised to $\frac{46}{50} = 92$ per cent, and the quantity of work which the machine will accomplish will be increased in the ratio of 92 : 74.

80. Mechanical Powers. Machines, however complicated or complex, are largely composed of a few simple machines long known as the "mechanical powers." As usually given they are the Lever, Wheel and Axle, Inclined Plane, Screw, and Pulley.

81. Experiments with the Lever.
Experiment 3. Support a lever, as in Fig. 51, so that there shall be unequal arms. Move W until the lever is balanced in a horizontal position. Suspend (say) seven balls from the short arm (say) one space from the fulcrum. Then from the other arm suspend a single ball from such a place (in this case seven equal spaces from the fulcrum) that it will balance the seven balls. There is now equilibrium between the two forces. Suppose the smaller force to be increased a little and to produce motion, what mechanical advantage (*i.e.* intensity of force or speed) would be

EXPERIMENTS WITH THE LEVER. 83

gained by the use of the machine? What is the ratio of gain, the small additional force being neglected? How does this ratio compare with the ratio between the length of the two arms? For convenience we call the distance of the point of application of the force from the fulcrum the *force-arm*, and the distance of the resistance from the fulcrum the *resistance-arm*.

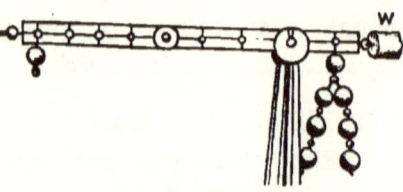

FIG. 51.

Suppose the small additional force to be applied to the short arm, what mechanical advantage would be gained? What would be the ratio of gain?

While the general law of machines (§ 78) is always applicable, its application is not always convenient, since, for example, it necessitates putting the machine in motion in order to measure s and s' (the distances traversed, respectively, by the points of application of the force and resistance in the same time), an operation which would be very difficult and tedious in many cases. Hence a *special law*, one in which the relation between the ratio of gain and the ratio between certain dimensions of the machine is stated, is often more convenient in practice. For example, in our experiment with the lever we discover that $r:f::$ force-arm : resistance-arm, *i.e. the force and resistance vary inversely as the lengths of their respective arms.* Compare this special law with the general law. Place the fulcrum at other points in the lever, and thereby vary the length of the arms, and verify by numerous experiments the special law of levers.

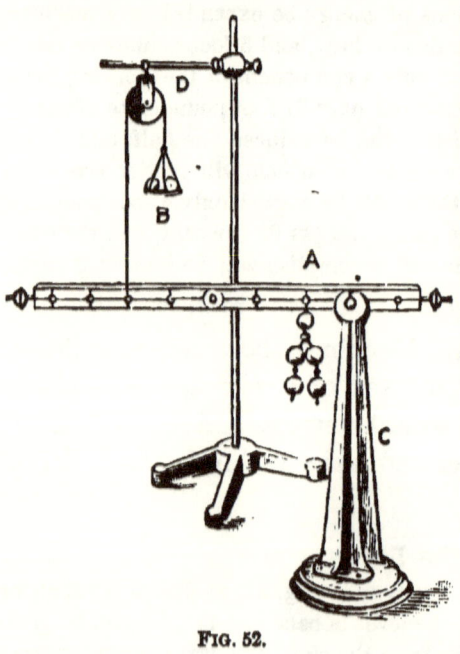

FIG. 52.

Experiment 4. By means of a pulley, D, so arrange that both f and r may be on the same side of the fulcrum (Fig. 52). First, place in the

84 MOLAR DYNAMICS.

pan weights sufficient to produce equilibrium in the machine (for example, in this case, one ball). Then suspend weights at some point, as A, and place other weights in the pan to counterbalance these. Verify the law of levers. If A be the resistance, what mechanical advantage is gained? What is the ratio of gain? If B be the resistance, what mechanical advantage is gained?

82. Wheel and Axle. The wheel and axle consists of two cylinders having a common axis, the larger of which is called the *wheel*, and the smaller the *axle*, as A and C (Fig. 53). The

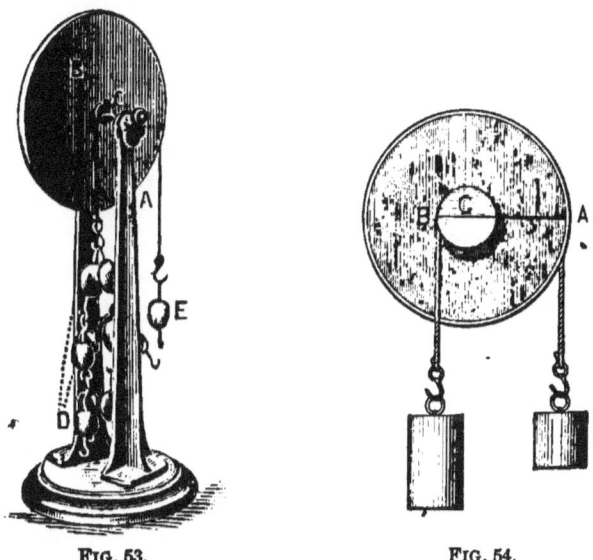

FIG. 53.   FIG. 54.

wheel may be moved by the hand or by a string with a weight attached to it. The wheel is often replaced by a crank, as in the windlass, or by a spoke, as in the capstan, and is thus employed in hoisting apparatus, such as cranes, derricks, etc.

The wheel and axle will be seen (Fig. 54) to be only a modification of the lever, which, unlike the latter, may be continuous in its operation. C is the fulcrum, the radius C A is the force-arm, and the radius C B the resistance-arm. The laws pertaining to this machine are virtually the same as those of the lever. For example, when the force f is applied to the wheel and the resistance r is at the axle,

$$f : r = \frac{1}{R \text{ (radius of wheel)}} : \frac{1}{R' \text{ (radius of axle)}}.$$

83. Inclined Plane. Any plane surface not horizontal or vertical, known as an *inclined plane*, may be used as a simple machine for gaining intensity of force; *e.g.* a plank resting with one end on a cart body and the other on the ground, a hillside, or a road-grade. The *gradient* is the quantity of rise per horizontal foot, or it is the ratio of the vertical rise to the horizontal distance.

When a body is pressed against a hard, smooth surface, the resistance offered by the surface is at right angles to the

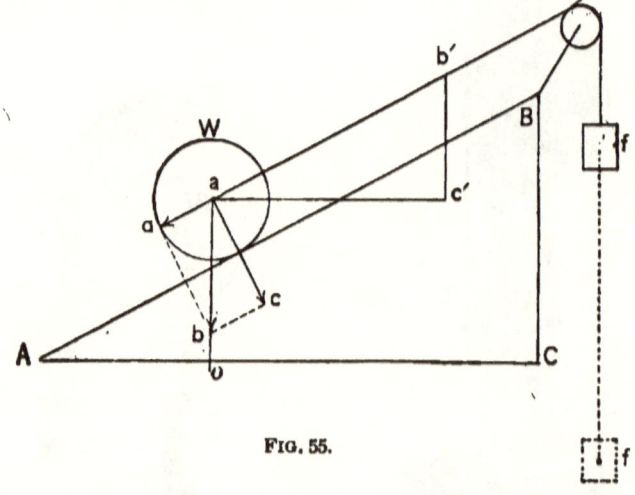

Fig. 55.

surface. A body, *e.g.* a sphere, may be supported on a horizontal surface, for the weight acting downward is counteracted by the upward reaction of the plane. But since on an inclined plane the reaction is not vertically upward, a body cannot rest on it without the aid of another force.

The mechanical advantage of this machine depends on the principle of the resolution of a force into its components. Let A B (Fig. 55) be an inclined plane whose gradient is $\frac{BC}{AC}$. Let a be the center of mass of the weight W (technically called the *load*). The line of direction of the load is along the vertical a o, but the pressure exerted upon the plane is in the direction a c, and the reaction of the plane is in the direction c a. We may take any length along the vertical as a b to represent the load W.

Draw b c parallel to A B to meet a c. Complete the parallelogram a d b c with a b as its diagonal. The force a b is thereby resolved into two forces, a c representing the pressure upon the plane, and a d representing an unbalanced force tending to move W along the plane B A. It is evident that to produce equilibrium, *i.e.* to support the body on the plane, A B, a force equal to a d, but opposite in direction, must be employed. Now it may be proved geometrically that the triangles a d b and B C A are similar ; *i.e.*

$$d\,a : a\,b :: C\,B : B\,A.$$

But d a represents the force f necessary to produce equilibrium, while a b represents the load or resistance r; C B represents the hight h, and B A the length l, of the inclined plane. Therefore,

$$f : r :: h : l, \text{ or } f = \frac{h}{l}r.$$

Hence, *a given force acting parallel to the direction of inclination of an inclined plane will support a weight as many times greater than itself as the length of the inclined plane is greater than its vertical hight.* Corollary : with a given length of inclined plane, the greater its vertical hight, *i.e.* the steeper it is, the greater f must be.

84. The screw is a variety of inclined plane, as may be shown by winding a triangular piece of paper around a cylinder, *e.g.* a lead pencil (Fig. 56). The hypotenuse will form a spiral about the cylinder resembling the threads of a screw.

In actual practice the screw consists of two parts : (1) a convex grooved cylinder, or *screw*, S (Fig. 57), which turns within (2) a hollow cylinder, or *nut*, N. The concave surface of the latter is cut with a thread corresponding to the thread of the screw. The force is employed either to turn the screw within an immovable nut, or to turn the nut

THE SCREW.

about a fixed screw. In either case the force is usually applied to a lever or wheel fitted either to the screw or to the nut.

During a single rotation of the screw or nut, the load or resistance is moved a distance equal to the vertical distance between the correspond-

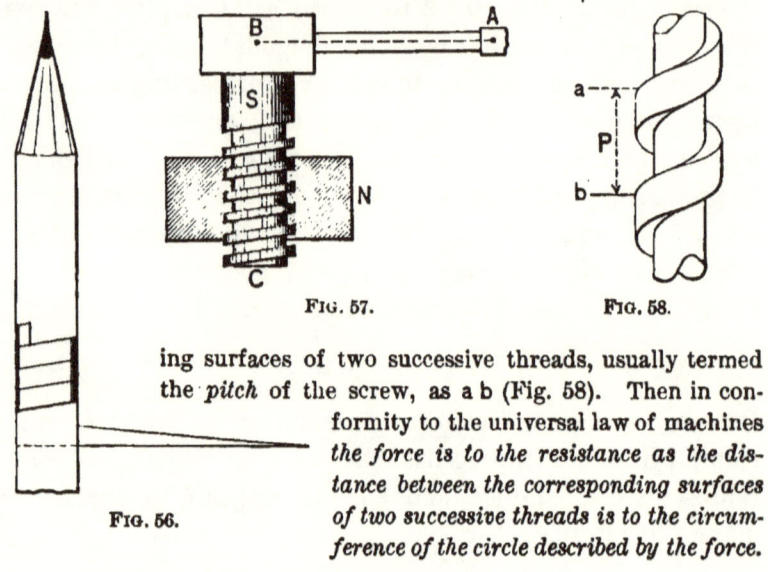

FIG. 56. FIG. 57. FIG. 58.

ing surfaces of two successive threads, usually termed the *pitch* of the screw, as a b (Fig. 58). Then in conformity to the universal law of machines *the force is to the resistance as the distance between the corresponding surfaces of two successive threads is to the circumference of the circle described by the force.*

EXERCISES.

1. (*a*) When is a machine said to gain intensity of force? (*b*) When is it said to gain speed?
2. (*a*) How is intensity of force gained by the use of a machine? (*b*) How is speed gained by the use of a machine?
3. (*a*) What is mechanical advantage? (*b*) Give a rule by which

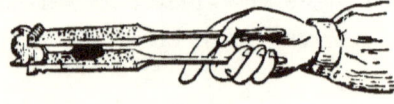

FIG. 59.

the mechanical advantage that may be gained by any machine may be calculated.

4. Energy is applied to a machine at the rate of 250 foot-pounds per minute, and it transmits 200 foot-pounds per minute. What is its efficiency?

5. (*a*) What advantage is gained by a nut-cracker (Fig. 59)? (*b*) What is the ratio of gain?

6. (*a*) What advantage is gained by cutting far back on the blades of shears near the fulcrum (Fig. 60)? Why? (*b*) Should shears for

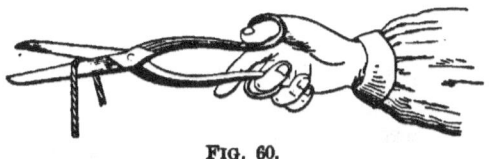

FIG. 60.

cutting metals be made with short handles and long blades, or the reverse? (*c*) What is the advantage of long blades?

7. The arm is raised by the contraction (shortening by muscular force) of the muscle A (Fig. 61), which is attached at one extremity to the shoulder and at the other extremity, B, to the fore-arm, near the elbow. (*a*) When the arm is used, as represented in the figure, to raise

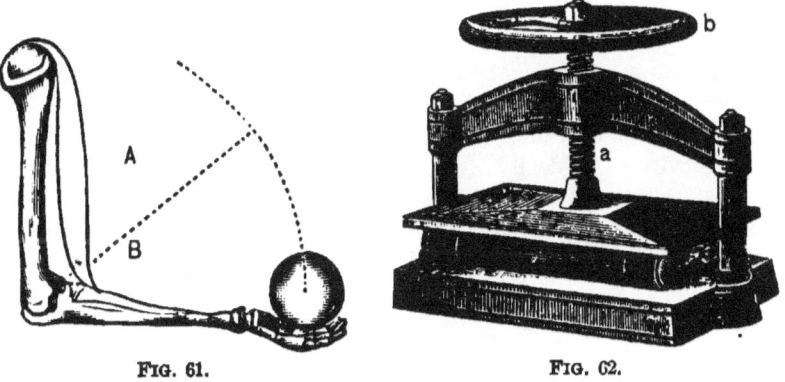

FIG. 61. FIG. 62.

a weight, what kind of machine is it? (*b*) What mechanical advantage is gained by it? (*c*) How can the ratio of gain be computed? (*d*) For which purpose is the arm adapted, to gain intensity of force or speed?

8. Is work done when the moment of the force applied to a lever is equal to the moment of the resistance? Why?

9. Suppose the screw in the letter-press (Fig. 62) to advance ¼ inch at each revolution, and a force of 25 pounds to be applied to the circumference of the wheel b, whose diameter is 14 inches; what pressure would be exerted on articles placed beneath the screw?

10. Two weights, of 5 k. and 20 k., are suspended from the ends of a lever 70 cm. long. (*a*) Where must the fulcrum be placed that they may balance? (*b*) What will be the pressure on the prop?

EXERCISES.

11. (*a*) A skid 12 feet long rests with one end on a cart at a hight of 3 feet from the ground. What force will roll a barrel of flour weighing 200 pounds over the skid into the cart? (*b*) What amount of work will be required?

12. (*a*) Draw a line to represent an inclined plane. Find what is the least force that will prevent a ball weighing 96 pounds from rolling down the plane. (*b*) Find the pressure which the ball will exert upon the plane.

13. An iron safe on trucks, weighing 2 tons, is prevented from rolling down an inclined plane by a force of 250 pounds. What is the ratio of the length of the plane to its hight?

14. If the circumference of an axle (Fig. 63) be 60 cm., and the point of application of the force applied to the crank travel 240 cm. during each revolution, what force will be necessary to raise a bucket of coal weighing 40 k.?

FIG. 63.

15. Through how many meters must the force act to raise the bucket from a cavity 10 m. deep?

16. The truck (Fig. 64) is a lever; the fulcrum is at the axis M of the wheels. A B represents the line of direction of the load, *i.e.* the direction in which the resistance acts; and C D represents the direction in which

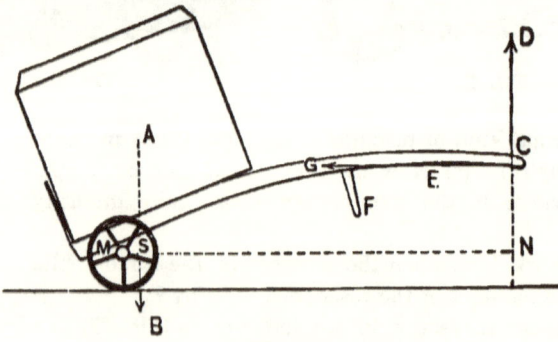

FIG. 64.

a force acts to produce equilibrium in the load in its present position. (*a*) What represents the force-arm? (*b*) What represents the resistance-arm? (*c*) The force required to support the load is what part of the load? (*d*) Would greater, or less, force be required if it were applied at

90 MOLAR DYNAMICS.

E instead of C? Why? (e) How may the load be supported without any force applied to the lever, the legs not touching the ground? (f) Would its equilibrium in this position be stable or unstable? Why? (g) Suppose the feet F to rest upon the ground, how would the pressure of the load be distributed between the feet and wheels? (h) Which is better suited for moving heavy burdens, a wheelbarrow or a truck? Why? (i) Suppose that C D represents the supporting force and C G the force employed in moving the load, how would the intensity and direction of the single force that accomplishes both results be found?

FIG. 65.

17. What must be the diameter of a wheel in order that a force of 20 pounds applied at its circumference may be in equilibrium with a resistance of 600 pounds applied to its axle, which is 3 inches in diameter?

18. (a) Where is the fulcrum in a claw-hammer (Fig. 65)? (b) What is the ratio of the mechanical advantage gained by means of it?

19. In its technical meaning, a "perpetual motion machine" is not a machine that will run indefinitely, but a machine which *can do work indefinitely without the expenditure of energy.* Show that such a machine is impossible.

In the arrangement shown in Fig. 66 three simple levers are combined so as to form one *compound lever.* The supporting force is applied at A; the resistance applied to this simple lever at B is identical with the force applied at A', and so on. Now

$$f = l \times \frac{\text{continuous product of resistance-arms}}{\text{continuous product of force-arms}}.$$

A combination of levers similar to this may be seen in scales used for weighing very heavy bodies, such as the so-called platform "hay-scales," in which a comparatively small weight counterbalances the heavy load.

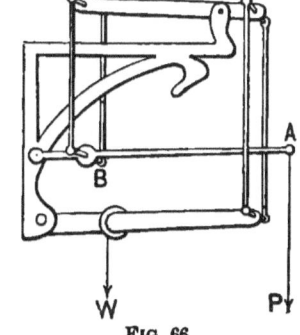

FIG. 66.

Fig. 67 represents a train of wheels in gear. A train of wheels being analogous to a compound lever, the mechanical advantage gained is obviously the ratio of the continued product of the radii of the wheels to the continued product of the radii of the axles.

20. Suppose the lengths of the arms of the several levers in Fig. 66 bear the following relations to each other: 5 : 1, 4 : 1, and 3 : 1; what force applied at A will support 180 pounds at W?

21. (*a*) In what sense may machines be "labor-saving"?
(*b*) In what sense is no machine "labor-saving"?

22. If the arms of the lever of a hydrostatic press be 3 feet and ½ foot, and the diameters of the plungers be 3 inches and 3 feet, a force of 50 pounds will produce what pressure?

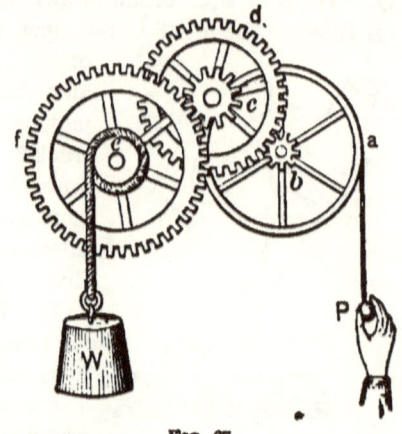

FIG. 67.

23. Suppose that the radii of the wheels a, d, and f (Fig. 67) are, respectively, 20 inches, 16 inches, and 24 inches, and the radii of their axles are, respectively, 2 inches, 4 inches, and 6 inches; how great advantage may be gained by the compound machine?

SECTION XIII.

SOME PROPERTIES OF MATTER DUE TO MOLECULAR FORCES.

85. Cohesion and Tenacity. According to the theory of the constitution of matter (§ 3) the molecules of every mass are in ceaseless motion, hitting and rebounding from one another. This tends to drive the molecules apart. In gaseous masses the molecules move without restraint; hence gaseous bodies always tend to expand.

In solids and liquids the molecules are held under the action of a very powerful attractive force, called *cohesion*, which prevents their separation except under the action of considerable external force. It is the force which resists an effort tending to break, tear, or crush a body. The *tenacity* or *tensile strength* of solids and liquids, *i.e.* the resistance which they offer to being pulled apart, is due to this force. It is usually greater in solids than in liquids, and is entirely wanting in a true gas.

86. Elasticity. *Elasticity is that property in virtue of which a solid tends to recover its size and shape, and a fluid its size, after deformation.* Solids are remarkable for high rigidity. A perfectly rigid solid is one which, when a force is applied to it in any way, suffers no strain before breaking. No body is absolutely rigid, though some bodies are approximately so. If the stress between the molecules in opposition to the distorting force continue constant, regardless of the time the strain is kept up, and restore the body to its normal condition immediately on the removal of the distorting force, without any permanent strain or "set," the body is said to be *perfectly elastic*. All fluids are perfectly elastic, and a few solids are approximately so, such as ivory, steel, and glass.

If a solid have little or no tendency to recover its size and shape after distortion, it is said to be *plastic* or *inelastic*. Such substances are putty, wet clay, and dough. A great number of substances are elastic when the distorting forces are small, but break or receive a "set" when these forces are too great. They are said to be elastic "within certain limits," called the *limits of elasticity*. If strained beyond those limits, they become more or less plastic. Hence, the springs of a buggy sometimes become set from bearing a too heavy load, and lose permanently much of their elasticity ; *i.e.* they become in a degree plastic.

87. Viscosity.
Experiment 1. Support in a horizontal position, by one of its extremities, a stick of sealing wax, and suspend from its free extremity an ounce weight, and let it remain in this condition several days. At the end of the time the stick will be found permanently bent. Should an attempt be made to bend the stick quickly, it will be found to be quite brittle.

The experiment shows that sealing wax possesses fluidity, freedom of motion of its molecules around one another, in a small degree. Resistance to deformation, due to the friction of the molecules of a body in sliding over one another, is called *viscosity*. Bodies that slowly suffer continuous and permanent deformation under the action of a continuous stress are said to be viscous. A lump of pitch in course of time loses its sharpness of outline and flows down hill of its own weight. It is very viscous. Cold molasses is quite viscous, but as its temperature is raised its viscosity diminishes and it becomes more and more plastic or mobile. A perfectly rigid solid is one of infinite viscosity. A *perfect fluid* is a fluid which possesses no viscosity. Gases are viscous to some extent and are therefore *imperfect fluids*.

Bodies surrounded by air have on their surfaces an adherent film of air. When they move, this film rubs against the surrounding air, and thus their movements are retarded by friction in the air. To the viscosity of the air is due in part the retardation of the velocity of falling bodies.

88. Hardness. *Hardness* is resistance to abrasion or scratching.

To enable us to express degrees of hardness, the following table of reference is generally adopted:

MOHR'S SCALE OF HARDNESS.

1. Talc.
2. Gypsum (or Rock-Salt).
3. Calcite.
4. Fluor-Spar.
5. Apatite.
6. Orthoclase (Feldspar).
7. Quartz.
8. Topaz.
9. Corundum.
10. Diamond.

By comparing a given substance with the substances in the table, its degree of hardness can be indicated approximately. Thus, "$H=7$" means that the body is about as hard as quartz.

89. Malleability; Ductility. Solids which possess that kind of fluidity which renders them susceptible of being rolled or hammered out into sheets are said to be *malleable*. Most metals are highly malleable. Gold may be hammered so thin as to be transparent, or to a thickness of one three-hundred-thousandth of an inch. Most substances that are malleable are also susceptible of being drawn out into fine threads, *e.g.* wire of different metals. Such substances are said to be *ductile*. Platinum has been drawn into wire .000165 inch thick, or so fine as to be scarcely visible to the unaided eye.

90. Cohesion in Liquids. Clean glass is wet by water. If a glass plate be dipped into water and then withdrawn, a layer of water clings to the glass. When the glass is withdrawn, water is torn from water, and not glass from water. This shows that the attraction of the molecules of water for one another is weaker than the attraction between glass and water. Or if, to save words, we call the attraction between the solid and the liquid *adhesion*,[1] then we may say that the *cohesion* between the molecules of the water is weaker than the *adhesion* between the glass and the water.

Clean glass is not wet by clean mercury, which shows that the adhesion between glass and mercury is not so great as the cohesion in mercury. Generally speaking, a solid is wet by a liquid when the adhesion of the solid to the liquid is greater than the cohesion of the liquid, and is not wet when the cohesion is greater than the adhesion.

91. Tension. When a rubber band is stretched, it is said to be in a state of *tension*, and there exists between its molecules a contractile or resilient stress which tends to restore the body to its normal condition. A rubber balloon inflated with compressed air is in a state of tension in every direction.

---

[1] There is no reason for supposing that adhesion is a different force from cohesion.

SURFACE TENSION.   95

92. Surface Tension and Surface Viscosity. Every liquid behaves as if a thin film forming its external layer were ever in a state of tension, or were exerting a constant effort to contract. This superficial film is tough or hard to break as compared with the interior mass. This property is called *surface viscosity*. It is not within the scope of this book to explain how the molecular forces produce this result; it must suffice to call attention to the peculiar condition, with reference to mutual attractions, of those molecules which compose the surface film. In the interior of a liquid body each molecule is surrounded by other similar molecules, and is acted upon equally in all directions. At a free surface the molecules can be acted upon only by others lying internal to them; the outcome of this condition is that it tends to reduce the free surface to the least possible area. This tendency of a liquid surface to contraction means that it acts like an elastic membrane, equally stretched in all directions, and by a constant tension.

Experiment 2. Form a soap-bubble at the orifice of the bowl of a tobacco-pipe, and then, removing the mouth from the pipe, observe that tension of the two surfaces (exterior and interior) of the bubble drives out the air from the interior and finally the bubble contracts to a flat sheet. The viscosity of a free surface of a solution of soap in water is greater than that of pure water; hence, its greater adaptability to the formation of bubbles.

As a consequence of surface tension, *every body of liquid tends to assume the spherical form*, since the sphere has less surface than any other form having equal volume. In bodies of large mass the distorting forces due to gravity are generally sufficient to disguise the effect; but in bodies of small mass, *e.g.* drops of liquids, and soap-bubbles, it is apparent.

93. Capillary Phenomena. If a glass rod be thrust vertically into water so as to leave a part projecting into the air, the surface of the water does not meet the rod at right angles, but is turned up so as to form a very small angle with the surface of the glass, as A C B (Fig. 68). Here the three substances, water, glass, and air, are brought in contact, and there are in operation a triplet of tensions (for the surfaces of all bodies tend to contract), the resultant of which is a force which pulls the water up against the glass wall. On the other hand, if mercury, glass, and air be brought in contact, the relation between the triplet of forces becomes

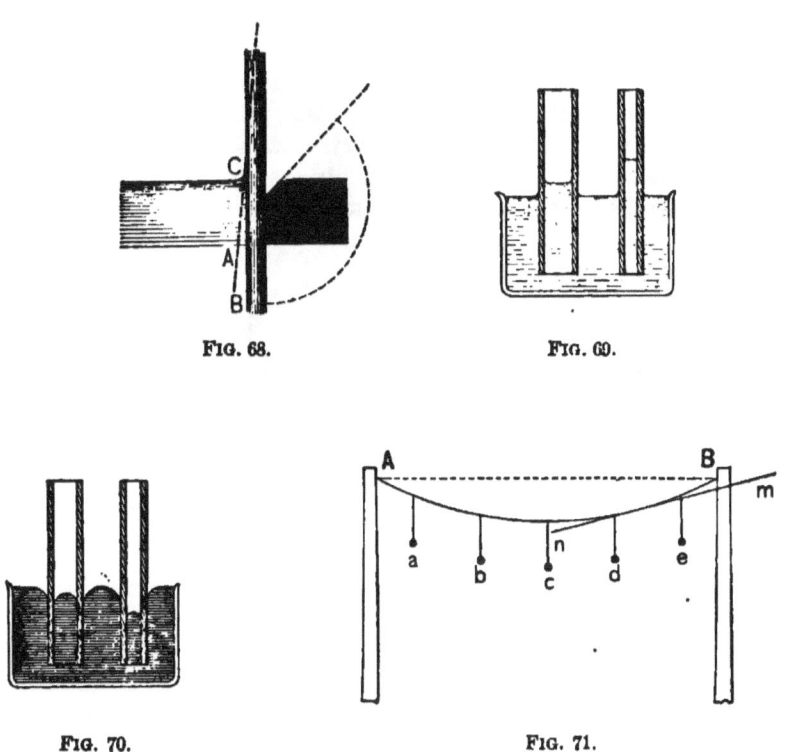

FIG. 68.   FIG. 69.

FIG. 70.   FIG. 71.

so changed as to cause the mercury to meet the glass at a very large angle, about 135°.

If glass tubes (Fig. 69) of capillary (hair-like) bore be thrust into water, the water will rise in the bores considerably above the general level outside. If similar tubes (Fig. 70) be thrust into mercury, the mercury within the bores will be depressed below the surface outside. Phenomena of this kind are called *capillary phenomena*. The surfaces of the liquids inside the bores are curved, the surface of water being concave and that of mercury convex. The size of the bores of the tubes is greatly exaggerated in order to show this more plainly. The concavity

CAPILLARY PHENOMENA. 97

and convexity of these interior surfaces are a necessary consequence of the angles of contact with which these liquids meet glass.

It remains to explain the elevation and depression of the column of liquid in the tube. This may be done in part by analogy. Let A B (Fig. 71) represent a clothesline suspended slackly between two posts. From this line hang by strings small stones, a, b, c, etc. If the hempen line become wet, as in a rain, it contracts and straightens, as shown by the dotted line A B. In other words, the contractile force which is exerted obliquely (*e.g.* n m, Fig. 71) is resolvable into two forces, one of which is horizontal and the other is vertically upward; the latter tends to elevate the stones. In a similar manner the curved surfaces of water and mercury tend to contract and become flat. In the case of the water surface (which is concave) the contractile force tends to elevate the pendent liquid; but in the case of the mercury surface (which is convex) the tendency is to produce depression. On the nature of the curvature depends the direction in which the contractile force acts on the pendent liquid. Now it is evident that water will be drawn up by this contractile force until the weight of the column balances this force; and mercury will be depressed until the force is balanced by the pressure of the mercury outside the tube. Capillary phenomena are, therefore, phenomena of surface tension. The phenomena of capillary action are well shown by placing various liquids in U-shaped glass tubes having one arm reduced to a capillary size, as A and B in Fig. 72. Mercury poured into A assumes convex surfaces in both arms, but does not rise as high in the small arm as it stands in the large arm. Pour water into B, and all the phenomena are reversed. Fig. 73 shows the forms that the surfaces of water and mercury take when contained in the same glass tube.

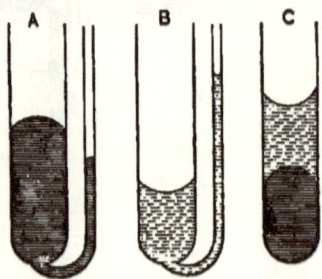

FIG. 72.   FIG. 73.

94. Laws of Capillary Action. The following laws may be verified by experiment:

I. *Liquids rise in tubes when they wet them, and are depressed when they do not.*

II. *The elevation or depression varies inversely as the diameter of the bore.*

CHAPTER III.

DYNAMICS OF FLUIDS.

SECTION I.

TRANSMISSION OF PRESSURE.

95. Law of Hydrostatic and Pneumatic Transmission of Pressure. That branch of science which treats of liquids in a state of equilibrium or rest is called *hydrostatics;* that branch which treats of liquids in motion is called *hydrokinetics;* and that branch which treats of the dynamics of air and other gases is called *pneumatics.* With the exception of phenomena occasioned by difference in compressibility and expansibility, liquids and gases are subject to the same laws and may be treated together, in so far as they are alike, under the common term *fluid.*

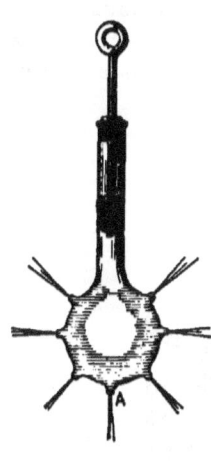

FIG. 74.

Experiment 1. Fill the glass globe and cylinder (Fig. 74) with water, and thrust the piston into the cylinder. Jets of water will be thrown not only from that aperture, A, in the globe toward which the piston moves and the pressure is exerted, but from all the apertures.

It thus appears not only that external pressure is exerted upon that portion of the liquid that lies in the path of the force, but that it is transmitted equally to all parts and in all directions.

When pressure is exerted upon a solid, on account of its rigidity it is incapable of transmitting the pressure in other directions than that in which it is pressed. But fluids, on

TRANSMISSION OF PRESSURE.

account of the mobility of their molecules, are incapable of resisting a change of shape when acted upon at any point by a force, and hence any force applied to a fluid body must be transmitted by the fluid in every direction. Consequently, every portion of the interior walls of the containing vessel with which the fluid is in contact is subjected to pressure.

A pressure exerted on a fluid enclosed in a vessel is transmitted undiminished to every part of that vessel; and the total pressure exerted on the interior of the vessel is equal to the area of the interior multiplied by the pressure per unit of area.

The pressure exerted by a fluid upon the vessel containing it is normal to the walls of the vessel. Fluid pressure is expressed by stating the force exerted on a *unit area,* as 2 lbs. per sq. in., 5 g. per cm.², etc.

Experiment 2. Fig. 75 represents a section of an apparatus called (from the number of uses to which it may be put) the *seven-in-one apparatus.* A is a hollow cylinder closed at one end. B is a tightly

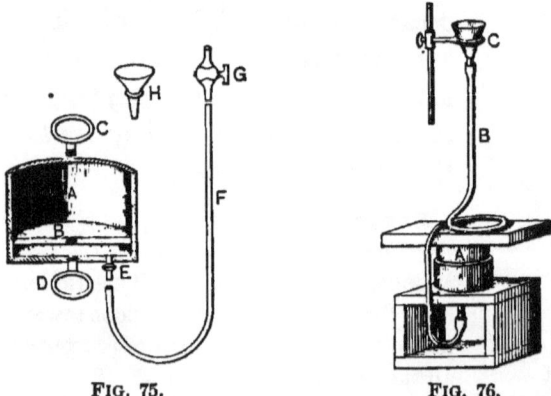

FIG. 75. FIG. 76.

fitting piston which may be pushed into or drawn out of the cylinder by the detachable handle C when screwed into the piston. D is another handle permanently connected with the closed end of the cylinder. E is a nipple, opening into the space below the piston. To this may be attached a thick-walled rubber tube F. G is a stop-cock, and H is a funnel, either of which may be inserted at will into the free end of the tube.

Support the seven-in-one apparatus with the open end upward, force the piston in, place on it a block of wood, A (Fig. 76), and on the block a heavy weight. Attach one end of the rubber tube B (12 feet long) to the apparatus, and insert a funnel, C, in the other end of the tube. Raise the latter end as high as practicable, and pour water into the tube. Explain how the few ounces of water standing in the tube can exert a pressure of many pounds on the piston, and cause it to rise together with the burden that is on it.

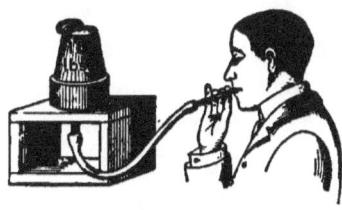

FIG. 77.

Experiment 3. Remove the water from the apparatus, place on the piston a 16-pound weight, and blow (Fig. 77) from the lungs into the apparatus. Notwithstanding that the actual pushing force exerted through the tube by the lungs probably does not exceed a few ounces, the slight increase of pressure caused thereby, when exerted upon the (about) 26 square inches of surface of the piston, causes it to rise together with its burden.

96. The Hydrostatic Press. Closely allied to the seven-in-one apparatus is the *hydrostatic press*. Water drawn from

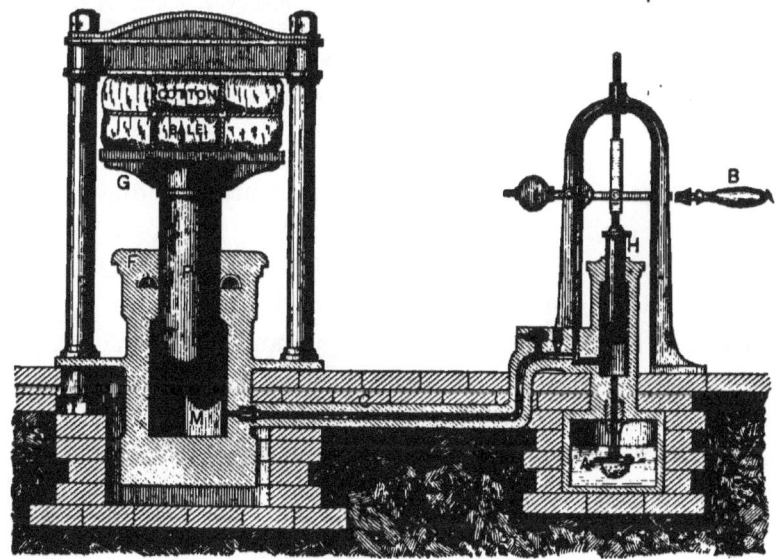

FIG. 78.

a reservoir, A (Fig. 78), by a suction and force-pump worked by a lever, B, is forced along the tube C into the cylinder

M. This cylinder contains a plunger, P, which works watertight in the collar F. The plunger carries a plate, G, upon which are placed objects to be pressed. The water forced into the cylinder exerts upon the plunger a total upward pressure which is as many times greater than the downward pressure exerted upon the liquid through the plunger H as the area of the cross section of the plunger P is times greater than the area of the cross section of the plunger H. To obtain the entire theoretical gain of force that may be obtained by this machine the ratio of the cross sections of the plungers is multiplied by the ratio of the two arms of the lever B.

The pressure that may be exerted by these presses is enormous. The hand of a child can break a strong iron bar. But observe that, although the pressure exerted is very great, the upward movement of the plunger P is very slow. In order that the plunger P may rise 1 cm., the plunger H must descend as many centimeters as the area of the cross section of P is times the area of the cross section of H. The disadvantage arising from slowness of operation is insignificant, however, when we consider the great advantage accruing from the fact that one man can produce as great a pressure with the press as several hundred men can exert without it. The modern engineer finds it a most efficient machine whenever great resistances are to be moved through short distances.

97. Pascal's Principle. Fluids exert pressure due to their weight. Imagine a vessel filled with shot; the upper layer of shot will press upon the layer next beneath with a force equal to its weight, the second upon the third with a force equal to the sum of the weights of the first two, and so on. You therefore conclude that the pressure exerted upon the successive layers will be exactly *proportional to their depths*. In like manner and for the same reason *the pressure at different points in a liquid is proportional to the depth.*

Since shot possess a certain degree of mobility or freedom of motion around one another, their weight will cause to some extent a lateral pressure against one another and against the walls of the containing vessel. In consequence of the extreme mobility of the molecules of fluids the downward pressure due to gravitation at any point in a fluid gives rise to an equal pressure at that point in all directions. Hence the so-called *Pascal's principle: At any point in a fluid at rest the pressure is equal in all directions.*

Thus, let a, b, c, etc. (Fig. 79), represent imaginary surfaces, and the arrow-heads the direction of pressûre exerted at points in these surfaces at equal depths in a liquid. The pressures exerted at these several points are equal.

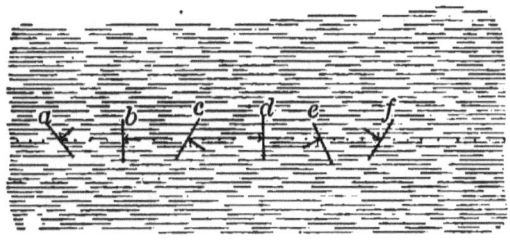

FIG. 79.

The truth of this principle is obvious, for if there be any inequality of pressure at any point, the unbalanced force will cause particles at that point to move, which is contrary to the supposition that the fluid is *at rest*. Conversely, when there is motion in a body of fluid it is evidence of an inequality of pressure.

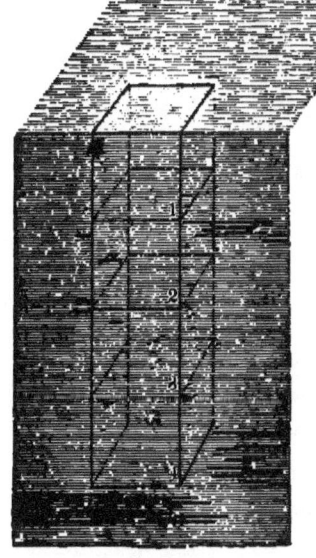

FIG. 80.

98. Methods of Calculating Liquid Pressure. Conceive of a square prism of water (Fig. 80) in the midst of a body of water, its upper surface coinciding with the free surface of the liquid. Let the prism be 4 cm. deep and 1 cm. square at the end; then the area of one of its ends is 1 cm.^{2}, and the volume of the prism is 4 cc. Now the weight of 4 cc. of water is 4 g., and hence this prism must exert a downward pressure of 4 g. upon an area of 1cm.^{2} But at the same depth the pressure in all directions is the same; hence, generally, the pressure at any depth in water may be taken approximately

RULES FOR CALCULATING LIQUID PRESSURE. 103

as one gram per square centimeter for each centimeter of depth (= 1,000 k. per m.² for each meter of depth; or, since the weight of water is about 62.3 pounds per cubic foot, the pressure is 62.3 pounds per square foot for each foot of depth). To determine the pressure at any given depth in any other liquid, the water pressure at the given depth must be multiplied by the specific density (see Appendix) of the liquid.

99. Rules for Calculating Liquid Pressure against the Bottom and Sides of a Containing Vessel. *The total pressure due to gravity on any portion of the horizontal bottom of a vessel containing a liquid is equal to the weight of a column of the same liquid whose base is the area of that portion of the bottom pressed upon, and whose hight is the depth of the water in the vessel.*

Evidently, the lateral pressure at any point of the side of a vessel depends upon the depth of that point; and, as depth at different points of a side varies, *to find the total pressure upon any portion of a side of a vessel, find the weight of a column of liquid whose base is the area of that portion of the side, and whose hight is the average depth of that portion.*

100. The Surface of a Liquid at Rest is Level. By jolting a vessel, the surface of a liquid in it may be made to assume the form seen in Fig. 81. Can it retain this form? Take two molecules of the liquid at the points a and b, on the same level. The total downward pressures upon a and b are in the ratio of their respective depths, c a and d b. But since the pressure at a given depth is equal in all directions, c a and d b represent the lateral pressures at the points a and b, respectively. But d b is greater than c a; hence, the molecules a and b, and those lying in a straight line between them, are acted upon by two unequal forces in opposite directions. Hence, the liquid cannot remain at rest in

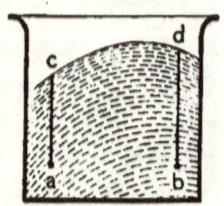

FIG. 81.

the position assumed, and there will, therefore, be a movement of molecules in the direction of the greater force, toward a, till there is equilibrium of forces, which will occur only when the points a and b are equally distant from the surface; or, in other words, *there will be no rest till all points in the surface are on the same level.*

This fact is commonly expressed thus: "Water seeks its lowest level." In accordance with this principle, water flows down an inclined plane, and will not remain heaped up. An illustration of the application of

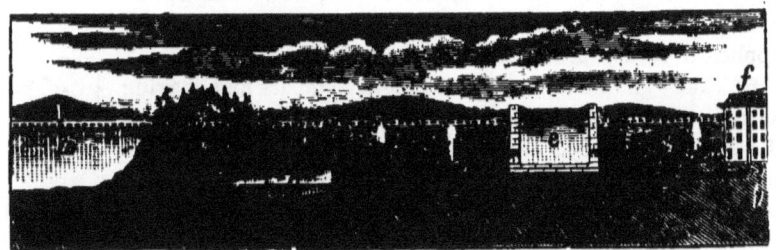

FIG. 82.

this principle, on a large scale, is found in the method of supplying cities with water. Fig. 82 represents a modern aqueduct, through which water is conveyed from an elevated pond or river, a, beneath a river, b, over a hill, c, through a valley, d, to a reservoir, e, from which water is distributed by service-pipes to the dwellings, f, in a city.

EXERCISES.

1. The areas of the bottoms of vessels A, B, C, and D (Fig. 83) are equal. The vessels have the same depth, and are filled with water. (*a*) Which vessel contains the most water? (*b*) On the bottom of which vessel is the pressure equal to the weight of the water which it contains? (*c*) How does the pressure upon the bottoms of vessels A, B, and C compare with the weight of the water in them, respectively?

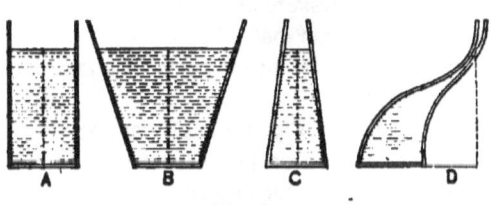

FIG. 83.

2. A cubic foot of water weighs about 62.3 pounds. Suppose that the area of the bottom of each vessel is 100 square inches, and the depth is 14 inches, what is the pressure on the bottom of each?

3. The bottom of vessel A is square. What is the total pressure against one of its vertical sides?

4. Let A (Fig. 84) be a cubical tank whose inside dimension is 20 inches. Leading from its side is a tube whose top is 80 inches above the interior top surface of the tank. (a) What mass of water will the tank (not including the tube) contain? (b) Suppose the tank and tube to be filled with water, what pressure will be exerted upon the bottom of the tank? (c) What upon the top of the tank? (d) What upon one of its sides?

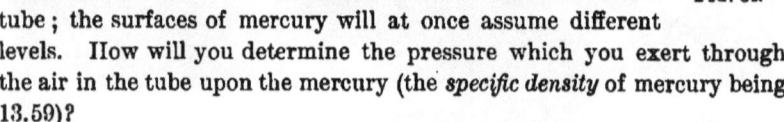

5. Suppose that the area of the end of the large piston of of a hydrostatic press is 100 square inches, what must be the area of the end of the small piston that a force of 100 pounds applied to it may produce a pressure of 2 tons?

6. Take a glass U-tube (Fig. 85) about 40 inches high, having a stout rubber tube, a, attached, and containing mercury with the surfaces at the same level in both arms. Blow into the tube; the surfaces of mercury will at once assume different levels. How will you determine the pressure which you exert through the air in the tube upon the mercury (the *specific density* of mercury being 13.59)?

FIG. 84.

7. (a) Suck air from a. What happens to the mercury? (b) How may you determine the diminution of pressure which you produce by suction?

8. Take a similar tube containing water instead of mercury, connect it with a gas jet and turn on the gas. How would you determine how much greater (or less) its pressure is than that of the atmosphere?

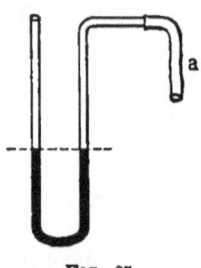

9. How great is the hydrostatic pressure in fresh water at the depth 50 feet?

10. (a) A house is supplied with water by a system of pipes from a distant reservoir, as is customary in cities. What data should you require in order to compute the pressure at any point in the pipe? (b) How much greater is the pressure at a point in the pipe in the cellar than at another point in the attic? (c) Is the pressure in the pipe the same when water is running from a faucet in the house as when the water is at rest?

FIG. 85.

11. Suppose that the aqueduct at point b (Fig. 82) is 150 feet below the level of the pond, what is the pressure at this point (expressed in tons per square foot) tending to burst the walls of the aqueduct?

106 MOLAR DYNAMICS.

SECTION II.

ATMOSPHERIC PRESSURE.

101. Introduction. We live at the bottom of an exceedingly rare and elastic ocean of air, called the *atmosphere*. Every molecule in this gaseous ocean is drawn towards the earth's center by gravitation, and the atmosphere is thus

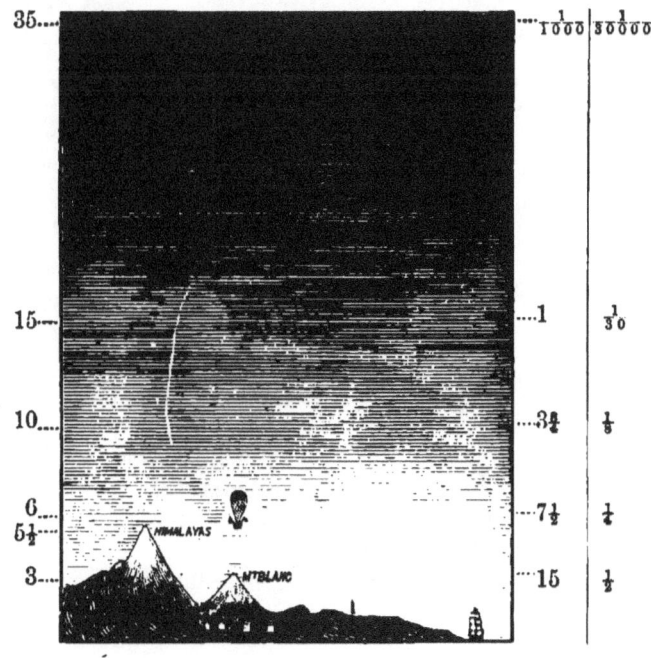

FIG. 80.

bound to the earth by this force, just as is the liquid ocean. Evidently the pressure in the atmosphere due to its weight increases with the depth; or, since in our position we are more accustomed to speak of *hight* in the atmosphere, decreases with the hight. The pressure does not diminish regularly with the hight as in an ocean of incompressible

ATMOSPHERIC PRESSURE.

fluid. Air is very compressible; the lower strata of the atmosphere, which sustain the weight of the upper strata, are much compressed, and are therefore relatively very dense. The density of the air diminishes more rapidly than the hight above sea level increases. Owing to this fact, the greater part of the mass of the atmosphere is within three and a half miles of the sea level. Above this hight the air is much rarified and vanishes, as it were, very gradually into airless space.[1]

FIG. 87.

Experiment 1. Fill (or partly fill) a tumbler with water, cover the top closely with a card or writing-paper, hold the paper in place with the palm of the hand, and quickly invert the tumbler (Fig. 87). How is the water supported?

FIG. 88.

Experiment 2. Force the piston A (Fig. 88) of the seven-in-one apparatus quite to the closed end of the hollow cylinder, and close the stop-cock B. Try to pull the piston out again. Why do you not succeed? Hold the apparatus in various positions, so that the atmosphere may press down, laterally, and up, against the piston. You discover no difference in the pressure which it receives from different directions.

Atmospheric pressure is exerted all over the outside of a body, so that usually it is not noticeable. But if, as in the case of the piston, pressure be removed from one side of a body, the other side experiences the whole unbalanced pressure.

[1] The shading in Fig. 86 is intended to indicate roughly the variation in the density of the air at different elevations above sea level. The figures in the left margin show the hight in miles; those in the first column on the right, the corresponding average hight of the mercurial column in inches; and those in the extreme right, the density of the air compared with its density at sea level. If the aerial ocean were of a density equal to that at sea level, its hight would be a little less than 5 miles. Only the highest peaks of the Himalaya Mountains would rise above it. If an opening were made in the earth 35 miles in depth below sea level, the density of the air at the bottom would be greater than that of water.

102. How Atmospheric Pressure is Measured.

Experiment 3 (preliminary). Take a U-shaped glass tube (Fig. 89), half fill it with water, close one end with a thumb, and tilt the tube so that the water will run into the closed arm and fill it; then restore it to its original vertical position. Why does not the water settle to the same level in both arms?

FIG. 89.

Let Fig. 90 represent a U-shaped glass tube closed at one end, about 34 inches in hight, and having a bore of 1 square-inch section. The closed arm having been filled with mercury, the tube is placed with its open end upward, as in the cut. The mercury in the closed arm sinks about 2 inches to A and rises 2 inches in the open arm to C; but the surface A is 30 inches higher than the surface C. This can be accounted for only by the atmospheric pressure. The column of mercury B A, containing 30 cubic inches, is an exact counterpoise for a column of air of the same diameter extending from C to the upper limit of the atmospheric ocean, — an unknown hight.

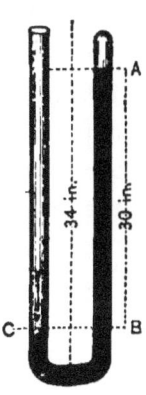

FIG. 90.

The weight of the 30 cubic inches of mercury in the column B A is 14.7 pounds. Hence the weight of a column of air of 1 square-inch section, extending from the surface of the sea to the upper limit of the atmosphere, is about 14.7 pounds. But in fluids gravitation causes equal pressure in all directions. Hence, *at the level of the sea, all bodies are pressed upon in all directions by the atmosphere with a force of about* 14.7 *pounds per square inch*, or about one ton per square foot.

103. Standard Pressure.
Many physical operations require a standard pressure for reference. The standard generally

adopted is the pressure exerted by a column of pure mercury at 0° C. and 76 cm. (29.922 inches) high, which is about the average hight of the barometric column at sea level in latitude 45°. The pressure corresponding to this hight is 1033.3 grams per square centimeter, or 14.7 pounds per square inch.

A pressure of 14.7 pounds per square inch is quite generally adopted by engineers as a unit of pressure, and is called an *atmosphere*. Physicists, however, generally measure gaseous pressure in terms of *mm.* of mercury at 0° C.; that is, the hight in *mm.* of mercury that the given pressure would sustain in the vacuum tube.

104. The Barometer. The hight of the column of mercury supported by atmospheric pressure is quite independent of the area of the surface of the mercury pressed upon; hence, the apparatus is more conveniently constructed in the form represented in Fig. 91.

A straight tube about 34 inches long is closed at one end and filled with mercury. The tube is inverted, with its open end tightly covered with a finger, and this end is inserted into a vessel of mercury. When the finger is withdrawn, the mercury sinks until there is equilibrium between the downward pressure of the mercurial column A B and the pressure of the atmosphere. The empty space at the top of the tube is called a *Torricellian*[1] *vacuum*. An apparatus designed to measure atmospheric pressure is called a *barometer* (pressure-measurer).

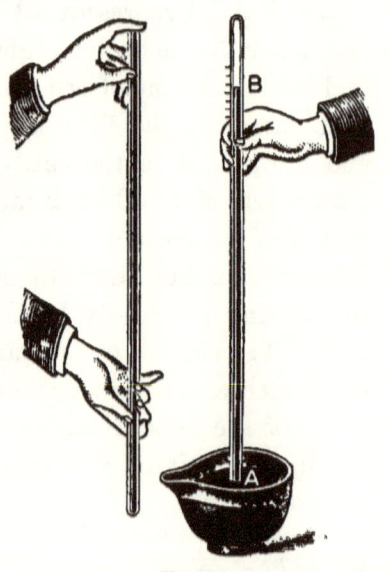

FIG. 91.

---

[1] The first barometer was constructed by Torricelli, a Florentine, in 1643.

A common and inexpensive form of barometer is represented in Fig. 92. Beside the tube and near its top is a scale, graduated in inches or centimeters, indicating the hight of the mercurial column. For ordinary purposes this scale needs to have a range of only three or four inches, so as to include the maximum fluctuations of the column.[1]

Fluctuations in barometric pressure are of hourly occurrence. Some of the many conditions which influence atmospheric pressure are *changes in temperature, humidity of the air, and currents in the atmospheric ocean.*[2]

105. The Fortin Barometer. We will suppose the scale of a barometer to be fixed so as to indicate correctly the hight of the surface of mercury in the tube above that in the cistern at a time, for instance, when this distance is 30 inches. A point on the surface of the mercury in the cistern in this case is called technically the *zero point*. Now, should the mercury in the tube fall to 29 and the mercury in the cistern remain at zero, then the scale reading would indicate correctly the barometric hight. But the mercury does not remain at zero, but rises a little (less as the diameter of the cistern is greater); consequently, the scale reading is too great. When the mercury in the tube is higher than 30 all the readings will be too small. Evidently, then, the mercury in the cistern must be brought to zero at every observation in order to eliminate this error. This is easily accomplished in the Fortin barometer. The bottom of the cistern of this barometer is of pliable leather, resting on

FIG. 92.

---

[1] At the Central Station in Boston, Feb. 8, 1895, the mercury fell to 28.01 in., the lowest on record at this station. The highest point ever recorded at this station was 30.97 in., on Dec. 1, 1887.

"A variation of two inches in barometric hight is attended by a corresponding variation in the hight of tides at the place of about three feet." — G. H. DARWIN.

[2] The barometer is sometimes called a "weather-glass," chiefly because its scale frequently bears the words *fair, rainy, storm,* etc. These words are very objectionable, since they are totally misleading from a meteorological point of view. To form a forecast of the weather of much value, a barometer, a thermometer, and a hygrometer must be consulted, and one must be familiar with the laws which govern the relations between atmospheric pressure, temperature, moisture, etc.

a thumb-screw, A (Fig. 93). Projecting from the tube inside of the cistern is a little pointer, B, of colored glass. The lower end of this pointer, called the *fiducial point*, corresponds to the zero point. The level of the mercury in the cistern must be set to this point by raising or lowering the cistern base by the adjusting screw, before taking a reading. A sliding piece, C (Fig. 94), furnished with a vernier, can be slid along the tube so as to enable one to read with great accuracy.

In refined scientific researches it is necessary to make suitable allowances for expansion and contraction of the mercury attending changes in temperature, hence a very sensitive thermometer is attached to the barometer.

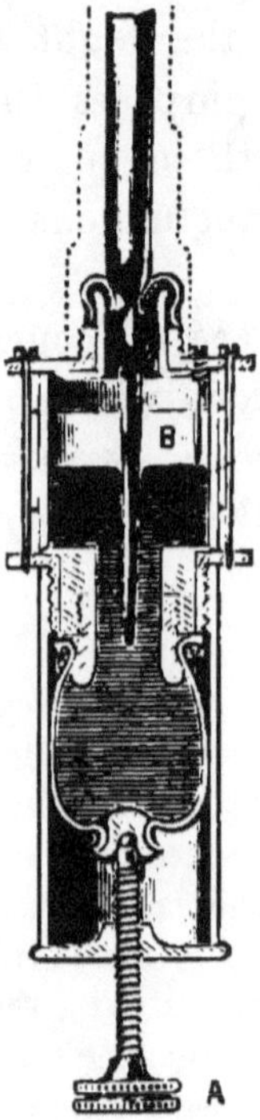

106. Barometric Measurement of Hights. Since atmospheric pressure varies with the hight above sea level, it is evident that changes in elevation may be determined from changes of pressure as indicated by the barometer. For example, the hight of a mountain may be ascertained from barometric readings made at the same time on the summit and at sea level. For moderate hights the barometric column falls at a nearly uniform rate of one inch for every 900 feet of ascent.

FIG. 93.

FIG. 94.

EXERCISES.

1. The average barometric hight in the city of Denver is 24.5 in. What is the average atmospheric pressure in this city, expressed in pounds per square inch?

2. When the barometric column stands at 748 mm., what is the atmospheric pressure, expressed in grams per cm.^{2}?

3. What is the hight of the barometric column when the pressure of the air is 1045 g. per cm.^{2}?

4. Suppose that, on a day when the pressure of the air is 756 mm., air is exhausted from the receiver of an air-pump until the mercury in a barometer placed in the receiver stands at 30 mm., what per cent of the air has been removed?

SECTION III.

RELATION BETWEEN THE DENSITY, VOLUME, AND PRESSURE OF A BODY OF GAS.

107. Elasticity of Gases. *The elasticity of all fluids is perfect.* By this is meant that the force exerted in expansion is equal to the force used in compression; and that, however much a fluid is compressed, it will always completely regain its former volume when the pressure is removed. Hence the barometer, which measures the compressing force of the atmosphere, also measures at the same time the elastic force of the air. A so-called *vacuum gauge* (Fig. 95) is simply a short mercury barometer, — short because it is seldom required to make measurements except in tolerably high vacua, where the mercurial column is correspondingly low. For instance, this apparatus, placed under the receiver of an air-pump from which air is exhausted, will measure the elastic force of the air in the receiver. This known, the degree of exhaustion is readily determined.

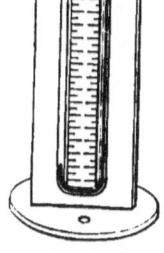

FIG. 95.

108. Boyle's (or Mariotte's) Law.

Experiment. Take a bent glass tube (Fig. 96), the short arm being closed, and the long arm, which should be at least 34 inches (85 cm.) long, being open at the top. Pour mercury into the tube till the surfaces in the two arms are at the same level A B. The body of air to be experimented with is in the short arm between A and C. The dimensions of this body can vary only in hight; hence its hight, H, may represent its volume. Measure H (*i.e.* the distance between A and C) and regard the number of inches (or centimeters) as representing the volume, V. Its pressure, P, evidently

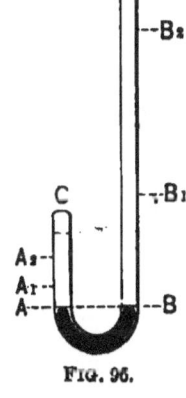

FIG. 96.

is the same as that of the atmosphere at the time. Consult a barometer, and ascertain the hight of the barometric column; represent this hight by P. Pour a little mercury into the tube; the mercury rises (say) to A_1 and B_1. Measure the vertical distance between A_1 and C; this number represents the volume, V_1, of the body of air now. Measure the vertical distance between A_1 and B_1; this number represents the increase in pressure, which, added to P, gives its present pressure, P_1.

Now pour more mercury into the tube, so that it will rise to (say) A_2 and B_2. Determine as before the new volume, V_2, and the new pressure, P_2. So continue to add mercury a third and a fourth time, and get new values for the volume, V_3 and V_4, and for the pressure, P_3 and P_4. Arrange the results as follows:

$V = \ldots$ $P = \ldots$ $V \times P = \ldots$
$V_1 = \ldots$ $P_1 = \ldots$ $V_1 \times P_1 = \ldots$
$V_2 = \ldots$ $P_2 = \ldots$ $V_2 \times P_2 = \ldots$
   etc.   etc.

It will be found that the series of products in the last column are approximately equal (due allowance being made for errors in measurement, etc.); consequently, the product of the volume of a body of gas multiplied by its pressure is constant, and the volume varies inversely as its pressure. Hence the (Boyle's) law:

The volume of a body of gas at a constant temperature varies inversely as its pressure, density, and elasticity.[1]

EXERCISES.

1. If the volume of a certain body of gas be 500 cc. when its pressure is 800 g. per cm.^{2}, what is the volume of the same body when its pressure is 1200 g. per cm.^{2}.

2. If a body of air whose volume is 1 m.^{3} and pressure is 760 mm. expands and occupies 4.5 m.^{3}, what will be its pressure?

3. A bubble of air liberated at a depth of 2 meters in water has a volume of 3 cm.^{3}. What will be its volume when it has risen 1 meter?

---

[1] For many years after the announcement of this law it was believed to be rigorously correct for all gases; but more recently more precise experiments have shown that it is approximately but not rigidly true for any gas. There is a limit beyond which this law does not hold. This limit is soonest reached with those gases, like carbon-dioxide, chlorine, etc., that are most readily liquefied. A gas conforms more nearly to Boyle's law, in proportion as it is farther, as regards both pressure and temperature, from its liquefying point. As a gas approaches this point its density increases more rapidly than its elasticity.

114 MOLAR DYNAMICS.

4. Air is rarefied in the receiver of an air-pump so that the difference in level of the two surfaces of mercury in the vacuum gauge is .2 mm. What is the elastic force of the remaining air, expressed in grams per cm.^{2}?

5. A mass of air occupies 80 cc. when the pressure is 100 mm. What space will it occupy when the pressure is 150 mm.?

6. A mass of air occupies 160 cc. when the pressure is 760 mm. What must be the pressure that it shall occupy only 50 cc.?

SECTION IV.

PUMPS AND SIPHONS.

109. The Air-pump. The air-pump is used to rarefy air in a closed vessel. Fig. 97 will illustrate its operation. R is a glass *receiver* within which the air is to be rarefied; B is a hollow cylinder of brass, called the *pump-barrel;* the plug P, called a *piston*, is fitted to the interior of the barrel, and can be moved up and down by the handle H; s and t are valves. A valve acts on the principle of a door intended to open

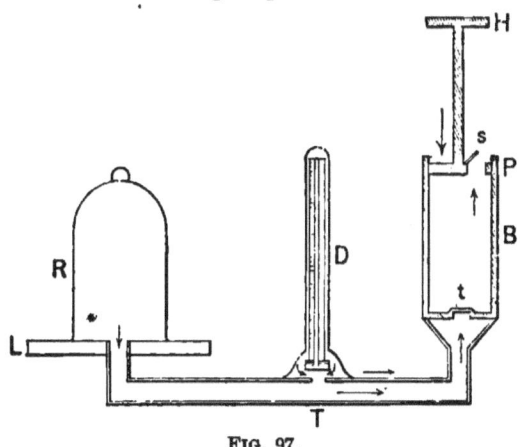

FIG. 97.

or close a passage. If you walk against a door on one side, it opens and allows you to pass; but if you walk against it on the other side, it closes the passage and stops your progress.

Suppose the piston to be in the act of descending; the compression of the air in B closes the valve t, and opens the valve s, and the enclosed air escapes. After the piston reaches the bottom of the barrel, it begins its ascent. This would cause a vacuum between the bottom of the barrel and the ascending piston (since the unbalanced pressure of the outside air immediately closes the valve s), but the pressure of the air in the receiver R opens the valve t and fills this space. As the air in R expands, it becomes rarefied and exerts less pressure. The external pressure of the air on R, being no longer balanced by the pressure of the air within, presses the receiver firmly upon the plate L. Each repetition of a double

LIFTING-PUMP FOR LIQUIDS. 115

stroke of the piston results in the removal of a portion of the air remaining in R. The air is removed from R by its own expansion.

However far the process of exhaustion may be carried, the receiver will always be filled with air, although it may be exceedingly rarefied. The operation of exhaustion is practically ended when the pressure of the air in R becomes too feeble to lift the valve t, unless the apparatus be so constructed that the valves are opened and closed by mechanical action. It is obvious that if s and t opened downward instead of upward, then as

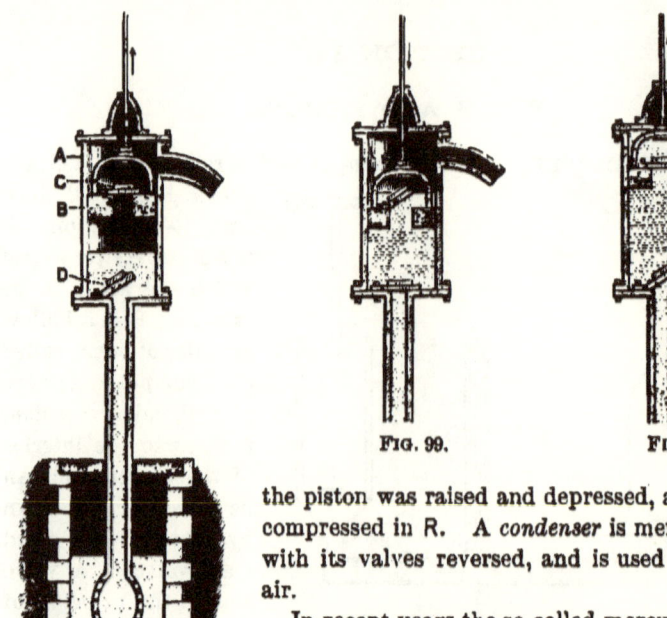

FIG. 99. FIG. 100.

FIG. 98.

the piston was raised and depressed, air would be compressed in R. A *condenser* is merely a pump with its valves reversed, and is used to condense air.

In recent years the so-called mercury air-pump has largely displaced the pump described above, since it is capable of producing a much greater rarefaction. In brief, it makes use of the Torricellian vacuum, such as is formed in the top of a barometer tube.

110. Lifting-pump for Liquids. The common *lifting-pump* is constructed like the barrel of an air-pump. Fig. 98 represents the piston B in the act of rising. As the air is rarefied below it, water rises in consequence of atmospheric pressure on the water in the well, and opens the lower valve D. Atmospheric pressure closes the upper valve C in the piston. When the piston is pressed down (Fig. 99), the lower valve closes, the upper valve opens, and the water between the bottom of the barrel and the piston passes through the upper valve above the piston.

When the piston is raised again (Fig. 100), the water above the piston is raised and discharged from the spout.

111. Force-pump. In this pump the ordinary piston with valve is replaced by a solid cylinder of metal, B (Fig. 101), called the *plunger*.

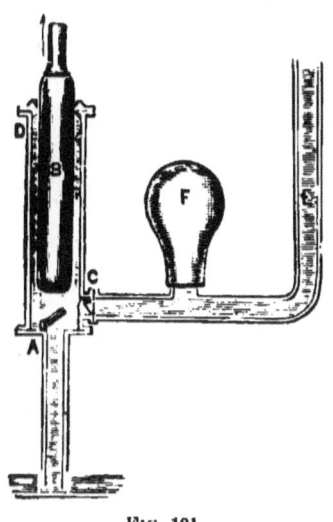

FIG. 101.

This passes through a stuffing box, D, in which it fits air-tight. Valves opening upward and outward are placed at A and C, respectively. When the plunger is raised, A opens and C closes, and water is raised into the barrel by atmospheric pressure. When the plunger descends, A closes and C opens, and the water is forced up through the pipe E. An air-dome, F, is usually connected with these pumps to regulate the pressure so as to give through the delivery pipe a very steady stream. This dome contains air. When the plunger descends it forces water into the dome and compresses the air within. As soon as the down stroke of the piston ceases, the valve C closes, and the compressed air in the dome forces the water out through E in a continuous stream.

112. Siphon. Take two vessels, A and B (Fig. 102), containing water (or other liquid). Let the surface of the liquid in one vessel be lower than the surface in the other. Bend a tube, C C, of any kind (*e.g.* rubber or glass) into the form of the letter U, fill it with some of the same liquid, cover the ends with your fingers, invert the tube, dip the ends of the tube into the liquids and remove the fingers. Liquid will flow from the vessel in which the liquid has a higher level into the other vessel.

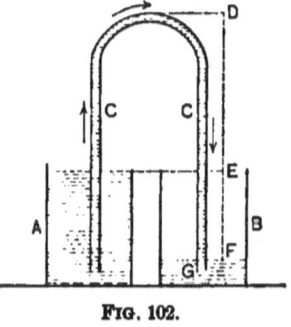

FIG. 102.

How does the atmospheric pressure upon the surfaces of the liquids in the two vessels compare? The flow of liquid shows the existence of an unbalanced force. How do you account for the unbalanced force? How great is this unbalanced force? As the liquids in the two vessels approach the same level, does this unbalanced force change in magnitude? What will happen when the liquids in the two vessels reach the same level? If vessel B were removed, would the liquid flow from vessel A? A tube used in this manner for transferring a liquid through the agency of atmospheric pressure is called a *siphon*.

SECTION V.

THE BUOYANT FORCE OF FLUIDS.

113. The Principle of Archimedes. Suppose d c b a (Fig. 103) to be a cubical block of marble immersed in a liquid. It is obvious that the difference between the upward pressure against the surface c b and the downward pressure on the surface d a is the weight of a column of liquid, e c b o, less the weight of a column of liquid, e d a o, which is a column of liquid, d c b a (e c b o − e d a o = d c b a). But a column of liquid, d c b a, has precisely the volume of the solid submerged. Therefore, *a body is buoyed up by a fluid in consequence of the unequal pressures upon its top and bottom at their different depths, and the amount of the buoyancy is the weight of a volume of that fluid equal to the volume of the immersed body.*

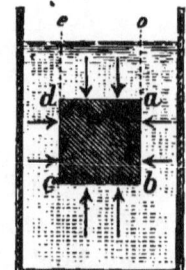

FIG. 103.

This principle, commonly called *the Principle of Archimedes*, from the name of the discoverer, may be thus stated : *a body immersed in a fluid is buoyed up by a force equal to the weight of the fluid displaced.*

Experiment 1. Suspend from one arm of a balance beam a cylindrical bucket, A (Fig. 104), and from the bucket a solid cylinder, B, whose volume is exactly equal to the capacity of the bucket ; in other words, the latter would just fill the former. Counterpoise the bucket and cylinder with weights.

Place beneath the cylinder a tumbler of water, and raise the tumbler until the cylinder is completely submerged. The buoyant force of the water destroys the equilibrium. Pour water into the bucket. When it becomes just even full, the equilibrium is restored.

Now it is evident that the cylinder immersed in the water displaces its own volume of water, or just as much water as fills the bucket. But the bucket full of water is just sufficient to restore the weight lost by the submersion of the cylinder. What principle does this experiment illustrate?

FIG. 104.

A floating body (as a cork on water) *sinks until it displaces a mass of the fluid equal to its own mass, or until it reaches a depth where the upward pressure of the fluid is equal to its own weight.*

Experiment 2. Place a baroscope (Fig. 105), consisting of a scale-beam, a small weight, and a hollow brass sphere, under the receiver of an air-pump, and exhaust the air. In the air the weight and sphere balance each other; but when the air is removed, the sphere sinks, showing that in reality it is heavier than the weight. In the air each is buoyed up by the weight of the air it displaces; but as the sphere displaces more air, it is buoyed up more. Consequently, when the buoyant force is withdrawn from both, their equilibrium is destroyed.

The *absolute weight of a body* is its weight in a vacuum. How much greater is this weight than the weight of the body in air?

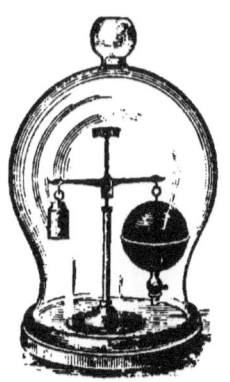

FIG. 105.

The density of the atmosphere is greatest at the surface of the earth. A body free to move cannot displace more than its own weight of a fluid; therefore a balloon, which is a large bag filled with a gas many

ARCHIMEDES.
[From bust in National Museum, Naples.]

times lighter than air at the sea level, will rise till the weight of the balloon, together with its car and cargo, equals the weight of the air displaced.

SECTION VI.

DENSITY, SPECIFIC DENSITY, AND SPECIFIC GRAVITY.

114. Terms Defined. The *density* of a substance at any temperature is the mass per unit of volume of the substance at that temperature. Thus, the density of water at 4° C. is one gram per cubic centimeter, and the density of cast iron at the same temperature is about 7.12 grams per cubic centimeter. The mean density of a body is found by dividing its mass by its volume.

The *specific density* of a substance is the number which expresses *how many times denser* the substance is than some standard substance. The *specific gravity* of a substance is *the ratio of the weight of a body of that substance to the weight of an equal volume of some standard.* The standard adopted for solids and liquids is distilled water at some definite temperature (in scientific work at 4° C.). Evidently the number which expresses the specific density of a substance and the number which expresses the specific gravity of the same substance are identical, and both are abstract numbers.

115. Formulas for Specific Density and Specific Gravity. Let D represent the density of any given substance, and D' the density of water, and let W and W' represent, respectively, the weights of equal volumes of the same substances; then, by definition,

(1) $\dfrac{\text{Density of given substance}}{\text{Density of water}} = \dfrac{D}{D'} = Sp.\ D.$, or

(2) $\dfrac{\text{Weight of a given volume of the substance}}{\text{Weight of equal volume of water}} = \dfrac{W}{W'} = Sp.\ G.$

116. Experimental Methods of Finding the Specific Gravity of Substances. (1) *Solids.*

The Principle of Archimedes is commonly made use of in determining the specific gravity of solids.

Experiment 1. From a hook beneath a scale-pan (Fig. 106) suspend by a fine thread a small portion of the solid substance whose specific gravity is to be found, and weigh it, while dry, in the air. Then immerse the body in a tumbler of water (see that it is completely submerged), and weigh it in water. The loss of weight in water is evidently W', *i.e.* the weight of the water displaced by the body; or, in other words, the weight of a body of water having the same volume as that of the specimen. Apply the formula (2) for finding the specific gravity.

Fig. 106.

Experiment 2. Take a piece of sheet lead 1 inch long and ½ inch wide, weigh it in air and then in water, and find its loss of weight in water. Weigh in air a piece of cork or other substance that floats in water; then fold the lead-sinker, place it astride the string just above the specimen, completely immerse both, and find their combined weight in water. Subtract their combined weight in water from the sum of their weights in air; this gives the weight of water displaced by both. Subtract from this the weight lost by the lead alone, and the remainder is W', *i.e.* the weight of water displaced by the cork. Apply formula (2), as before.

(2) *Liquids.*

Experiment 3. Take a bottle that holds when filled a certain (whole) number of grams of water, *e.g.* 100 g., 200 g., etc. Fill the bottle with the liquid whose specific gravity is sought. Place it on a scale-pan (Fig. 107), and on the other scale-pan place a piece of metal, a, which is an exact counterpoise for the bottle when

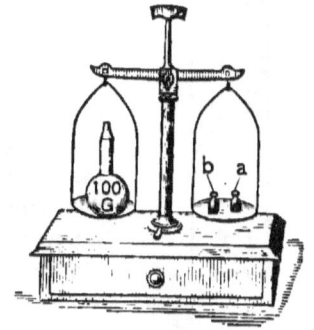

Fig. 107.

empty. On the same pan place weights b until there is equilibrium. The weights placed in this pan represent the weight W of the liquid in the bottle. The W' (*i.e.* the 100 g., 200 g., etc.) is usually etched on bottles constructed for this purpose. Apply formula (2).

117. The Densimeter. The principle of the *densimeter* (commonly called *hydrometer*) is based upon two facts: (1) a floating solid sinks until it displaces its own weight of the liquid in which it floats; (2) the volumes of two liquids displaced by the same floating solid vary inversely as their densities.

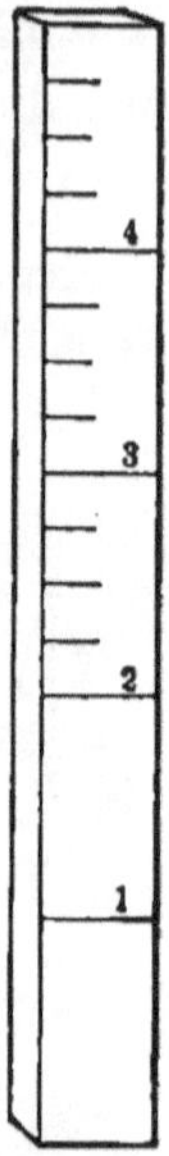

Experiment 4. Take a prism of paraffined wood (Fig. 108) ¼ inch square and 5 inches long, with a quarter-inch scale on one of its faces. It should be so loaded as to assume a vertical position and sink just 4 inches when placed in water. It displaces, therefore, a volume, V (¼ × ¼ × 4 = 1 cu. in.), of water. Place it in some liquid whose specific density is sought. It displaces a volume, V', of this liquid. Then $\dfrac{V}{V'}$ = the specific density of the given liquid. This experiment illustrates the principle on which the densimeter is based.

FIG. 108.

Instead of a prism of wood, a glass tube, A (Fig. 109), terminating in a bulb containing shot or mercury, is generally used. It has a scale of specific densities on the stem, so that no computation is necessary. The experimenter merely places it in the liquid to be tested, and reads the specific density at that point which is at the surface of the liquid.

The specific density of a gas is found by the application of the same principles as those employed in determining that of a liquid, but the operation is attended with peculiar difficulties (see author's "Principles of Physics"). For many purposes it is most convenient to employ hydrogen gas — the lightest gas — as a standard for gases. Then, assuming the density of hydrogen to be 1, that of air is 14.7, oxygen 16, etc. A cubic centimeter of hydrogen at 0° C. and at the barometric pressure of 760 mm. weighs at Paris 0.0000895682 g., and a cubic centimeter of pure dry air under the same conditions weighs 0.0012032 g.

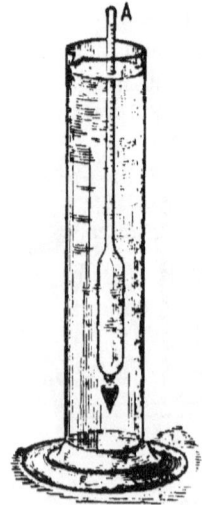

FIG. 109.

118. Miscellaneous Experiments.

Experiment 5. Find the volume of an irregularly shaped body, e.g. a stone. Find its loss of weight in water. Remember that the loss of weight is precisely the weight of the water it displaces, and that the volume of one gram of water is one cubic centimeter.

Experiment 6. Find the capacity of a test-tube, or of an irregularly shaped cavity in any body. Weigh the body; then fill the cavity with water and weigh again. As many grams as its weight is increased, so many cubic centimeters is the capacity of the cavity.

Experiment 7. Float a sensitive densimeter in water at about 60° F. (15° C.), and in other water at about 180° F. (82° C.). Which water is denser?

EXERCISES.

1. Can you by placing the neck of a bottle in your mouth suck liquid out of the bottle? Explain.

2. What is the weight of a cubic foot of beechwood? (Consult the Table of Specific Densities in the Appendix.)

3. Into what space must you compress 30 cu. ft. of air that its elastic force may be made five times as great?

4. If when the barometer stands at 760 mm. a cubic meter of air be forced into a vessel whose capacity is 1000 cc., what pressure will be exerted upon its interior walls?

5. Why do ironclad vessels float in water?

6. A block of ice weighing 500 grams floats on water. (a) What volume of water does it displace? (b) What volume of ice is out of water?

7. Will ice float or sink in alcohol? (See Table of Specific Densities in the Appendix.)

8. Give the density and specific density of gold, cork, and alcohol.

9. The effective weight of a stone in water is 56 grams; its weight in air is 112 grams. (a) What is the volume of the stone? (b) What is its density?

10. How many cubic centimeters of dry air at 760 mm. and at 0° C. weigh as much as one cubic centimeter of water at 4° C.?

11. If 4 cu. ft. of a body have a mass of 180 pounds, what is its specific gravity?

12. How much will one k. of copper weigh in water?

13. What does a piece of lead $20 \times 10 \times 5$ cm. weigh?

14. How much does a cubic foot of gold weigh?

15. A solid body weighs 10 pounds in air and 6 pounds in water. (a) What is the weight of an equal volume of water? (b) What is its specific gravity? (c) What is the volume of the body?

EXERCISES.

16. A thousand-gram bottle filled with sea-water requires in addition to the counterpoise of the bottle 1026 grams to balance it. (a) What is the specific gravity of sea-water? (b) What is the quantity of salt, etc., dissolved in 1000 grams of sea-water?

17. A piece of cork floating on water displaces 2 pounds of water. What is the weight of the cork?

18. In which would a hydrometer sink farther, in milk or in water?

19. What metals will float in mercury?

20. (a) Which has the greater specific density, water at 10° C. or water at 20° C.? (b) If water at the bottom of a vessel could be raised by application of heat to 20° C., while the water near the upper surface had a temperature of 10° C., what would happen?

21. A block of wood weighs 550 grams; when a certain irregularly shaped cavity is filled with mercury the block weighs 570 grams. What is the capacity of the cavity?

22. In which is it easier for a person to float, in fresh water or in sea-water? Why?

23. Fig. 110 represents a beaker graduated in cubic centimeters. Suppose that when water stands in the graduate at 50 cc., a pebble-stone is dropped into the water, and the water rises to 75 cc. (a) What is the volume of the stone? (b) How much less does the stone weigh in water than in air? (c) What is the weight of an equal volume of water?

24. If a piece of cork be floated on water in a graduate and displace (i.e. cause the water to rise) 7cc., what is the weight of the cork?

25. You wish to measure out 50 g. of sulphuric acid. To what number on a beaker graduated in cubic centimeters will that correspond?

26. A measuring beaker contains 35 cc. of ether. What is the weight of the ether?

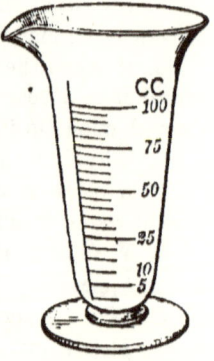

FIG. 110.

27. If 15 g. of salt be dissolved in 1 liter of water without increasing the volume of the liquid, what will be the specific density of the solution?

28. A mass whose weight in air is 30 g. weighs in water 26 g. and in another liquid 27 g. What is the specific density of the other liquid?

29. Find the specific gravity of wax from the following data: weight of a given mass of wax in air is 80 g.; wax and sinker displace 102.88 cc. of water; sinker alone displaces 14 cc.

30. A boat displaces 25 m.^{3} of water. How much does it weigh?

31. What mass of alcohol can be put into a vessel whose capacity is 1 liter?

CHAPTER IV.

MOLECULAR DYNAMICS. — HEAT.

SECTION I.

THEORY OF HEAT.

In the preceding pages the theory of heat has been several times anticipated; we are now better qualified to judge of its validity.

119. Energy of Mechanical Motion Convertible into Heat.

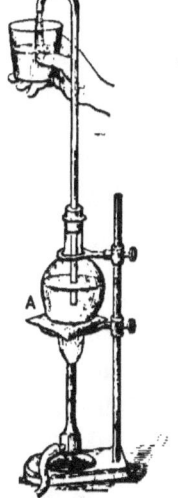

FIG. 111.

Experiment 1. Place a tenpenny nail upon a stone, and hammer it briskly; it soon becomes too hot to be handled with comfort, and we may conceive that if the blows were rapid and heavy enough, it might soon become red hot. Rub a desk with your fist, and your coat sleeve with a metallic button; both the rubbers and the things rubbed become heated.

You observe that in every case heat is generated at the expense of work or mechanical energy; *i.e. mechanical energy destroyed becomes heat.* When the brakes are applied to the wheels of a rapidly moving railroad train, its energy is converted into heat, much of which may be found in the wheels, brake-blocks, and rails.

120. Heat Convertible into Mechanical Energy.

Experiment 2. Take a thin glass flask, A (Fig. 111), half fill it with water, and fit a cork air-tight into its neck. Perforate the cork, insert a glass tube bent as indicated in the figure, and extend it into the water. Apply heat to the flask; soon the liquid rises in the tube and flows from its upper end.

KINETIC THEORY OF HEAT. 125

Here heat produces mechanical motion, *i.e.* it does work in raising a mass in opposition to gravitation. Every steam engine is a *heat engine, i.e.* the power of steam is due to its heat. The steam which leaves the cylinder of an engine, after it has set the piston in motion, is cooler than when it entered.

It will be shown hereafter that in all cases when work is done by heat without waste or loss, the quantity of heat consumed is proportional to the mechanical work done; and, conversely, by the performance of a definite quantity of mechanical work an equivalent quantity of heat may be generated. In other words, there is a *definite quantitative relation between heat and mechanical work.*

If heat be consumed, and mechanical work thereby performed, we are justified in saying that *heat becomes transformed into mechanical energy;* and, conversely, if mechanical energy be expended and heat thereby produced, we may say that *mechanical energy has become transformed into heat.* This has lead to the idea that heat is a *form of energy.*

121. Kinetic Theory of Heat. A hammer descends and strikes an anvil. Its motion ceases, but the anvil is not sensibly moved; the only observable effect produced is heat. Instead of a motion of the hammer and anvil, there is now supposed to be an increased *vibratory* motion of the *molecules* that compose the hammer and anvil — *simply a change from mass motion to molecular motion.* Of course this latter motion is invisible. According to the modern view, *heat is but a name for the energy of vibration of the molecules of a body*, or, briefly, HEAT IS MOLECULAR KINETIC ENERGY.[1] A body is heated by having the motion of its molecules quickened, and cooled by parting with some of its molecular motion.

---

[1] As late as the beginning of the nineteenth century heat was generally regarded as an "igneous *fluid*," sometimes called "caloric." Experiments performed by Count Rumford, Joule, and others have demonstrated the falsity of this view and have led to the adoption of the *kinetic theory.*

SECTION II.

SOURCES OF HEAT.

122. Mechanical Energy a Source of Heat. As heat is energy, so *all heat originates in some form of energy, i.e. by the transformation of some other form of energy into heat.*

In the preceding section it was shown that heat may be generated at the expense of molar motion, *i.e.* mass motion checked usually results in increased molecular motion. By friction, by compression, by percussion, or by any process by which mass motion is arrested, heat is mechanically generated.

123. Chemical Union a Source of Heat.

Experiment. Take a glass test-tube half full of cold water, and pour into it one fourth its volume of strong sulphuric acid. The liquid almost instantly becomes so hot that the tube cannot be held in the hand.

When water is poured upon quicklime, heat is rapidly developed. The invisible oxygen of the air combines with the constituents of the various fuels, such as wood, coal, oils, and illuminating gas, and gives rise to what we call *combustion*, by which a large amount of heat is generated. In all such cases the heat is generated by the combination or clashing together of molecules of substances that have an affinity (*i.e.* an attraction) for one another.

The energy possessed by fuels and all substances which by chemical union generate heat is called the *potential energy of chemical separation*. Chemical affinity converts the potential energy of the molecules into kinetic energy of vibration, *i.e.* into heat.

124. The Sun as a Source of Energy. The sun is not only a direct source of heat but it is also the source, directly or indirectly, of very nearly all the energy employed by man in doing work. The growth of vegetation is maintained by solar light and heat. Our coal-beds, the results of the deposit of vegetable matter, are vast storehouses of the sun's energy rendered potential during the growth of the plants many ages ago. Animals feed upon vegetable matter and thereby appropriate solar energy. Every drop of water that falls to the earth and rolls its way to the sea, contributing its mite to the immense water-power of the earth, and every wind that blows, derives its power directly from the sun.

125. Origin of the Sun's Heat and Energy. This has been the subject of much speculation. The theory of combustion or any other form of chemical action has been abandoned by physicists. Two rival theories, *viz. the meteoric theory* and *the contraction or compression theory*, now offer about equal claims for credence. According to the former theory, the heat of the sun originates from the arrested motion of cosmical bodies, meteorites,[1] that fall into the sun. According to the latter theory, heat is generated by the falling in of the sun upon itself, or the contraction and compression due to the mutual attraction between its parts. It is perhaps not amiss to attribute the heat to both causes.

It is the conclusion of both Professor Langley and Lord Kelvin that the temperature of the sun's photosphere is about $8000°$ C. The temperature of the voltaic arc (§ 374) is about $4000°$ C.

SECTION III.

TEMPERATURE AND THERMOMETRY.

126. Temperature Defined. The words *warm, hot, cool, cold,* are associated in our minds with a series of sensations which correspond to a series of states of matter with respect to heat. These are all temperature terms, and refer to the *state* of an object with reference to heat. When the quantity of heat in a body increases, its temperature is said *to rise;* and when this diminishes, its temperature is said *to fall.*

---

[1] The intense heat of meteorites that fall to the earth is due to their arrested mass motion on entering the atmosphere. Their temperatures are almost instantly raised from the temperature of outer space (more than 200 degrees below zero) to a temperature above that of the melting point of all substances. Their speed before entering the atmosphere is enormous. In the outer space a meteor moves farther in one second than the fastest express train moves in an hour.

128   MOLECULAR DYNAMICS.

If body A when brought into contact with body B tend to impart heat to it, then A is said to have a higher *temperature* than B. *Temperature is the state of a body with reference to its tendency to communicate heat to, or receive heat from, other bodies.* The direction of the flow of heat determines which of two bodies has the higher temperature. If the temperature of neither body rises at the expense of the other, then both have the same temperature, and are said to be in *thermal equilibrium*.

Temperature depends on the average kinetic energy of the molecules. The temperature of a body increases proportionally to the mean square of the velocity of vibration of its molecules.

127. Construction of a Thermometer. A *thermometer* is an instrument for indicating temperature. It consists of a glass tube of uniform capillary bore, terminating at one end in a bulb, the bulb and a part of the tube being filled with mercury, and the space in the tube above the mercury being a partial vacuum. On the tube, or on a plate of metal behind the tube, is a scale to show the hight of the mercurial column.

If a thermometer be brought into intimate contact with a body whose temperature is sought, as, for instance, a liquid into which it is plunged, or the air in a room, the mercury in the tube rises or falls [1] until it reaches a certain point, at which it remains stationary. We then know that it is in thermal equilibrium with the surrounding body. Hence the *reading*, as it is called, of the thermometer indicates the temperature not only of the mercury, but also of the surrounding body.

128. Graduation of Thermometers. First, the bulb of a thermometer is placed in melting ice (Fig. 112) and allowed to stand until the surface of the mercury becomes stationary, when a mark is made upon the stem at that point, which indicates the *melting point*. Then the instru-

---

[1] The thermometer primarily indicates changes of volume; but as changes of volume in this case are caused by changes of temperature, it is commonly used for the more important purpose of indicating *temperature*.

ment is suspended in steam rising from boiling water (Fig. 113), so that all but the very top of the column is in the steam.

The bulb is placed in a metallic vessel, M, with a narrower upper part, A. This narrower part is surrounded by a larger part, B. By observing the arrows it is seen that steam surrounds the inner part, and thus prevents its cooling; it escapes by the tube D. The orifice of D is large enough to allow the steam to escape freely, and thus prevent a pressure inside the vessel greater than the atmospheric pressure. To guard against such a contingency a pressure gauge, m, is inserted in the vessel. The liquid in both arms of the gauge must be kept at the same level throughout the operation. The mercury rises in the stem of the thermometer until its temperature becomes the same as that of the steam, when it becomes stationary. A barometer is consulted, and due allowance for atmospheric pressure at the time having been made, a mark is placed on the stem to indicate the *boiling point*. This boiling point is the temperature of steam at a pressure of 760 mm. of mercury at 0° C. Then the space between the two points found is divided into a convenient number of equal parts called *degrees*, and the scale is extended above and below these points as far as is desirable.

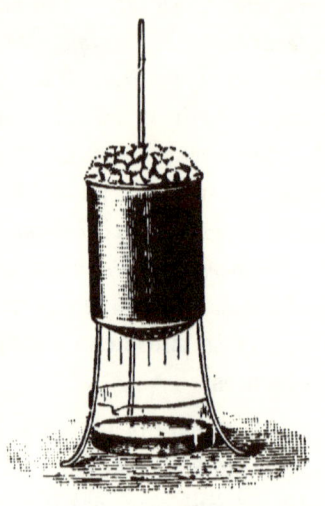

FIG. 112

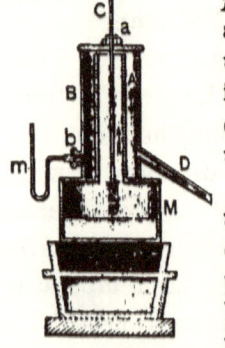

FIG. 113.

Two methods of division are adopted in this country (see a and b, Fig. 114): by one, the space is divided into 180 equal parts, and the result is called the *Fahrenheit* scale, from the name of its designer; by the other, the space is divided into 100 equal parts, and the resulting scale is called *Centigrade*, which means *one hundred steps*. In the Fahrenheit scale, which is generally employed in the United States for ordinary household purposes, the melting and boiling points are marked, respectively, 32° and 212°. The Centigrade scale, which is generally employed by scientists, has its melting and boiling points more conveniently marked, respectively, 0° and 100°. A temperature below 0° in either scale is indicated by a minus sign before the number. Thus,

— 12° F. indicates 12° below 0° (or 44° below the melting point of ice), according to the Fahrenheit scale.[1] The Fahrenheit and Centigrade scales agree at — 40°, but diverge both ways from this point.

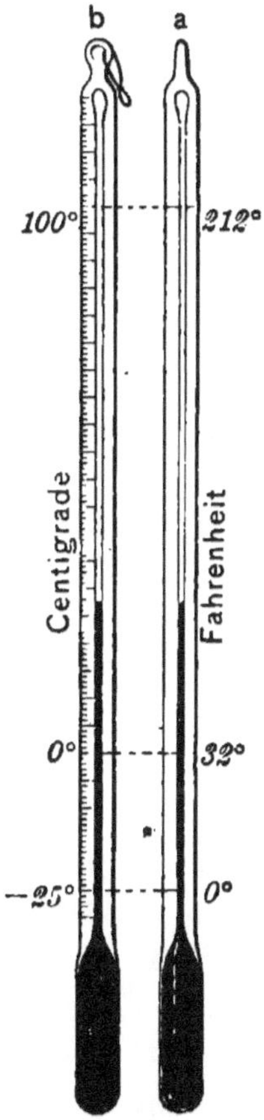

The expansion of gases is made the basis of the *standard scale of temperatures*, to which all other scales are referred for comparison and correction.

129. Conversion from One Scale to the Other. Since 100° C. = 180° F., 5° C. = 9° F., or 1° C. = $\frac{9}{5}$ of 1° F. Hence, to convert Centigrade degrees into Fahrenheit degrees, we multiply the number by $\frac{9}{5}$; and to convert Fahrenheit degrees into Centigrade degrees, we multiply by $\frac{5}{9}$. In finding the temperature on one scale that corresponds to a given temperature on the other scale, it must be remembered that the number that expresses the temperature on a Fahrenheit scale does not express the number of degrees above melting point, as it does on a Centigrade scale. For example, 52° on a Fahrenheit scale is not 52° above melting point, but (52° — 32° =) 20° above it.

Hence, to reduce a Fahrenheit reading to a Centigrade reading, *first subtract* 32 *from the given number, and then multiply by* $\frac{5}{9}$. Thus,

$$\tfrac{5}{9}(F. - 32) = C.$$

FIG. 114.

To change a Centigrade reading to a Fahrenheit reading, *first multiply the given number by* $\frac{9}{5}$, *and then add* 32. Thus,

$$\tfrac{9}{5} C. + 32 = F.$$

---

[1] The 0 of the Fahrenheit scale is about the lowest temperature that can be obtained by a mixture of snow and salt, and was formerly thought to represent the lowest temperature attainable. The first practical thermometer was constructed (1620) by Drebel, a Dutch physician. This was improved (1749) by Fahrenheit of Danzig. Celsius of Upsala, Sweden, added (1742) the scale now known as Centigrade.

EXERCISES.

1. Express the following temperatures of the Centigrade scale in the Fahrenheit scale: 100°; 40°; 56°; 60°; 0°; —20°; —40°; 80°; 150°.

NOTE. In adding or subtracting 32°, it should be done *algebraically*. Thus, to change —14° C. to its equivalent in the Fahrenheit scale: $\frac{9}{5} \times (-14) = -25.2°$; —25.2° + 32° = 6.8°, the required temperature in the Fahrenheit scale. Again, to find the equivalent of 24° F. in the Centigrade scale: $24 - 32 = -8$; $-8 \times \frac{5}{9} = -4\frac{4}{9}$; hence, 24° F. is equivalent to —4.4° + C.

2. Express the following temperatures of the Fahrenheit scale in the Centigrade scale: 212°; 32°; 90°; 77°; 20°; 10°; —10°; —20°; —40°; 40°; 59°; 329°.

3. Explain the origin of the heat obtained by burning coal.

4. In what does the value of coal consist?

5. How does all heat originate?

6. (*a*) Give an illustration of a transformation of matter; (*b*) of a transformation of energy.

7. The absolute zero (§ 141) is —273° C.; what is this on the Fahrenheit scale?

SECTION IV.

CALORIMETRY.

130. Distinction between the Questions "How Hot" and "How Much Heat." The former, like the question "how sweet" when applied to a solution of sugar, is answered only relatively. The latter, like the question "how much sugar in the solution," is answered quantitively. Sweetness and temperature are independent of the mass of the body. A pint of boiling water is as hot as a gallon of the same; but the latter contains eight times as much heat. *Temperature depends on the average kinetic energy of the molecules. Quantity of heat is the product of the average kinetic energy of the molecules multiplied by the number of molecules.* The quantity of heat a body has depends, therefore, upon both its mass and its temperature. What is the meaning of the statement that the temperature of the air is 20° C.?

131. Thermal Units. A thermal unit is the quantity of heat required to produce a definite effect. The thermal unit generally adopted is the *calorie*, which is the quantity of heat necessary to raise one kilogram of water from 4° to 5° C. (Glazebrook). The thermal unit in the C.G.S. system is the *gram-calorie*, sometimes called the *smaller calorie*, which is the quantity of heat required to raise one gram of water from 4° to 5° C. The operation of measuring heat is called *calorimetry*.

132. Heat Capacity; Specific Heat. The expression *heat capacity* applied to a body refers to the quantity of heat necessary to raise the temperature of the body 1°. The expression *specific heat* is applied only to some particular *substance*, and refers to the quantity of heat [1] required to raise one kilogram of that substance from 4° to 5° C. It is apparent that *the specific heat of a substance is the heat capacity of 1 unit of mass of that substance.*

Experiment 1. Mix 1 k. of water at 0° with 1 k. at 20°; the temperature of the mixture becomes 10°. The heat that leaves 1 k. of water when it falls from 20° to 10° is just capable of raising 1 k. of water from 0° to 10°.

Experiment 2. Take (say) 300 g. of sheet lead, make a loose roll of it, and suspend it by a thread in boiling water for about five minutes, that it may acquire the same temperature (100° C.) as the water. Remove the roll from the hot water, and immerse it as quickly as possible in 300° g. of water at 0°, and introduce the bulb of a thermometer. Note the temperature of the water when it ceases to rise, which will be found to be about 3° (accurately 3.3° +). The lead cools very much more than the water warms. The temperature of lead falls about 33° for every degree an equal mass of water is warmed.

From the first experiment we infer that a body in cooling a certain number of degrees gives to surrounding bodies as

[1] Specific heat is also defined as the ratio of the quantity of heat required in order to raise a given mass of a substance through one degree to the quantity required in order to raise an equal mass of water through one degree. Numerically it is equal to the number of calories required to raise one kilogram of the substance one degree.

SPECIFIC HEAT. 133

much heat as it takes to raise its temperature the same number of degrees. From the second experiment we learn that the quantity of heat that raises 1 k. of lead from 3.3° to 100°, when transferred to water, can raise 1 k. of water only from 0° to 3.3°. Hence we conclude that *equal quantities of heat, applied to equal masses of different substances, raise their temperatures unequally.* (See Table of Specific Heat, Appendix.)

If equal masses of mercury, alcohol, and water receive equal quantities of heat, the mercury will rise 30°, and the alcohol nearly 2°, for every degree the water rises. From this we infer that, to raise equal masses of each of these substances 1°, 30 times as much heat is required for the water as for the mercury, and twice as much as for the alcohol. Since a given quantity of heat affects the temperature of a given mass of water less than it does that of an equal mass of mercury or alcohol, water is said to have greater specific heat than these substances. It is also apparent that a given mass of water in cooling imparts to surrounding bodies more heat than the same mass of mercury or of alcohol would impart in cooling the same number of degrees, and that the excess is in proportion to the difference in specific heat between the substances.

133. Method of Measuring Specific Heat. A known mass, m (in kilograms), of the substance of which the specific heat is required is heated, as in Experiment 2, to a known temperature t_1 (C.); then it is mixed with (or immersed in) a known mass of water, m_2, at a lower temperature, t_2, and as soon as thermal equilibrium is established throughout, the temperature of the mixture t is taken. Let s represent the specific heat of the substance sought. Then the quantity of heat lost by the substance is $m\,s\,(t_1 - t)$ calories; while that gained by the water is $m_2\,(t - t_2)$ calories. Now if no heat be lost during the operation, $m\,s\,(t_1 - t) = m_2\,(t - t_2)$, whence $s = \dfrac{m_2\,(t - t_2)}{m\,(t_1 - t)}$. For example, taking the quantities obtained in the experiment above, we find for lead (300 g. = .3 k.) $s = \dfrac{.3\,(3.3 - 0)}{.3\,(100 - 3.3)} = .034$ calorie.

134. Specific Heat of the Same Substances at Different Temperatures and in the Three States of Matter. The specific heat of solids and liquids usually increases slightly with the temperature, and diminishes with increase of density. The specific heat of water at 0°, 40°, and 80° is, respectively, 1, 1.003, and 1.0089 calories. Substances usually have a higher specific heat in the liquid state than in the solid or gaseous state. Thus, water has nearly double the specific heat of ice, and a little more than double the specific heat of steam.

135. Great Capacity of Water for Heat. Water requires more heat to warm it, and gives out more in cooling through a given range of temperature than any other substance except hydrogen. The quantity of heat that raises a kilogram of water from 0° to 100° C. would raise a kilogram of iron from 0° to 800° or 900° C., or above a read heat. Conversely, a kilogram of water in cooling from 100° to 0° C. gives out as much heat as a kilogram of iron gives out in cooling from about 900° to 0° C.

"The vast influence which the ocean must exert as a moderator of climate here suggests itself. The heat of summer is stored up in the ocean, and slowly given out during the winter. This is one cause of the absence of extremes in an island climate."

The high specific heat of water is utilized in heating buildings by hot water.

EXERCISES.

1. The specific heat of mercury is .033 caloric. Explain this statement.

2. What is the heat capacity of 90 k. of mercury?

3. If the heat yielded by 1 k. of water in cooling from 100° to 0° C. were employed in heating 100 k. of mercury initially at 20°, to what temperature would the mercury be raised?

4. A mass of 700 g. of copper at 98° C. put into 800 g. of water at 15° contained in a copper vessel whose mass is 200 g. raised the temperature of the water to 21°. Find the specific heat of copper.

5. A copper ball weighing 3 k., taken out of a furnace and plunged into 10 k. of water at 10° C., heated the water to 25°. Find the temperature of the furnace [s. h. of copper = .095]. *Ans.* 551.3° C.

6. 10 k. of a certain substance, at a temperature of 120° C., is mixed with 2 k. of water at 20° C. The temperature of the mixture is 25° C. What is the specific heat of the substance?

7. A platinum ball whose mass is 900 g. and whose temperature is 110° C. is dropped into 500 g. of water at 20°. To what temperature will it raise the water?

SECTION V.

EFFECTS OF HEAT. EXPANSION.

136. Experiments Illustrating Expansion of Solids, Liquids, and Gases.

Experiment 1. Take two brass tubes, one of a size that will just permit it to enter the bore of the other. Heat the smaller tube; it will not in its expanded state enter the other. Thrust the heated tube into cold water; its temperature falls, and it now enters the bore of the other tube. "Heat expands," but "cold" does not "contract." Cohesion, when a diminution of heat (which acts as a repellent force) permits, causes a solid or liquid body to contract. Cold is a term of *negation* signifying merely a greater or less deficiency of heat; it is not an entity, and hence it cannot be the direct cause of any phenomenon.

Experiment 2. Fig. 115 represents a thin brass plate and an iron plate of the same dimensions riveted together so as to form what is called a *compound bar*. Place the bar edgewise in a flame, dividing the flame in halves (one half on each side of the bar) so that both metals may be equally heated.

FIG. 115.

The bar, which at first was straight, is now bent, owing to the *unequal expansion* of the two metals on receiving *equal increments of temperature*.

Experiment 3. Fit stoppers tightly in the necks of two similar thin glass flasks (or test-tubes), and through each stopper pass a glass tube about 60 cm. long. The flasks must be as nearly alike as possible. Fill one flask with alcohol and the other with water, and crowd in the stoppers so as to force the liquids in the tubes a little way above the corks. Set the two flasks into a basin of hot water, and note that, at the instant the flasks enter the hot water, the liquids sink a little in the tubes, but quickly begin to rise, until, perhaps, they reach the top of the tubes and run over.

When the flasks first enter the hot water they expand, and thereby their capacities are increased; meantime, the heat has not reached the liquids to cause them to expand, consequently the liquids sink momentarily to accommodate themselves to the enlarged vessel. Soon the heat reaches the liquids, and they begin to expand, as shown by their rise in the tubes. The alcohol rises faster than the water. *Different substances, in both the solid and the liquid states, expand unequally on experiencing equal changes of temperature.*

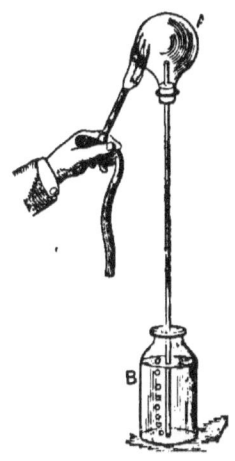

FIG. 116.

Experiment 4. Take a dry flask like that used in Experiment 3; insert the end of the tube in a bottle of colored water (Fig. 116), and apply heat to the flask; the enclosed air expands and comes out through the liquid in bubbles. After a few minutes withdraw the heat, keeping the end of the tube in the liquid; as the air left in the flask cools, its pressure decreases, and the water is forced by atmospheric pressure up the tube into the flask, and partially fills it.

137. Expansion-coefficients. The expansion which attends a rise of temperature depends not only upon the size of the body, and upon the number of temperature degrees through which it is heated, but upon a quantity peculiar to the substance itself, called its *expansion-coefficient.* The so-called *linear expansion-coefficient is the increase of unit-length per degree rise of temperature.*

Suppose that a rod of length l, at 0° C., be heated through t degrees, so that its length becomes l_1; then, representing the linear expansion-coefficient by c, we have

$$c = \frac{l_1 - l}{lt}, \text{ whence } l_1 = l\,(1 + ct).$$

The expression $1 + ct$, called the *expansion-factor,* is evidently the ratio of the final to the original length. Hence $l_1 = l\,(1 + ct)$; that is, multiplying the length of a solid at 0° C. by the expansion-factor gives its length at t degrees above zero. Conversely, dividing its length at $t°$ by the expansion-factor gives its length at 0°.

In the expansion of fluids we have to do only with increase of volume, called *volume* or *cubical* expansion. A *volume expansion-coefficient is the increase of unit volume per degree rise of temperature.* This is approximately 3 c, or three times the linear expansion-coefficient, and may be taken as such for most practical purposes. Likewise, the surface or superficial expansion-coefficient is approximately 2 c.

Not only do the expansion-coefficients of liquids and solids vary with the substance, but the coefficient for the same substance varies with the temperature, being greater at high than at low temperatures. Hence, in giving the expansion-coefficient of any substance it is customary to give the *mean* coefficient through some definite range of temperature, as from 0° to 100° C.

It is found that the expansion-coefficient of all gases is approximately the same as long as they remain true gases, but as they approach the vaporous state the coefficient changes rapidly.

138. Anomalous Expansion and Contraction. Water presents a partial exception to the general rule that matter expands on receiving heat and contracts on losing it. If a quantity of water at 0° C., or 32° F., be heated, it contracts as its temperature rises, until it reaches 4° C., or about 39° F., when its volume is least, and it therefore has its *maximum density*. If heated beyond this temperature it expands, and at about 8° C. its volume is the same as at 0°. On cooling, water reaches its maximum density at 4° C., and expands as the temperature falls below that point. The mass of one cubic decimeter of pure water at 4° C. is one kilogram (§ 6).

Water is said to have a negative expansion-coefficient between 0° and 4° C., or between 32° and 39.2° F. A few other substances, such as india rubber and iodide of lead, contract when heated, and have, therefore, negative coefficients.

EXERCISES.

1. How does rise of temperature affect the density of a body?

2. A rod of copper measures 10 feet at 0° C.; its length at 100° C. is 10.0159 feet. Find the linear expansion-coefficient of copper.

3. Two rods, one of brass, the other of wrought iron, are each 10 feet long at 0° C. Find their lengths at 100° C. (See Table of Expansion-coefficients in the Appendix.)

4. The volume of a leaden ball at 60° F. is 100 cu. in. Find its volume at the boiling point of water (coefficient of linear expansion of lead on F. scale = .0000157).

5. What is the length of the standard platinum meter rod at 80° C.?

6. (*a*) If the volume of a brass ball at 10° C. be 840 cc., what is its volume at 90°? (*b*) At which temperature is its density greater?

SECTION VI.

KINETIC THEORY OF MATTER. LAWS OF GASEOUS BODIES. ABSOLUTE TEMPERATURE.

139. Kinetic Theory of Matter. The theory that the molecules composing all bodies of matter are in perpetual relative motion is called the *kinetic theory of matter*. This theory claims that in gases the molecules are so far separated from one another that their motions are not generally influenced by molecular attractions. Hence, in accordance with the first law of motion, the molecules of gases move in *straight lines* and with uniform velocity until they collide with one another or strike against the walls of the containing vessel, when, in consequence of their perfect elasticity, they at once rebound and start on new paths.

140. Pressure of a Gas Due to the Kinetic Energy of its Molecules. Consider, then, what a molecular storm must be raging about us, and how it must beat against us and against every exposed surface. According to the kinetic theory, a

PRESSURE OF GASES. 139

gas exerts pressure upon the receptacle which confines it, in consequence of the incessant impacts of the molecules of the gas upon the surfaces against which the gas is said to press, the impulses following one another in such rapid succession that the effect produced cannot be distinguished from constant pressure. Upon the energy of these blows, and upon the number of blows per second, must depend the amount of pressure. But we have learned that on the kinetic energy of the individual molecules depends that condition of a gas called its *temperature;* so it is apparent that *the pressure of a given quantity of gas varies with its temperature.* Again, as at the same temperature the number of blows per second must depend upon the number of molecules in the unit of space, it is apparent that *the pressure varies with the density.*

141. Absolute Zero. The zeros on the thermometric scales which we have hitherto considered are provisional and arbitrary. Absolute zero is the temperature corresponding to a total absence of heat. At the *absolute zero* the molecules must be supposed to be *at rest*. At this temperature gases (if they may be called such) exert no pressure, and occupy no space save that which their molecules take up when closely packed together. The point of absolute zero is independent of the conventions of man. It is a point of absolute cold or total absence of heat, beyond which no cooling is conceivable.

The pressure in air increases or diminishes by .00367 = (about) $\frac{1}{273}$ of its pressure at 0° for each Centigrade degree of rise or fall of temperature, the volume being maintained constant. If air were a perfect gas, and could be cooled down to − 273° C. (− 459.4° F.), it would cease to exert pressure. The reason it would exert no pressure is that its particles would possess no kinetic energy, no motion. This is assumed, therefore, to be the absolute zero of temperature.

142. Absolute Temperature. Absolute temperature is that reckoned from the absolute zero, or $-273°$ C. *Temperatures measured from absolute zero are proportional to the pressure of a theoretically perfect gas of constant volume and density.*

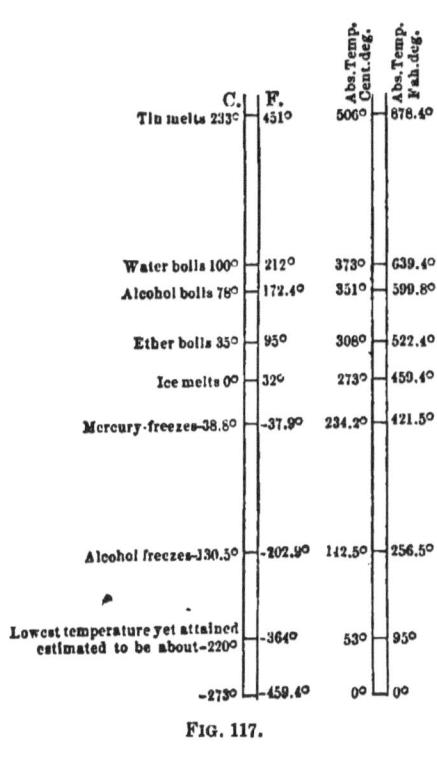

Fig. 117.

The absolute temperature (based on the above theory) of any body is found by adding 273 to its temperature as indicated by a Centigrade thermometer, or 459.4 to its temperature as indicated by a Fahrenheit thermometer. Figure 117 furnishes a comparative view of both the arbitrary and the absolute thermometric scales expressed in both C and F degrees.

143. Laws of Gaseous Masses. It follows, from the above discussion, that (1) *the volume of a given mass of gas at constant pressure is proportional to its absolute temperature;* i.e. at constant pressure $\dfrac{v \text{ (volume of a given mass of gas)}}{t \text{ (absolute temperature)}}$ remains constant. This is called the *Law of Charles*.

(2) *The pressure of a given mass of gas whose volume is kept constant is proportional to its absolute temperature.*

Boyle's law states that (3) *at a constant temperature the volume of a given mass of gas is inversely proportional to its pressure;* i.e. *the product of its pressure and its volume is con-*

stant. Now, when both the pressure and the volume vary at the same time, it may be shown that (4) *the product of the pressure and the volume of a given mass of gas is proportional to its absolute temperature.* A gas is said to be *perfect* when it perfectly obeys these laws.

We may also state the fourth law as follows: the product of the pressure and the volume of a given mass of gas divided by its absolute temperature is a constant quantity, or

$$\frac{PV}{T} = C,$$

in which P = pressure, V = volume, T = absolute temperature of a given mass of gas, and C = a constant quantity, the value of which depends on the gas in question.

EXERCISES.

1. Find, in both Centigrade and Fahrenheit degrees, the absolute temperatures at which mercury boils and freezes.

2. At 0° C. the volume of a certain mass of gas under a constant pressure is 500 cc. (*a*) What will be its volume if its temperature be raised to 75° C.? (*b*) What will be its volume if its temperature become −20° C.?

3. If the volume of a mass of gas at 20° C. be 200 cc., what will be its volume at 30° C.? *Solution:* 20° C. is equivalent to (20 + 273) 293 abs. temp.; then 293 : 303 :: 200 : 206.8 cc. *Ans.*

4. To what volume will a liter of gas contract, if cooled from 30° C. to −15° C.?

5. One liter of gas under a pressure of one atmosphere will have what volume, if the pressure be reduced to 900 g. per square centimeter, while the temperature remains constant?

6. The volume of a certain mass of air at a temperature of 17° C., under a pressure of 800 g. per square centimeter, is 500 cc. What will be its volume at a temperature of 27° C., under a pressure of 1200 g. per square centimeter? *Solution:* 17° C. is equivalent to 290° abs. temp.; 27° C. is equivalent to 300° abs. temp. Then 290 : 300 :: 500 × 800 : x × 1200. Whence x = 344.8 cc. *Ans.*

7. If the volume of a mass of gas under a pressure of 1 k. per square centimeter at a temperature of 0° C. be 1 liter, at what temperature will

its volume be reduced to 1 cc. under a pressure of 200 k. per square centimeter? *Ans.* 54.6° abs. temp., or — 218.4° C.

8. If a cubic foot of coal gas at 32° F., when the barometer is at 30 in., have a mass of $\frac{1}{25}$ lb., what will be the mass of an equal volume at 68° F., when the barometer is at 29 in.?

SECTION VII.

EFFECTS OF HEAT. FUSION.

144. Changes of Properties in Solids Attending Change of Temperature. "Every known property of a piece of matter, except its mass, varies with variation of temperature." Inasmuch as heat tends to weaken cohesion, the rigidity and the tenacity of solids are generally lessened, and their plasticity is increased, by the addition of heat.

145. Fusion. Whether a given substance exist in a solid, liquid, or gaseous state depends upon its temperature and the pressure it is under. Solids exposed to heat generally liquefy or fuse. Some, like ice and tin, change their state abruptly; others, like glass and wrought iron, become plastic prior to liquefaction. The temperature at which a substance melts is called its *fusion point*. The fusion points of different substances vary greatly: that of alcohol (— 130.5° C.) and that of iridium (1950° C.) may be taken as extreme examples. (See Table of Melting Points, Appendix.)

Experiment and experience teach that (1) *the melting or solidifying point (they are approximately the same for the same substance) may vary widely for different substances, but for the same substance when under the same pressure the point is invariable.*

(2) *The temperature of a solid or a liquid remains constant at the melting point from the moment that melting or solidification begins until it ceases.*

HEAT OF FUSION.

Experiment 1. Put a lump of ice as large as your two fists into boiling water; when it is reduced to about ¼ its original size, skim it out. Wipe the lump, and place one hand on it and the other on a lump to which heat has not been applied; you will not perceive any difference in their temperatures. Under ordinary pressure ice cannot be made warmer than 0° C.

146. Heat of Fusion. The temperature of ice remains constant while melting, and, generally, heat imparted to a melting body affects its temperature very little if any. Furthermore, ice and other solids are not instantly converted into liquids on reaching the fusion point, but absorb a quantity of heat proportionate to their mass before fusion is accomplished. Inasmuch as none of the heat applied during melting raises the temperature of the body, the question arises, *What becomes of the heat applied to the body?* The thermo-dynamical theory furnishes the only satisfactory answer to this important question. The answer is, that about all the heat applied to a body during fusion is *consumed in doing internal work*, as it is called. The molecules that were held firmly in their places by molecular forces are, during fusion, moved from their places, and so work is done against these forces. Heat, the energy of motion, performs this work, and is thereby converted into *potential energy*. The heat which disappears in melting is called the *heat of fusion*.

If a large quantity of heat be required in order to effect the fusion of a body, it must be inferred that the amount of work done is proportionately great. It is fortunate that it does require much heat to melt moderately small masses of ice and snow, since otherwise on a single warm day in winter all the ice and snow would melt, creating most destructive freshets.

147. Measurement of the Heat of Fusion. Let it be required to find approximately the quantity of heat that disappears during the melting of one kilogram of ice. This quantity is most readily determined by the *method of mixtures*.

Experiment 2. Weigh out 200 g. of dry ice chips (dry them with a towel), whose temperature in a room of ordinary temperature may be safely assumed to be 0° C. Weigh out 200 g. of boiling water, whose temperature we assume to be 100° C. Pour the hot water upon the ice, and stir it until the ice is all melted. Test the temperature of the resulting liquid.

Suppose its temperature is found to be 10° C. It is evident that the temperature of the hot water in falling from 100° to 90° would yield sufficient heat to raise an equal mass of water from 0° to 10° C. Hence, it is clear that the heat which the water at 90° yields in falling from 90° to 10° — a fall of 80° — in some manner disappears. At this rate, had you used 1 k. of ice and 1 k. of hot water, the amount of heat lost would be 80 calories. Careful experiments, in which suitable allowances are made for loss or gain of heat by radiation, conduction, absorption by the calorimeter, etc., have determined that *80 calories of heat are consumed in melting 1 kilogram of ice.*

148. Transformation of Heat Reversible. As stated at the beginning of this chapter, work is transformable into heat, and, as stated on page 70, potential energy is transformed into kinetic energy "by the return of the particles to their original positions"; so when water freezes or any liquid is resolidified, the potential energy reappears as heat.

Water in freezing undergoes no change of temperature; hence, if heat be developed during the operation, it must become diffused or must be "given off" in order to allow the freezing to go on. As the diffusion is necessarily slow, so freezing must be slow; and this slow development of heat and its immediate dispersion accounts for the fact that we are seldom made conscious of the development of heat during solidification.

Farmers sometimes turn to practical use this well-known phenomenon. Anticipating a cold night, they carry tubs of water into cellars to be frozen. The heat generated thereby, although of a low temperature, is sufficient to protect vegetables which freeze at a lower temperature than water.

Heat disappears in the process of melting ice; and, paradoxical as it may seem, heat is generated by freezing water. By freezing one kilogram of water 80 calories of heat are generated, or a quantity sufficient to raise the temperature of one kilogram of water from 0° to 80° C. Heat of fusion of water is much greater than that of other liquids.

SECTION VIII.

EFFECTS OF HEAT CONTINUED. VAPORIZATION.

149. Evaporation; Ebullition. The process of converting a liquid into a vapor [1] is called *vaporization*. A comparatively slow vaporization, which takes place only at the exposed surface of a liquid, is called *evaporation*. A rapid process, which may take place throughout the liquid, but usually is most rapid at the point where heat is applied, is called *boiling*,[2] or *ebullition*. Liquids which evaporate readily are called *volatile liquids*.

150. Boiling Point Dependent upon Pressure. In evaporation, molecules fly from the surface of the liquid and mingle with the particles of the air, and drive only a certain small portion of them away. In boiling, the molecules which fly away from the surface drive all the air particles away a certain distance. Hence the vapor of a boiling liquid must exert a pressure at least as great as the atmospheric pressure. The greater the external pressure to be overcome, the greater must be the energy, *i.e.* the higher the temperature, of the vapor. When the vapor of a liquid exerts a pressure equal to that of the atmosphere, the liquid begins to boil, and the temperature at which this occurs is called the *normal boiling point* of that liquid.

Experiment 1. Half fill a glass flask with water. Boil the water over a Bunsen burner; the steam will drive the air from the flask. Withdraw the burner, quickly cork the flask very tightly, and plunge the flask into cold water, or invert the flask and pour cold water upon the part containing steam, as in Fig. 118; the water in the flask, though cooled several degrees below the usual boiling point, boils again violently. The

---

[1] A vapor is a gaseous state of a substance which at ordinary temperatures exists as a solid or liquid.

[2] Boiling is the agitation of a liquid produced by its own vapor, which, since it is lighter than the surrounding liquid, tends to rise rapidly to the surface.

application of cold water to the flask diminishes the pressure of the steam, so that the pressure upon the water is diminished, and the water boils at a temperature lower than its normal boiling point.

Experiment 2. Place a test-tube half filled with ether (Fig. 119) in a beaker containing water at a temperature of 60° C. Although the temperature of the water is 40° below its boiling point, it very quickly raises the temperature of the ether sufficiently to cause it to boil violently. Introduce a thermometer into the test-tube, and ascertain the boiling point of ether.

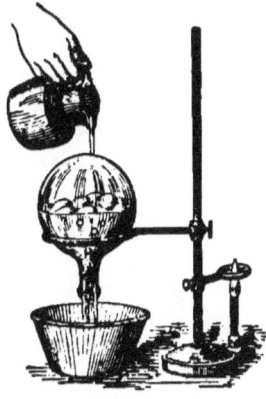

FIG. 118.

Experiment 3. In a beaker half full of distilled water suspend a thermometer so that the bulb will be covered by the water and yet be at least two inches above the bottom of the beaker. Apply heat to the beaker, and observe any changes of temperature which may occur, both before and after boiling begins. The mercury in the thermometer rises continuously until the water begins to boil, but soon after (*i.e.* as soon as thermal equilibrium between the mercury and water is established) it ceases to rise, thereby showing that the temperature of the water remains constant notwithstanding heat is constantly applied to it.

It is found that (1) *for a given pressure* (for example, that of the atmosphere at 760 mm.) *every liquid has a definite boiling point;* (2) *this boiling point remains constant after boiling has begun;* (3) *the boiling point of a liquid increases with the pressure.* A change of 27 mm. in the hight of a barometric column is attended with a change of about 1° C. in the boiling point of water boiled in an open vessel.

FIG. 119.

The boiling point of water varies with the altitude of places, in consequence of the change in atmospheric pressure. Roughly speaking, a difference of altitude of 533 ft. causes a variation of 1° F. in the boiling point. The measurement of hights by means of the boiling point is called *hypsometry*. A *hypsometer* is simply a convenient portable apparatus for boiling water, provided with a thermometer sensitive to (say) 0.01°.

EXERCISES.

1. What is the difference between a body that has heat and a body that has no heat?
2. What is the temperature of a body that has no heat?
3. What quantity of water at 90° C. is required to melt 800 g. of ice? *Ans.* 711 g.
4. 4.2 k. of steam at 100° C. enters a radiator in a room and is condensed into a liquid, and the liquid leaves the radiator at a temperature of 97° C. What quantity of heat is left in the room by this operation?
5. (a) How much heat is required to convert 4 k. of ice at $-18°$ C. into steam at 120° C? (b) How much of this heat is changed into "latent heat" (better, potential energy), and how much remains as sensible heat?
6. Let 900 g. of water at 85° C. be poured upon 200 g. of ice at 0° C.; what will be the temperature of the water after the ice is melted? *Ans.* 55° C.

151. Heat of Vaporization. Heat that is consumed in the process of vaporization is called *heat of vaporization*. The quantity of heat required to convert a gram of water at 100° C. into steam without altering its temperature (which is the same as the quantity of heat generated by the condensation of one gram of steam at 100°) is called the *heat of vaporization of water*.

Let it be required to find the heat of vaporization of water. Find the mass in grams of the glass beaker or calorimeter C (Fig. 120), and since it will receive a small portion of the heat generated by the condensation of the steam, find its *water equivalent* by multiplying its mass by the specific heat of glass (.177). Represent this quantity by m_1. Take in the calorimeter a certain known mass, M, of cold water at a known temperature, t. When water in the flask A begins to boil, introduce the end of the delivery tube B into the water in C. Screen the calorimeter from the heat of the lamp and flask by means of a board, D. The steam that passes through the tube is condensed on entering the cold water, and heats the water. When a considerable portion of the water in A has been vapor-

FIG. 120.

ized, the temperature t_1 of the water in C is taken again, and the contents of the calorimeter are again weighed. The increase m in the mass of water in C is the mass of steam which has been condensed. Let L be the heat of vaporization. Then the whole quantity of heat generated by the condensation of m grams of steam is Lm, and the quantity of heat imparted to the cold water in falling from 100° to $t_1°$ is $m(100 - t_1)$, or the total quantity of heat given to the calorimeter and its original contents is $Lm + m(100 - t_1)$. The heat required to raise the calorimeter and its original contents from t to t_1 is $(M + m_1)(t_1 - t)$. But these two quantities are equal, hence,

$$Lm + m(100 - t_1) = (M + m_1)(t_1 - t);$$

whence, $L = \dfrac{(M + m_1)(t_1 - t) - m(100 - t_1)}{m}$.

Careful experiments have determined that it requires 536 small calories of heat to convert one gram of water at 100° into steam at 100°, or 536 calories per kilogram, and when the process is reversed 536 calories per kilogram of steam are generated by the condensation.

When water is converted into steam, the larger portion of the heat which disappears is consumed in separating the molecules so far that molecular attraction is no longer sensible; a small portion — about $\frac{1}{13}$ — is consumed in overcoming atmospheric pressure. The amount of work done in boiling is very great, as is shown by the amount of heat consumed. Hence, it requires a long time for the water to acquire the requisite amount of heat. This is a protection against sudden and disastrous changes.

Steam is a most convenient vehicle for the conveyance of heat of vaporization, *i.e.* potential energy, from the boiler to distant rooms requiring to be heated. For example, for every kilogram of steam condensed in the pipes of the radiator, 536 calories, or heat enough to raise 5.36 kilograms (about 12 pounds) of ice-water to the boiling point, are generated.

152. Distillation.

Experiment 4. Vessel A (Fig. 121), called a *condenser*, contains a coil, B, called a *worm*, of copper tubing, terminating at one extremity at a. The other end of the tube b projects through the side of the vessel near its bottom. Near the top of the vessel projects another tube, F, called the *overflow*, with which is connected a rubber tube, H. This tube conveys the warm water which rises from the surface of the heated worm away to a sink or other convenient receptacle.

Take a glass flask, C, of a quart capacity, and fill it three fourths full of pond or bog water. Connect the flask by means of a glass delivery-tube with the extremity a of the worm. Heat the water in the flask; as soon as it begins to boil, commence siphoning cold water through a small tube, D, from an elevated vessel, E, into the con-

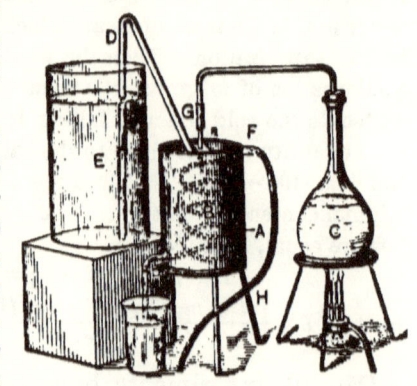

FIG. 121.

denser. Inasmuch as the worm is constantly surrounded with cold water, the steam on passing through it becomes condensed into a liquid, and the liquid (called the *distillate*) trickles from the extremity b into a receiving vessel. The distillate is clear, but the water in the flask acquires a yellowish brown tinge as the boiling progresses, due to the concentration of impurities (largely of vegetable matter) which are held in suspension and solution in ordinary pond water. The apparatus used is called a *still*, and the operation *distillation*.

When a volatile liquid is to be separated from water, — for example, when alcohol is separated from the vinous mash after fermentation, — the mixed liquid is heated to its boiling point, which is lower than that of water. Much more of the volatile liquid will be converted into vapor than of the water, because its boiling point is lower. Thus a partial separation is effected. By repeated distillations of the distillate, a 95 per cent alcohol is obtained.

Crude petroleum consists of a mixture of substances having various boiling points. The separation is effected by slowly distilling the mixture, with a thermometer in the path of the vapor; as the temperature of the vapor changes, the receptacles of the distillate are changed.

SECTION IX.

METHODS OF PRODUCING COLD ARTIFICIALLY.

153. Artificial Cold. A body becomes cold only by losing heat. As heat passes only from warmer to colder bodies, it is evident that the temperature of a body cannot fall below that of surrounding bodies — for example, below the temperature of other bodies in the same room — by the natural process of imparting heat to its neighbors. The temperature of a body, then, can be reduced below that of its neighbors only by some artificial means.

The fact that heat is consumed in the conversion of solids into liquids, and of liquids into vapors, is turned to practical use in many ways for the purpose of producing *artificial cold*. The following experiments will illustrate this process.

154. Heat Consumed in Dissolving; Freezing Mixtures.

Experiment 1. Prepare a mixture of 2 parts, by mass, of pulverized ammonium nitrate and 1 part of ammonium chloride. Take about 75 cc. of water (not warmer than 8° C.), and into it pour a large quantity of the mixture, stirring it while dissolving with a test-tube containing a little cold water. The water in the test-tube will be quickly frozen. A finger placed in the solution will feel a painful sensation of cold, and a thermometer will indicate a temperature of about − 10° C.

One of the most common freezing mixtures consists of 3 parts of snow or broken ice and 1 part of common salt. The affinity of salt for water tends to produce liquefaction of the ice, and the resulting liquid dissolves the salt, *both operations consuming heat*.

155. Heat Consumed in Evaporation. The heat consumed in vaporization is greater than that consumed in liquefaction; for example, in the case of water it is greater in the ratio of 536 : 80. Hence evaporation is the more efficient means of producing extremely low temperatures. Whatever tends to hasten evaporation tends to accelerate the reduction of temperature. *Rapidity of evaporation increases with the temperature, the extent of surface exposed, diminution of pressure, and the dryness of the atmosphere.* Other things being equal, the more volatile the liquid employed for evaporation, the more rapid the consumption of heat.[1]

Experiment 2. Fill the palm of the hand with ether; the ether quickly evaporates, and produces a sensation of cold.

Experiment 3. Place water at about 40° C. in a thin, porous cup, such as is used in electric batteries, and the same amount of water at the same temperature in a glass beaker of as nearly as possible the same size as the porous cup. Introduce into each a thermometer. The comparatively large amount of surface exposed by means of the porous vessel will so hasten the evaporation in this vessel, that, in the course of 10 to 15 minutes a very noticeable difference of temperature will be indicated by the thermometers in the two vessels.

In warm climates water is frequently kept in porous earthen vessels in order that its temperature may by evaporation be kept low enough to render it suitable for drinking.

156. Freezing Machines. The production of artificial cold has become an important industry. The impulse was given by the need of finding means of preserving meat in a fresh condition during its transit to foreign countries; also of preserving perishable articles of food, *e.g.* fish, eggs, butter, etc., in our storehouses.

The so-called *cold-air freezing machine* is driven by a steam engine, and its working parts are quite similar to those of a steam engine. In both there is a system of cylinders, pistons, and valves, and a working substance which undergoes alternately compression and expansion. The working substance in the freezing machine is air. Power furnished by the steam engine compresses air by the stroke of a piston in the compression cylinder. This cylinder is surrounded by a jacket in which cold

---

[1] Water may be frozen by its own evaporation in the receiver of an air-pump from which the air (and consequently the air pressure) is removed. By reducing the pressure on liquid air and thus causing rapid evaporation, Dr. Olszewski cooled helium, without liquefying it, to a temperature calculated to be $-264°$ C. under a pressure of one atmosphere. The same experimenter gives, as results of his investigations, the following: Under an atmospheric pressure, oxygen boils at $-164°$; specific gravity of the liquid 1.124 at $-181.4°$. Nitrogen melts at $-214°$, boils at $-194.4°$; specific gravity of the liquid .885 at $-194°$. Hydrogen boils at about $-240°$. In most instances the temperature is taken with a hydrogen thermometer.

water constantly circulates, so that the heat generated by the compression of the air is taken up by the cold water. Thus is obtained air at ordinary temperature but under high pressure.

If the air, when the pressure is removed, is allowed to expand into a vacuum, no work will be done by the air and (as proved by Joule) no change of temperature takes place. But in this machine the air in expanding drives a piston in an expansion cylinder and mechanical work is done at the expense of the heat of the expanding air. This piston is connected with the piston rod of the steam engine in such a way as to lighten the work of the latter.

By means of this expansion the air is readily cooled to $-50°$ F. Articles to be kept cool are placed in large chambers, the walls of which are double, the interspace being filled with charcoal, a non-conducting material. A stream of intensely cold air is injected into the chamber at each stroke of the piston of the expansion cylinder, and the temperature of the chamber is thus kept constantly near the freezing point.

The *ammonia machine* is another type of freezing machine. In this the frigorific effect is due first to the heat consumed by the vaporization of liquid ammonia which has been condensed under high pressure; and, secondly, as in the air machine, to the expansion of the vapor. In this machine the expansion of the vapor takes place in long coils of tubing placed in a bath of brine which has a low freezing point. From this bath the cold brine is driven by pumps through a system of tubes to places where refrigeration is required. In this manner ice is manufactured artificially in the summer season.

SECTION X.

HYGROMETRY.

157. Dew-point. *Hygrometry* treats of the state of the air with regard to the water vapor it contains. A given space, *e.g.* a cubic meter (it matters little whether there is air in the space or whether it is a vacuum), can hold only a limited quantity of water vapor. This quantity depends upon the temperature. The capacity of a space for water vapor increases rapidly with the temperature, being nearly doubled by a rise of $10°$ C. On the other hand, if air containing a given quantity of water vapor be cooled, it will continually approach and finally reach *saturation*, since the lower the temperature, the less the capacity for water vapor. It is evident that air saturated with vapor cannot have its temperature lowered without the condensation of some of the vapor into a liquid, which will appear, according to location and condition of objects within it, as *dew*, *fog*, or *cloud*. The temperature at which this condensation occurs is called the *dew-point* for air containing this proportion of water vapor. The dew-point may be defined as the *temperature of saturation* for the quantity of water vapor actually present in the air. The greater the quantity of water vapor present in the air, the higher is its dew-point. Capacity for water vapor depends upon temperature; dew-point depends upon the quantity of vapor present.

If the existing temperature be far above dew-point, it indicates that the air can contain much more vapor than there is in it at the time, and the air is said to be *dry* and to have a *low dew-point*. If the temperature of the air be little above dew-point, the air is said to be *humid* and to have a *high dew-point*, which means that it can hold but little more vapor. The sensation of dryness experienced, especially in rooms heated artificially, does not depend upon the *absolute* quantity of water vapor present per cubic foot.

The heat of a stove, for instance, *dries* the air of a room without destroying any of its water vapor. In such a room the lips, tongue, throat, and skin experience a disagreeable sensation of dryness, owing to the rapid evaporation which takes place from their surfaces. This should be taken as nature's admonition to keep water in the stove urns, and in tanks connected with furnaces.

The quantity of water vapor present in the air is expressed either (1) by the mass of vapor per unit of volume; or (2) by the ratio between the quantity actually present and that which would be present if the air were saturated at the temperature of observation. The latter is the more common and more useful method, and this ratio is called the *relative humidity*, or simply "humidity" of the air. It is expressed in percentages. Thus, "relative humidity = 75 per cent, or 0.75," denotes that the air contains three fourths the quantity of water vapor required to saturate it at the present temperature.

SECTION XI.

DIFFUSION OR TRANSFERENCE OF HEAT.

158. Three Processes of Diffusion. There is always a tendency to *equalization of temperature;* that is, heat has a tendency to pass from a warmer body to a colder, or from a warmer to a colder part of the same body, until there is an equality of temperature.

There are commonly recognized three processes of diffusion of heat — *conduction, convection,* and *radiation.*

159. Conduction.

Experiment 1. Place one end of a wire about 10 inches long in a lamp flame, and hold the other end in the hand. Heat gradually travels from the end in the flame toward the hand. Apply your fingers successively at different points nearer and nearer the flame; you find that the nearer you approach the flame the hotter the wire is.

The flow of heat through an unequally heated body, from places of higher to places of lower temperature, is called *conduction;* the body through which it travels is called the *conductor.* The molecules of the wire in the flame have their motion quickened; they strike their neighbors and quicken their motion; the latter in turn quicken the motion of the next; and so on, until some of the motion is finally communicated to the hand, and creates in it the sensation of heat.

FIG. 122.

Experiment 2. Fig. 122 represents a board on which are fastened, by means of staples, four wires: (1) iron, (2) copper, (3) brass, and (4) German silver. Place a lamp flame where the wires meet. In about a minute run your fingers along the wires from the remote ends toward the flame, and see how near you can approach the flame on each without suffering from the heat. Make a list of these metals, arranging them in the order of their conductivity.

You learn that some substances conduct heat much more rapidly than others. The former are called *good conductors*, the latter *poor conductors*. Metals are the best conductors, though they differ widely among themselves.

Experiment 3. Nearly fill a test-tube with water, and hold it somewhat inclined (Fig. 123), so that a flame may heat the part of the tube near the surface of the water. Do not allow the flame to touch the part of the tube that does not contain water. The water may be made to boil near its surface for several minutes before any change of the temperature at the bottom will be perceived.

FIG. 123.

Liquids, as a class, are poorer conductors than solids. Gases are much poorer conductors than liquids.

It is difficult to discover that pure, dry air possesses any conducting power. The poor conducting power of our clothing is due partly to the poor conducting power of the fibers of the cloth, but chiefly to the air which is confined by it. Loose garments, and garments of loosely woven cloth, inasmuch as they hold a large amount of *confined air*, furnish a good protection from heat and cold. Bodies are surrounded with bad conductors in order to *retain* heat when their temperature is above that of surrounding objects, and to *exclude* it when their temperature is below that of surrounding objects. In the same manner the confined air between double windows and double doors protects from cold.

160. Convection in Gases. Conduction takes place gradually and slowly at best from particle to particle, the body and its particles being relatively at rest. *Convection* takes place when the body moves, or when there is relative motion between its parts, the heat in either case being *conveyed* from one place to another.

Experiment 4. Cover a candle flame with a glass chimney (Fig. 124), blocking the latter up a little way so that there may be a circulation of air beneath. Hold smoking touch-paper near the bottom of the chimney; the smoke seems to be *drawn* with great rapidity into the chimney at the bottom; in other words, the office of the chimney is to create what is called a *draft* of air. Notice whether the combustion takes place any more rapidly with the chimney than without it.

Experiment 5. Place a candle within a circle of holes cut in the cover of a vessel, and cover it with a chimney, A (Fig. 125). Over an orifice in the cover place another chimney, B. Hold a row of smoking touch-paper over B. The smoke descends this chimney, and passes through the vessel and out at A. This illustrates the method often adopted to produce a ventilating draft through mines. Let the interior of the tin vessel represent a mine deep in the earth, and the chimneys two shafts sunk to opposite extremities of the mine. A fire kept burning at the bottom of one shaft will cause a current of air to sweep down the other shaft, and through the mine, and thus keep up a circulation of pure air through the mine.

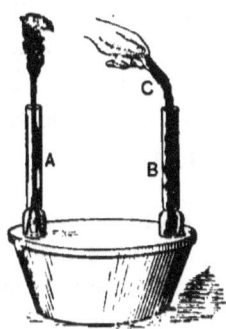

FIG. 124.   FIG. 125.

The cause of the ascending currents is evident. Air, on becoming heated, expands rapidly and becomes much rarer than the surrounding colder air; hence it rises, much like a cork in water, while cold air pours in laterally to take its place. In this manner winds are created. Sea and land breezes are convection currents.

161. Ventilation. Intimately connected with the topic *convection* is the subject (of vital importance) *ventilation*, inasmuch as our chief means of securing the latter is through the agency of the former. The chief constituents of our atmosphere are nitrogen and oxygen, with varying quantities of water vapor, argon, carbon dioxide gas, ammonia gas, nitric acid vapor, and other gases. The atmosphere also contains in

VENTILATION. 157

a state of suspension varying quantities of small particles of free carbon in the form of smoke, microscopic organisms, and dust of innumerable substances. All of these constituents except the first four are called *impurities*. Carbon dioxide is the impurity that is usually the most abundant and most easily detected; so it has come to be taken as the measure of the purity of the atmosphere, though not itself the most deleterious constituent. Its chief harm arises from its diluent effect upon the life-giving oxygen. Pure outdoor air contains about 4 parts of carbon dioxide by volume in 10,000. If the quantity rise to 10 parts, the air becomes unwholesome.

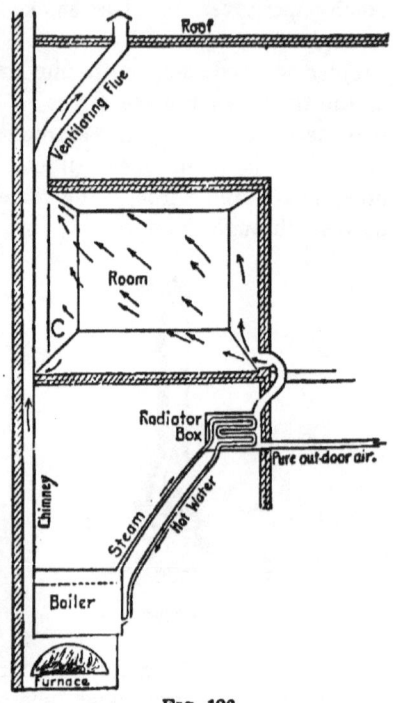

FIG. 126.

The quantity of fresh air introduced into an occupied room should be great enough to dilute the impurities till they are harmless. An adult makes about 18 respirations per minute, expelling from his lungs at each expiration about 500 cc. of air, over 4 per cent of which is carbon dioxide. At this rate, about 9 cubic decimeters of air per minute become unfit for respiration; and to dilute this sufficiently, good authorities say that about 100 times as much fresh air is needed; or, for proper ventilation, about *a cubic meter of fresh air per minute is needed for each person;* i.e. in British measures, 2,000 cubic feet per hour.

Fig. 126 represents a scheme for heating a room by steam, and ventilating it by convection. Steam is conveyed by a pipe from the boiler to a radiator box just beneath the floor of the room. The air in the box becomes heated by contact with and radiation (§ 163) from the coil of pipe in the box, and rises through a passage opening into the room by means of a register near the floor at C, a supply of pure air being kept up by means of a tubular passage opening into the box from the outside of the building. Thus the room is furnished with *pure warm* air, which, mingling with the impurities arising from the respiration of its occupants, serves to dilute them and render them less injurious. At the same time the warm and partially vitiated air of the room passes through the open ventilator A into the ventilating flue and escapes, so that in a moderate length of time a nearly complete change of air is effected. No system of ventilation dependent wholly on convection is adequate properly to ventilate crowded halls; air is too viscous and sluggish in its movements. In such cases ventilation should be assisted by some mechanical means, such as a blower or fan, driven by steam or water power.

162. Convection in Liquids.

Experiment 6. Fill a small (6-ounce) thin glass flask with boiling hot water colored with a teaspoonful of ink, stopper the flask, and lower

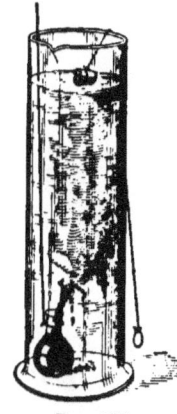

FIG. 127.

it deep into a tub, pail, or other large vessel filled with cold water (Fig. 127). Withdraw the stopper, and the hot, rarer, colored water will rise from the flask, and the cold water will descend into the flask. The two currents passing into and out of the neck of the flask are easily distinguished. The colored liquid marks distinctly the path of the heated convection currents through the clear liquid, and makes clear the method by which heat, when applied at the bottom of a body of liquid, becomes rapidly diffused through the entire mass, notwithstanding that liquids are poor conductors.

By similar convection currents the warming of buildings by hot water is effected. Water heated in a boiler in the basement rises through pipes to the radiators in the rooms above; there it gives heat to the air of the room, and, after being thus cooled, it returns by other pipes leading from the radiators to the boiler. Ocean currents, *e.g.* the gulf stream, are convection currents. The warmer portions of the waters flow away from the tropical toward the polar latitudes, while at greater depths the cold waters of high latitudes flow back toward the tropics.

Liquids are also cooled by convection currents. When the air above the surface of a pond, for instance, is cooler than the surface water, the latter gives heat to the former, cools, becomes denser, and sinks. Meanwhile, the warmer and rarer water below rises, and in this way the entire body is kept at an approximately uniform temperature until it reaches $4°$ C., at which point convection ceases.

163. Radiation. In *radiation* a hotter body loses heat, and a colder body is warmed, through the transmission of undulatory motion in a medium called the ether, *which is not itself heated thereby*. It is neither a mass nor a molecular transference of heat; in fact, heat itself is not transferred by radiation at all. Heat generates radiation (ether-waves) at one place, and the body which obstructs these waves transforms the energy of their motion, or, as it is commonly called, *radiant energy*, into heat. In this manner the earth is heated by the sun, though no heat passes between them. In this manner radiant energy passes through glass and slabs of ice without heating them much, since they offer little obstruction to the passage of ether-waves. All bodies emit radiant energy, and there is an exchange of energy between bodies by radiation going on at all times. This mode of transmission of energy is the most important of all, and will be treated fully in a future chapter.

SECTION XII.

THERMO-DYNAMICS.

164. Thermo-dynamics Defined. *Thermo-dynamics treats of the relation between heat and mechanical work.* One of the most important discoveries in science is that of the *equivalence of heat and work;* that is, that *a definite quantity of mechanical work, when transformed without waste, yields a definite quantity of heat; and, conversely, this heat, if there be no waste, can perform the original quantity of mechanical work.*

165. Transformation, Correlation, and Conservation of Energy. The proof of the facts just stated was one of the most important steps in the establishment of the grand twin conceptions of modern science : (1) that *all kinds of energy are so related to one another that energy of any kind can be transformed into energy of any other kind,* — known as the doctrine of CORRELATION OF ENERGY; (2) that *when one form of energy disappears, its exact equivalent in another form always takes its place, so that the sum total of energy is unchanged,*— known as the doctrine of CONSERVATION OF ENERGY.

These two doctrines are admirably summarized by Maxwell as follows : " *The total energy of any body or system of bodies is a quantity which can neither be increased nor diminished by any mutual action of these bodies, though it may be transformed into any of the forms of which energy is susceptible.*" Since all bodies of matter in the universe constitute a system, it

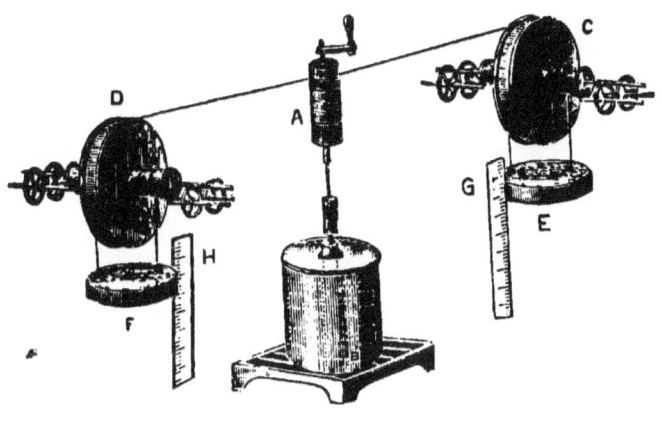

FIG. 128.

follows from the above that *the sum total of energy in the universe is a constant quantity.* Neither creation nor annihilation of energy is possible through any agency known to man. These doctrines constitute the corner stones of modern physical science. Chemistry teaches that there is a conservation of matter, *i.e.* that matter is neither creatable nor annihilable through any known natural agency or process.

166. Joule's Experiment. The experiment to ascertain the "mechanical value of heat," as performed by Dr. Joule of England, was conducted about as follows :

A copper vessel, B (Fig. 128), was provided with a paddle wheel (indicated by the dotted lines) which rotated about a vertical axle, A. The axle was rotated by the weights E and F, the cord of each being so arranged that each weight, in falling, rotated the axle in the same direction. By turning the crank above A the weights are raised to any desired hight measured on the scales G and H.

The resistance offered by the water to the motion of the paddles was the means by which the mechanical energy of the weights was converted into heat, which raised the temperature of the water. Taking two bodies whose combined mass was, *e.g.*, 80 k., he raised them a measured distance, *e.g.* 53 m. high; by so doing 4240 kgm. of work were performed upon them, and consequently an equivalent amount of energy was stored up in them, ready to be converted, first into that of mechanical motion, then into heat. He took a definite mass of water to be agitated, *e.g.* 2 k., at a temperature of 0° C. After the descent of the weights, the water was found to have a temperature of 5° C.; consequently the 2 k. of water must have received 10 calories of heat (careful allowance being made for all losses of heat), which is the number of calories that is equivalent to 4240 kgm. of mechanical energy; *or one calorie is equivalent to 424 kgm. of mechanical energy.*

In other words, *to produce a quantity of heat required to raise 1 kilogram of water through 1° C., 424 kilogrammeters of mechanical energy must be consumed.* What the experiment really shows is that whenever a certain quantity of mechanical energy is converted into heat, the number of thermal units produced is always proportional to the mechanical energy consumed, or to the work done.

167. Mechanical Equivalent of Heat. As a converse of the above it may be demonstrated by actual experiment that the quantity of heat required to raise 1 k. of water from 0° to 1° C. will, if converted into work, raise a 424 k. weight 1 m.

high, or 1 k. weight 424 m. high. In terms of the British system, the same fact is stated as follows : The quantity of heat that will raise the temperature of 1 pound of water from 60° to 61° F. will raise 772.55 pounds 1 foot high. The quantity, 424 kgm., is called the *mechanical equivalent of one calorie, or Joule's equivalent*. Or we may say that one calorie is the *thermal equivalent* of 424 kgm. of work.

SECTION XIII.

THERMO-DYNAMICS CONTINUED. — STEAM ENGINE.

168. Description of a Steam Engine. A steam engine is a machine in which the elastic force of steam is the motive agent. Inasmuch as the elastic force of steam is entirely due to heat, *the steam engine is properly a heat engine;* that is, it

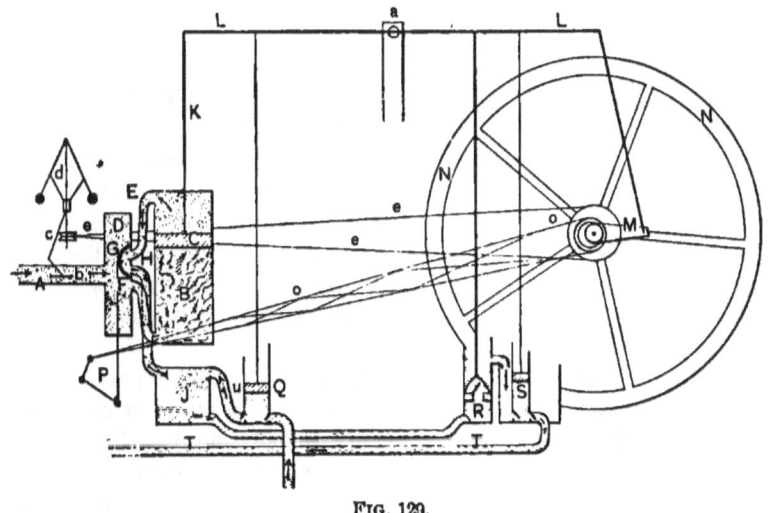

Fig. 129.

is a machine by means of which heat is continuously transformed into work, or the energy of mass motion. The modern steam engine consists essentially of an arrangement by which steam from a boiler is conducted to each side of a piston alternately; and then, having done its work in driving the piston to and fro, is discharged from each side alternately, either into the air or into a condenser.

The diagram in Fig. 129 will serve to illustrate the general features and the operation of a steam engine. The details of the various mechanical contrivances are purposely omitted, so as to present the engine as nearly as possible in its simplicity.

In the diagram, A represents the *steam pipe* through which steam passes from the boiler to a small chamber, D, called the *valve-chest*. In this chamber is a *slide-valve*, G, which, as it is moved to and fro, opens and closes alternately the passages leading from the valve-chest to the *cylinder* B, and thus admits the steam alternately each side of the *piston* C. When one of these passages is open, the other is always closed.

In the position of the valve represented in the diagram, the passage F is open, and steam entering the cylinder at the bottom drives the piston upward. At the same time the steam on the upper side of the piston escapes through the passage E and the exhaust-port H. While the piston rises, the valve closes the passage F leading from the valve-chest and opens the passage E into the same, and thus the order of things is reversed.

Motion is communicated by the piston through the *piston rod* K and the *walking beam* L L to the *crank* M, and by this means the shaft carrying a *fly-wheel*, N N, is rotated. Connected with an eccentric on the shaft is the rod o o, which by a simple device communicates a to-and-fro motion to the slide-valve G. Motion is also communicated by the band e e to the *governor* d, which is caused to rotate. If the speed of the parts, due to undue pressure of steam, become excessive, the centrifugal tendency will cause the arms of the governor to rise and thereby close the *throttle-valve* b and thus shut off steam.

The *fly-wheel* is a large, heavy wheel, having the larger portion of its mass located near its circumference; it serves as a reservoir of energy, which is needed to make the rotation of the shaft and all other machinery connected with it uniform, so that sudden changes of velocity resulting from sudden changes of the driving power or resistances may be avoided.

When the exhaust steam escapes through H into the air the engine is said to be a *high-pressure* or a *non-condensing* engine; when it is led to a condensing chamber (as J in the figure) and there condensed by a spray of cold water for the purpose of reducing the back pressure of the atmosphere, the engine is called a *low-pressure* or a *condensing* engine.

This water must be pumped out of the condenser by a special pump, R, called technically the *air-pump*; thus a partial vacuum is maintained. The advantage of such an engine is obvious, for if the exhaust-pipe, instead of opening into a condenser, communicate with the outside air, as in the *non-condensing engine*, the steam is obliged to move the piston constantly against a resistance arising from atmospheric pressure of 15 pounds for every square inch of the surface of the piston. But in the condensing engine a large portion of the pressure on the exhaust side of the piston is removed and an equivalent portion of the pressure on the steam side is utilized and made to do useful work. In well-proportioned condensing apparatus the pressure on the exhaust side may be reduced 90 per cent, so that the moving piston instead of working against a resistance of 15 pounds meets with a resistance of only 1.5 pounds per square inch.

169. Steam Gauge. An instrument called a steam gauge is connected with the boiler. It measures the excess of the pressure of the steam at any instant above the atmospheric pressure. The absolute pressure of the steam (*i.e.* measured from zero) is the pressure indicated by the steam gauge *plus* the pressure of the atmosphere at the time.

170. Power of a Steam Engine. The horse-power of a steam engine is calculated by means of the following formula:

(Mean effective pressure in pounds per square inch on piston × area of piston in square inch × length of stroke in feet × number of strokes per minute) ÷ 33,000.

The steam engine, with all its merits and with all the improvements which modern mechanical art has devised, is an exceedingly wasteful machine. The best engine that has been constructed utilizes less than 15 per cent of the heat energy generated by the combustion of the fuel.

EXERCISES.

1. Why does the temperature of steam suddenly fall as it moves the piston?

2. What do you understand by a ten horse-power steam engine?

EXERCISES. 165

3. Upon what does the power of a steam engine depend?

4. The area of a piston is 500 square inches, and the average unbalanced steam pressure is 30 pounds per square inch. What is the total effective pressure? Suppose that the piston travels 30 inches at each stroke, and makes 100 strokes per minute, 40 per cent being allowed for wasted energy, what power does the engine furnish, estimated in horse-powers?

5. Can ice at 0° C., and under ordinary atmospheric pressure, have its temperature raised? Explain.

6. Find the resulting temperature (C.) of the following mixtures:
   (a) 5 k. of snow at 0° with 25 k. of water at 28°.
   (b) 4 k. of ice at $-10°$ with 30 k. of water at 50°.
   (c) 10 k. of iron at 200° with 2 k. of ice at 0°.

7. How many thermal units are required to change 5 k. of ice at $-10°$ C. into water at 10°?

8. If 30 g. of steam at 100° C. be passed into 400 g. of ice-water at 0° C., what will be the temperature of the mixture?

9. A building is heated by hot-water pipes. How does heat get from the furnace of the boiler to a person in the building?

10. A building is heated by steam pipes. How does heat get from the furnace to objects in the building?

11. A rod of copper at 0° C. measures 10 feet; its length at 100° C. is 0.191 inch greater. Find the coefficient of expansion of copper.

12. A silver rod at 0° C. is 10 feet long. Find its length at 100° C.

13. A kettle contains 2 k. of water at 60° C. How much heat must be supplied to vaporize all the water?

14. In what state is heat when it is not manifest to the senses?

15. Which plays the greater part in the heating of a room, convection or conduction?

16. Why are woolen blankets good both for keeping a person warm in winter and for preserving ice in summer?

17. Explain the benefit derived from double windows and double doors in cold climates.

18. At what temperature will a liter hold the greatest quantity of water?

19. Explain how a piece of iron hammered on an anvil becomes hot.

20. What temperature of the arbitrary Centigrade scale corresponds to 450° of the absolute scale?

21. (a) Why do metals and marble in a cold room feel colder than wood and flannel? (b) Why do the latter in a hot oven feel cooler than the former?

CHAPTER V.

ENERGY OF MASS VIBRATION. — SOUND-WAVES.

SECTION I.

STUDY OF VIBRATIONS AND WAVES.

The subjects of Sound-waves and Light-waves, which we are about to study, have two important characteristics in common that distinguish them from the subjects already studied. First, each of them affects its peculiar organ of sense, the ear or the eye. Secondly, both originate in vibrating bodies, and reach us only by the intervention of some medium capable of being set in vibration.

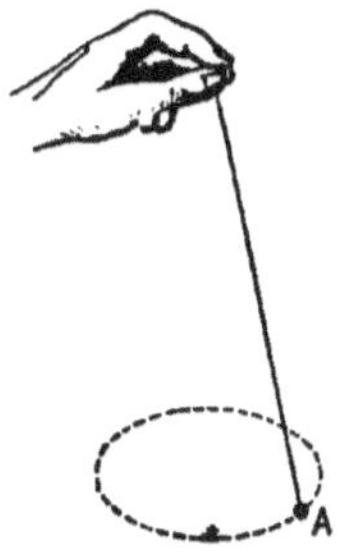

171. Simple Harmonic Motion. The motion of a lead bullet (Fig. 130) suspended by a thread and set swinging in a circular path in a horizontal plane is practically uniform. When viewed from above its path appears circular, when viewed obliquely its path appears elliptical, and when the eye is placed on the same level with it the motion appears to be in a straight line, M N (Fig. 131). The apparent motion of the bullet, when viewed from this position, is the projection of uniform circular motion on a straight line. While the bullet passes points 1, 2, 3, etc., lines drawn from these points normal to the line M N intersect it in points I, H, C, etc. While the bullet moves over equal spaces 1–2, 2–3, etc., in equal

FIG. 130.

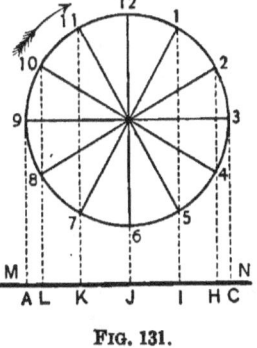

FIG. 131.

intervals of time, the speed of the point of intersection is variable, as shown by the unequal spaces I H, H C, etc., traversed by this point in equal intervals of time. Such a motion as that, whether in a line, an ellipse, or a circle, executed in equal intervals of time, is called *simple harmonic motion*, and it is the kind of motion with which we have to deal chiefly in the study of sound and light.

It is the kind of motion executed by the vibrating prongs of a tuning fork, and, generally, by the vibrating parts of all musical instruments. It is the kind of motion into which the particles of air are thrown when the atmosphere is traversed by sound-waves, and that is set up in the ether (§ 211) when it is traversed by light-waves or electrical waves.

The time occupied by a particle in executing a single complete harmonic motion, *i.e.* from A to C and back, is called the *period* of a complete vibration. The extent of the vibration on either side of the middle point, as A J or J C, is called the *amplitude* of the vibration.

172. Direction of Vibration.

Experiment 1. Grasp one end of a small rod or yardstick in a vise, pull the free end to one side, and set it in vibration. Pluck a string of a piano or violin. Note that the motions of all the bodies which thus far we have caused to vibrate are at *right angles to their length*. These are called *transverse vibrations*.

Experiment 2. Suspend a spiral spring or an elastic cord with a small weight attached at the lower end; lift the weight, and, dropping it, notice that the spring or cord vibrates *lengthwise*. This is a case of *longitudinal vibration*. There may also be *torsional vibrations;* for example, children often amuse themselves by producing these by twisting a curtain cord and tassel.

173. Wave-motion.
Imagine a series of particles moving with harmonic motion in parallel straight lines A, B, C, D, etc. (Fig. 132), so that each succeeding particle begins to move a definite time later than the preceding one. Suppose, for example, that the period of vibration is one second, and that each particle begins to move $\frac{1}{15}$ of a second later than the

preceding one. Draw a circle whose radius represents the amplitude of vibration, and divide it into twelve equal parts. Let the particle in line A be at a (*i.e.* at 4 in the circle); then the particle in line B will be $\frac{1}{12}$ of a period behind, that is, at b; the particle in C will be at c, etc. Join points a, b, c, etc., and a curve (represented by the thick line) is formed. In $\frac{3}{12}$ second the particle in A will have moved from a to a' (in the circle from 4 to 7); b will have moved to b', etc. Join these

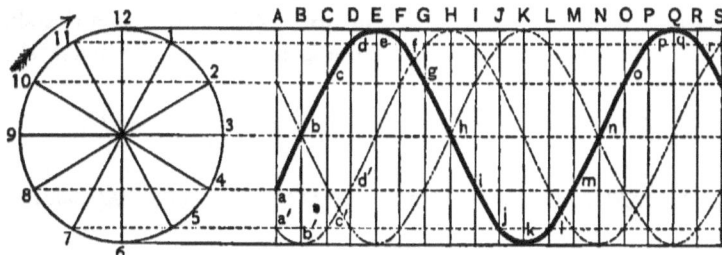

FIG. 132.

points, and the dotted curve a', b', c', d', etc., will be formed. The crest has moved from E to H; the wave-form is moving from A toward S. In $1\frac{2}{3}$ second, or one period, the particles will be in their original positions, but the crest of the wave will have moved from E to Q. This distance, or the distance from any particle to the next particle that is in a similar position in its path and is moving in the same direction (*e.g.* b to n, c to o) is called a *wave-length*. The wave-length may also be defined as *the distance the wave travels in one period*.

Since the wave travels one wave-length in a period, to determine the velocity with which the wave travels, it is only necessary to determine the number of wave-lengths the wave passes over in a period of time. Thus,

$$v = l \times n,$$

in which v, l, and n represent, respectively, velocity, wave-length, and number of wave-lengths.

174. Graphical Method of Studying Vibrations.

Experiment 1. Attach, by means of sealing-wax, a bristle or a fine wire to the end of one of the prongs of a large steel fork (like a tuning fork, but larger) called a *diapason*. Set the fork in vibration, and quickly draw the point of the bristle lightly

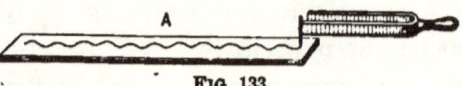

FIG. 133.

over smoked glass (A, Fig. 133). A beautiful wavy line will be traced on the glass, each wave corresponding to a vibration of the prong when vibrating as a whole.

Next, tap the fork, near its stem, on the edge of a table, and trace its vibrations on a smoked glass as before. You will generate a similar set of waves, but running over these is another set, of much shorter

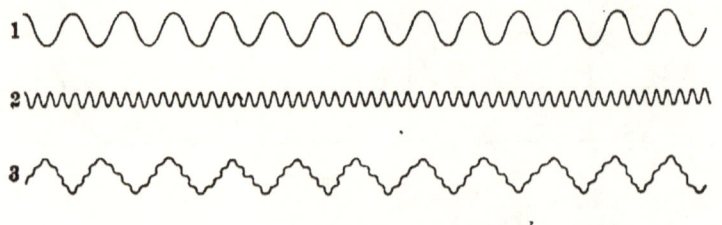

Fig. 134.

period, like the lower line of Fig. 134, showing that the prong vibrates, not only as a whole, but in parts. The serrated wavy line produced represents the resultant of the combined vibrations, and may be called a *complex wave-line*.

The vibration frequency of a fork may easily be found by means of an apparatus called a *vibrograph*. One of the tines of the fork a (Fig.

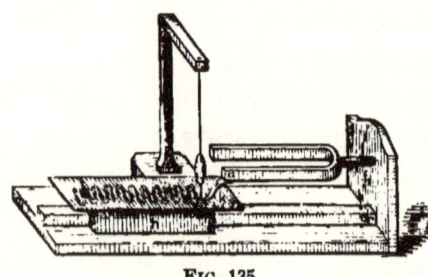

FIG. 135.

135) has a small elastic indicator attached to its extremity. The sharp point of this indicator touches a smoked glass plate, k, below. Above the glass plate is suspended a pendulum with a heavy bob. Beneath the bob is another indicator which just grazes the glass as it passes the lower part of its arc. The experimenter first finds the exact fraction of a second occupied by the pendulum in making one complete or double vibration. The fork is

then put in vibration and the block h, carrying the glass plate, is drawn along beneath the style, which marks upon the glass a wave-line. At the same time the pendulum is set swinging and allowed to traverse the plate width-wise three times, making, with its indicator, three lines athwart the wave-line. Now the interval of time between the instants when the first and the third of these lines are made is the period of one complete vibration. The number of vibrations which the fork made in this interval may be determined from the sinuous curved line intervening between the lines made by the pendulum. The number of vibrations made by the fork in a certain fraction of a second having been ascertained in this manner, the vibration number per second is calculated therefrom.

175. How Vibratory Motion, i.e. a Wave, is Propagated through an Elastic Medium.

Experiment 2. Fig. 136 represents a spring brass wire wound into the form of a spiral, about 12 feet long. Attach one end to a cigar box, and fasten the box to a table. Hold the other end of the spiral firmly in

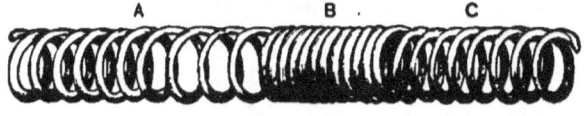

FIG. 136.

one hand, and with the other hand insert a knife blade between the turns of the wire, and quickly rake it for a short distance along the spiral toward the box, thereby crowding closer together for a little distance (B) the turns of wire in front of the hand, and leaving the turns behind pulled wider apart (A) for about an equal distance. The crowded part of the spiral may be called a *condensation*, and the stretched part a *rarefaction*. The condensation, followed by the rarefaction, runs with great velocity through the spiral, strikes the box, producing a sharp thump; is reflected from the box to the hand, and from the hand again to the box, producing a second thump; and by skillful manipulation three or four thumps will be produced in rapid succession. If a piece of twine be tied to some turn of the wire, it will be seen, as each wave passes it, to receive a slight jerking movement forward and backward in the direction of the length of the spiral.

The effect of applying force with the hand to the spiral spring is to produce in a certain section, B, of the spiral a crowding together of the turns of wire, and at A a separation; but the elasticity of the spiral instantly causes B to expand, the effect of which is to produce a crowding together of the turns of wire in front of it, in the section C, and thus a forward movement of the condensation is made. At the same time, the expansion of B causes a filling up of the rarefaction at A, so that this section is restored to its normal state. This is not all; the folds in the section B do not stop in their swing when they have recovered their original position, but, like a pendulum, swing beyond the position of rest, thus producing a rarefaction at B, where immediately before there was a condensation. Thus a forward movement of the rarefaction is made, and thus a pulse or wave is transmitted with uniform velocity through a spiral spring or any *elastic* medium.

A wave cannot be transmitted through an inelastic soft brass spiral. *Elasticity is essential in a medium, in order that it may transmit waves composed of condensations and rarefactions; and the greater the elasticity in a medium, the greater the facility and rapidity with which it transmits waves.*

176. Fluids as Media of Wave-motion.

Experiment 3. Arrange apparatus as shown in Fig. 137. The whole is filled with water, except the glass bulb A, and a portion of the glass tube connected with it. Glass tubes C and D are connected by a rubber tube, E, 10 or 12 feet long. The top of the thistle tube N is covered with thin sheet rubber.

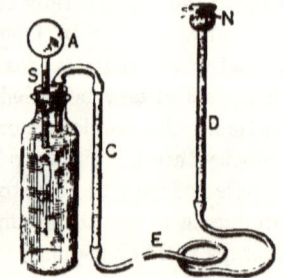

Fig. 137.

Tap with the end of a finger the diaphragm N; the water column in S is instantaneously thrown into oscillations. The observer should note the promptness with which the water in S responds to any impulse given the diaphragm. The impulse is transmitted by wave-motion with a velocity of about 1435 meters (4708 feet) per second.

Experiment 4. Place a candle flame at the orifice a of the long tin tube A (Fig. 138), and strike the table a sharp blow with a book near the orifice b. Instantly the candle flame is quenched. The body of air in the tube serves as a medium for transmission of motion to the candle.

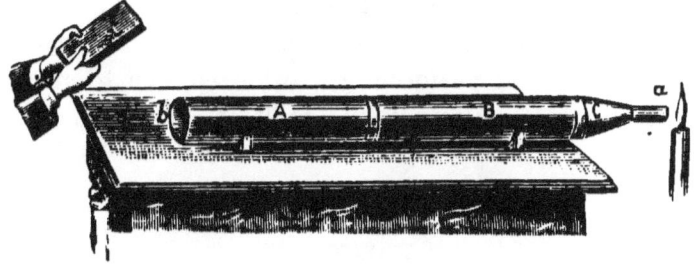

FIG. 138.

Is the motion transferred that of a current of air through the tube (a miniature wind), or is it a vibratory motion? Burn touch-paper [1] at the orifice b, so as to fill this end of the tube with smoke, and repeat the last experiment.

Evidently, if the body of the air be moved along through the tube, the smoke will be carried along with it. The candle is blown out as before, but no smoke issues from the orifice a. It is clear that there is no translation of material particles from one end to the other — nothing like the flight of a rifle bullet. The candle flame is struck by something like a *pulse* of air, not by a *wind*.[2]

Air is a fluid, and therefore has only volume elasticity. The only waves it can propagate are waves composed of compressions and rarefactions. There are two important distinctions between these waves and waves of water, or waves sent along a cord when one end is shaken: (1) the former consist of condensations and rarefactions; the latter of elevations and depressions; (2) in the former the vibrations of the particles are in the same line with the path of the wave,

[1] To prepare touch-paper, dissolve about a teaspoonful of saltpeter in a half-teacupful of hot water, dip unsized paper in the solution, and then allow it to dry. The paper produces much smoke in burning, but no flame.

[2] If a membrane be tied tightly over the orifice b and a sudden blow be given it (*e.g.* by snapping it with a finger), the vibratory character of the motion communicated through the tube is well shown by the fact that the flame is first driven from the orifice a and immediately afterward drawn toward it.

and hence they are called *longitudinal* vibrations; in the latter the vibrations take place in planes at some angle to the path of the wave, and are therefore called *transverse* vibrations. As an air-wave advances, every individual particle concerned in its transmission performs a short excursion to and fro in the direction of a straight line radiating from the source of the waves.

SECTION II.

SOUND AND SOUND-WAVES.

177. Sound and Sound-waves Defined. *Sound is a sensation caused usually by air-waves beating upon the organ of hearing.*[1]

Sound-waves are waves in any medium (usually air) *that are capable of producing the sensation of sound.*

If we could see the air as it is traversed by sound-waves, we should see spherical shells of condensed air alternating with shells of rarefied air. The condensed portions correspond to localities of greater pressure, while the rarefied portions represent localities of smaller pressure. When there is an increase of pressure on the drumhead of the ear it is pushed in, and when the pressure becomes less the drumhead springs back to its former position.

178. Sound-waves Originate in Mass-vibration. A body vibrating in an elastic medium, *e.g.* in air, does not necessarily produce sound-waves; in other words, not all waves are sound-waves. For example, the energy of the vibrations may be insufficient, or the vibrating body may be so small (or the medium so rare) that it cuts through the medium without condensing it sufficiently to produce audible effects.

---

[1] As commonly used, the term sound is ambiguous, being applied to both a sensation and the physical cause of the sensation. With sound as a sensation we have little to do, as this is a physiological rather than a physical phenomenon. No more appropriate name than sound-wave can be applied to the physical agent with which we are to deal; it suggests at once the reality, and is not suggestive of some vague mysterious *thing* shot through space. Sound is a condition, not a thing.

Experiment 1. Strike a bell or a glass bell-jar, and touch the edge with a small ivory ball suspended by a thread; you not only *hear* the sound, but, at the same time, you *see* a tremulous motion of the ball, caused by a motion of the bell. Touch the bell gently with a finger, and you feel a tremulous motion. Press the hand against the bell; you stop its vibratory motion, and at that instant the sound ceases. Watch the strings of a piano, guitar, or violin, or the tongue of a jew's-harp, when sounding. You can *see* that they are in motion.

As a bell while sounding possesses no peculiar property except motion, it has nothing to communicate but motion. The vibrations of a sonorous body cannot be communicated to the ear unless there be a continuous intervening medium of some kind.

Experiment 2. Lay a thick tuft of cotton wool on the plate of an air-pump, and on this, face downward, place a loud-ticking watch, and cover with the receiver. Notice that the receiver, interposed between the watch and your ear, greatly diminishes the sound, or interferes with the passage of *something* to the ear. Partially exhaust the air by a few strokes of the pump, and listen; the sound is more feeble, and continues to grow less and less distinct as the exhaustion progresses, until either no sound can be heard when the ear is placed close to the receiver, or an extremely faint one, as if coming from a great distance. The removal of air from a portion of the space between the watch and your ear prevents the passage of sound-waves to your ear. Let in the air again, and the sound is again heard.

179. Solids and Liquids are Capable of Transmitting Sound-waves.

Experiment 3. Place one end of a long pole on a cigar box, and apply the stem of a vibrating tuning fork to the other end; the sound-vibrations will be transmitted through the pole to the box, and a sound will be given out by the box, as though that, and not the tuning fork, were the origin of the sound.

Experiment 4. Place the ear to the earth, and listen to the rumbling of a distant carriage; or put the ear to one end of a long stick of timber, and let some one gently scratch the other end with a pin.

SECTION III.

VELOCITY OF SOUND-WAVES.

180. Velocity of Sound-waves Dependent on Elasticity and Density of Medium. *The velocity with which a particle of an elastic medium vibrates, and therefore the velocity of propagation in the medium (i.e. the velocity of a sound-wave), is directly proportional to the square root of the elasticity of the medium, and inversely proportional to the square root of its density.* The relation of these quantities is shown in the formula

$$V \propto \sqrt{\frac{e}{d}}.$$

If the elasticity and density of the medium vary alike, and in the same direction, it is evident that the velocity of the sound-wave is unaffected. Hence the velocity of a sound-wave is unaffected by barometric hight, or elevation above sea level. Temperature, however, affects only the density of air. Elevation of temperature of the air diminishes the density of the air, and therefore tends to increase the speed of the sound-wave. The velocity of a sound-wave is greatest in the direction of the wind. Velocity of sound-waves is very nearly independent of pitch and intensity.

The greater density of solids and liquids, as compared with gases, tends, of course, to diminish the velocity of sound-waves; but their greater incompressibility more than compensates for the decrease of velocity occasioned by the increase of density. As a general rule, solids are more incompressible than liquids; hence sound-waves generally travel faster in the former than in the latter. For example, sound-waves travel in water about four times as fast as in air, and in iron and glass sixteen times as fast.

The velocity of sound-waves in air at 0° C. is 332.4 m. (nearly 1091 feet) per second. The increase of velocity per degree C. is .608 m. (23.9 inches) per second.

176   MOLAR DYNAMICS.

SECTION IV.

ENERGY OF SOUND-WAVES. LOUDNESS.

181. Energy of Sound-waves Depends on the Amplitude of Vibration.
Fix your attention upon a particle of air as a sound-wave passes it. At a certain point of its vibratory excursion its velocity is at its maximum. Now since the energy of a moving particle varies as the square of its velocity, the *intensity* of the impact which it is capable of producing upon the ear is *proportional to the square of this maximum velocity*.

It is also clear that if the amplitude of vibration of a particle be doubled while its period remains constant, its velocity is doubled, and therefore its energy is increased fourfold. Hence, (1) *measured mechanically, the energy of a sound-wave is proportional to the square of the amplitude of the vibration of particles,* or, *it is proportional to the square of the maximum velocity of the vibrating particles.*

Loudness of sound refers to the intensity of a sensation. We have no standard of measurement for a sensation, so we are compelled to measure the energy of the sound-wave, knowing at the same time that *loudness is not proportional to this energy.*

182. Energy of Sound-waves Depends upon the Density of the Medium. In the experiment with the watch under the receiver of the air-pump (p. 174), the sound grew feebler as the air became rarer. Aëronauts are obliged to exert themselves more to make their conversation heard when they reach great hights than when in the denser lower air. In diving-bells persons are obliged to speak in undertones. In a rare medium, either a vibrating body sets in motion fewer particles during a single vibration, as in the case of the partially exhausted receiver, or, as in the case of the hydrogen gas, it sets in motion particles of less mass than in a dense medium; consequently it parts with its energy more slowly, and the sound is weaker.

(2) *The energy of gaseous sound-waves increases with the density of the medium in which they are produced.*

183. Energy of Sound-waves Depends on Distance from their Source. It is a matter of everyday observation that the loudness of a sound diminishes very rapidly as the distance from the source of the waves to the ear increases. As a sound-wave advances in an ever-widening sphere, a given quantity of energy becomes distributed over an ever-increasing surface; and as a greater number of particles partake of the motion, the individual particles receive proportionately less energy; hence it follows, — as a consequence of the geometrical truth that "the surface of a sphere varies as the square of its radius," — that (3) *the energy of a sound-wave varies inversely as the square of the distance from the source.* This is known as the Law of Inverse Squares.

184. Speaking-tubes.

Experiment. Place a watch at one end of the long tin tube (Fig. 138), and the ear at the other end. The ticking sounds very loud, as though the watch were close to the ear.

Long tin tubes, called *speaking-tubes*, passing through many apartments in a building, enable persons at the distant extremities to carry on conversation in a low tone of voice, while persons in the various rooms through which the tube passes hear little or nothing. The reason is that the sound-waves which enter the tube are prevented from expanding, consequently the energy of the sound-waves is not affected by distance, except as it is wasted by friction of the air against the sides of the tube, and by internal friction due to the viscosity of the air.

178   MOLAR DYNAMICS.

SECTION V.

CHANGES IN DIRECTION OF PROPAGATION OF SOUND-WAVES.

185. Reflection. So long as sound-waves are not obstructed in their motion they are propagated in the form of concentric spheres; but when they meet with an obstacle, they follow the general law of elastic bodies; that is, they return upon themselves, forming new concentric waves, called reflected waves, which seem to emanate from a second center on the other side of the reflecting body. This phenomenon is called the *reflection of sound-waves*.

A (Fig. 139) represents a vibrating particle or a sonorous center from which emanates a series of waves. P Q represents an obstacle with a flat surface turned toward the waves. Take, for example, the incident wave M C D N, emitted from the center A; the corresponding reflected wave is represented by the arc C K D of a circle whose center a is as far beyond the obstacle P Q as A is in front of it.

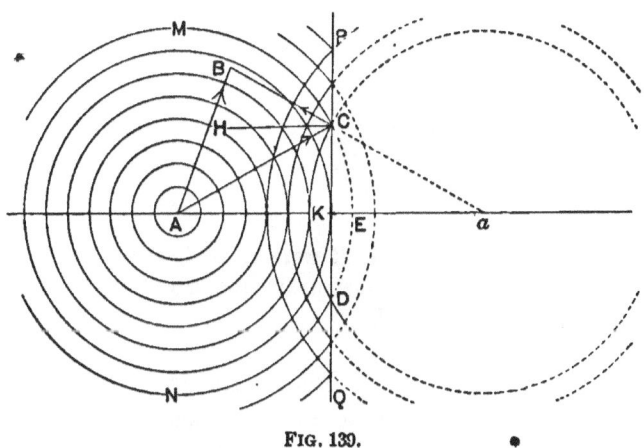

FIG. 139.

Join any point, C, of the reflecting surface to the sonorous center, and the line A C represents one of an infinite number of directions in which energy is transmitted by a sound-wave. Such a line may conveniently be called a *sound-ray*. Let fall the line H C normal to the surface at the point of incidence C. The angle A C H is called the *angle of incidence*. The ray A C after reflection takes the direction C B, which is a prolongation of a C. The angle B C H is called the *angle of reflection*. An observer at B receives sound-waves not only directly from A in the line A B, but also from C in the line C B. Hence he hears two sounds, one (to speak in common parlance) proceeding from point A, and the other from point C. The latter travels from A to C and from C to B, a longer distance than A B, and is therefore heard later than the former. If the interval of time between their arrivals at B be greater than about a fifth of a second, the ear is able to separate the two sensations and the latter appears as an *echo*. If the interval of time be too short, then only a single and perhaps somewhat blurred and indistinct sound is heard. The latter phenomenon is usually called *resonance*. Such an effect is experienced frequently when a person speaks in a large hall.

If the obstacle P Q present a concave surface, the wave-front after reflection will be less convex, and may become plane or even concave, according to the degree of the concavity of the reflector and the position of the sounding body.

186. Sound-waves Reflected by Concave Mirrors.

Experiment. Place a watch at the focus A (Fig. 140) of a concave mirror, G. At the focus B of another concave mirror, H, place the large opening of a small tunnel, and with a rubber connector attach the bent glass tube C to the nose of the tunnel. The extremity D being placed in the ear, the ticking of the watch can be heard very distinctly, as though it were somewhere near the mirror H. Though the mirrors be many feet apart, the sound will be louder at B than at an intermediate point E.

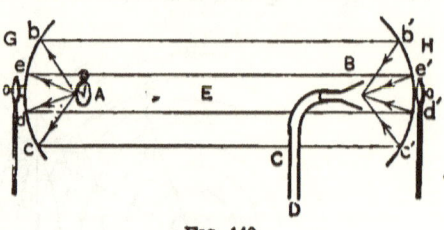

FIG. 140.

How is this explained? Every air particle in a certain radial line, as A c, receives and transmits motion in the direction of this line; the last particle strikes the mirror at c, and, being elastic, bounds off in the direction c c', communicating its motion to the particles in this line. At c' a similar reflection gives motion to the air particles in the line c′ B. In consequence of these two reflections, all divergent sound-rays, as A d, A e, etc., that meet the mirror G are there rendered parallel, and afterwards rendered convergent at the mirror H. The practical result of the concentration of this scattering energy is that a sound of great intensity is heard at B. The points A and B are called the *foci* of the mirrors. The front of the wave as it leaves A is convex, in passing from G to H it is plane, and from H to B it is concave. If you fill a large circular tin basin with water, and strike one edge with a knuckle, circular waves with concave fronts will close in on the center, heaping up the water at that point.

Long " whispering galleries" have been constructed on this principle. Persons stationed at the foci of the concave ends of the long gallery can carry on a conversation in a whisper which persons between cannot hear. The external ear is a wave-condenser. The hand held concave behind the ear, by increasing the reflecting surface, adds to its efficiency.

SECTION VI.

REËNFORCEMENT OF SOUND-WAVES; INTERFERENCE OF SOUND-WAVES.

187. Reënforcement of Sound-waves.

Experiment 1. Set a diapason in vibration; unless it is held near the ear, you can scarcely hear the sound. Press the stem against a table; the sound rings out loud, but the waves seem to proceed from the table.

When only the fork vibrates, the prongs, presenting little surface, cut their way through the air, producing very slight condensations, and consequently waves of little intensity. When the fork rests upon the table, the vibrations are communicated to the table; the table with its larger surface throws a larger mass of air into vibration, and thus greatly intensifies the sound-waves. The strings of the piano, guitar, and violin owe as much of their loudness of sound to their elastic sounding-boards as the fork does to the table.

188. Reënforcement by Bodies of Air; Resonators.

Experiment 2. Take a glass tube, A (Fig. 141), 16 inches long and 2 inches in diameter; thrust one end into a vessel of water, C, and hold over the other end a vibrating diapason, B, that makes (say) 256 vibrations in a second. Gradually lower the tube into the water, and when it reaches a certain depth, *i.e.* when the column of air o c attains a certain length, the sound becomes very loud; as the tube is lowered below this point, the sound rapidly dies away.

Columns of air, as well as sounding-boards, serve to *reënforce* sound-waves. The instruments which enclose the columns of air are called *resonators*. Unlike sounding-boards they can respond loudly to only one tone, or to a few tones of widely different pitch.

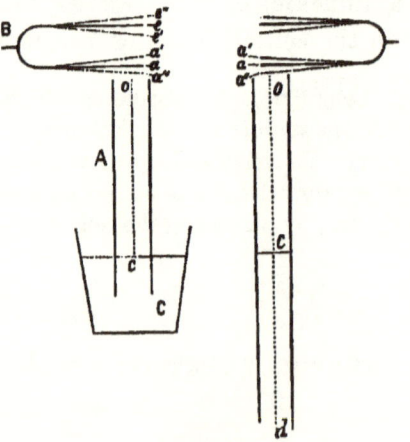

Fig. 141.

How is this reënforcement effected? When the prong a moves from one extremity of its arc a' to the other a", it sends a condensation down the tube; this condensation, striking the surface of the water, is reflected by it up the tube. Now suppose that the front of this reflected condensation should just reach the prong at the instant it is starting on its retreat from a" to a'; then the reflected condensation will combine with the condensation formed by the prong in its retreat to make a greater condensation in the air outside the tube. Again, the retreat of the prong from a" to a' produces in its rear a rarefaction, which also runs down the tube, is reflected, and reaches the prong at the instant it is about to return from a' to a", and to cause a rarefaction in its rear; these two rarefactions moving in the same direction conspire to produce an intensified rarefaction. The original sound-waves thus combine with the reflected, to produce

182 MOLAR DYNAMICS.

resonance; but this can happen only when the like parts of each wave coincide each with each; for if the tube were somewhat longer or shorter than it is, it is plain that condensations and rarefactions would meet in the tube and tend to destroy each other.

The loudness of sound of all wind instruments is due to the resonance of the air contained within them. A simple vibratory movement at the mouth or orifice of the instrument, scarcely audible in itself (such as the vibration of a reed in reed pipes, or a pulsatory movement of the air, produced by the passage of a thin sheet of air over a sharp wooden or metallic edge, as in organ pipes, flutes, and flageolets, or more simply still by the friction of a gentle stream of breath from the lips, sent obliquely across the open end of a closed tube), is sufficient to throw the large body of air enclosed in the instrument into vibration, and the sound thus reënforced becomes audible at long distances.

Experiment 3. Attach a rose gas-burner, A (Fig. 142), to a metal gas-pipe about 1 m. in length, and connect this by a rubber tube with a gas-nipple. Light the gas at the rose burner, and you will hear a low rustling noise. Remove the conical cap from the long tin tube (Fig. 138), support the tube in a vertical position, and gradually raise the burner into the tube; when it reaches a certain point not far up, the body of air in the tube will catch up the vibrations, and give out deafening sound-waves that will shake the walls and furniture in the room.

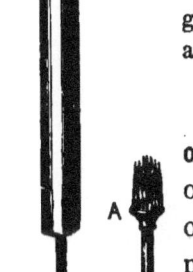
FIG. 142.

189. Measuring Wave-lengths and the Speed of Sound-waves. Experiments like that described on p. 181 enable us readily to measure the length of the wave produced by a fork whose vibration number is known, and also to measure the velocity of sound-waves. It is evident that if a condensation generated by the prong of the fork in its forward movement from a' to a'' (Fig. 141) meet with no obstacle, its front, meantime, will traverse the distance o d, or twice the distance o c; hence the length of the condensation is the distance o d. But a condensation is only one half of a wave, and the passage of the prong from a′ to a″ is only one half of a vibration; consequently the distance o d is one half of a wave-length, and the distance o c is one fourth of a wave-length. The measured distance of o c in this case is about 13.13 inches; hence, the length of wave produced by a C′-fork making 256 vibrations in a second is about (13.13 inches × 4 =) 52.5 inches = 4.38 feet. And since a wave from this fork travels 4.38 feet in $\frac{1}{256}$ of a second, it will travel in an entire second (4.38 feet × 256 =) 1121 feet. The distance o c varies with the temperature of the air.

It is evident that the three quantities expressed in the formula

$$\text{wave-length} = \frac{\text{velocity}}{\text{number of vibrations}}$$

bear such a relation to one another that if any two be known, the remaining quantity can be computed. It will further be observed that *with a given velocity the wave-length varies inversely as the number of vibrations;* i.e. the greater the number of vibrations per second, the shorter the wave-length.

190. Interference of Sound-waves.

Fig. 143.

Experiment 4. Hold a vibrating diapason over a resonance jar, as in Fig. 143. Roll the diapason over slowly in the fingers. At certain points a quarter of a revolution apart, when the diapason is in an oblique position with reference to the edge of the jar as represented in the figure, the reënforcement from the tube almost entirely disappears, but it reappears at the intermediate points. That is, there are four intervals in the space around the fork where the two series of waves generated by the two tines interfere to produce mutual destruction. These are called technically *the cones of silence*. Return to the position where there is no resonance, and, without touching the diapason, enclose in a loose roll of paper the prong farthest from the tube, so as to prevent the sound-waves produced by that prong from passing into the tube; the resonance resulting from the vibrations of the other prong immediately appears.

Two sound-waves may combine to produce a sound louder or weaker than either alone would produce, or may even cause silence. This combination of sound-waves to produce a louder or weaker sound is called *interference*.

191. Forced and Sympathetic Vibrations.

Experiment 5. Suspend from a frame several pendulums, A, B, C, etc. (Fig. 144). A and D are each 3 feet long, C is a little longer, and B and E are shorter. Set A in vibration; slight impulses will be communicated through the frame to D, and cause it to vibrate. The vibration-period of D being the same as that of A, all the impulses tend to accumulate motion in D, so that it soon vibrates through arcs as large as those of A. On the other hand, C, B, and E, having different rates of vibration from that of A, will at first acquire a slight motion, but soon their vibrations will be in opposition to those of A, and then the impulses received from A will tend to destroy the slight motion they had previously acquired.

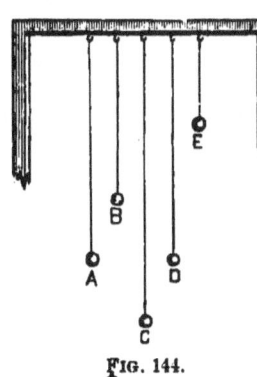

FIG. 144.

Experiment 6. Press down gently one of the keys of a piano so as to raise the damper without making any sound, and then sing loudly into the instrument the corresponding note. The string corresponding to this note will be thrown into vibrations that can be heard for several seconds after the voice ceases. If another note be sung, this string will respond only feebly.

Raise the dampers from all the strings of the piano by pressing the foot on the right-hand pedal, and sing strongly some note into the piano. Although all the strings are free to vibrate, only those will respond loudly that correspond to the note you sing, *i.e.* those that are capable of making the same number of vibrations per second as are produced by your voice.

So the pulses or waves that traverse the air between the vocal organs and the strings, so gentle that only the sensitive organ of the ear can perceive them, become great enough to bend the rigid steel wires when the energy of their blows, dealt at the rate of perhaps 512 in a second, accumulates. The large number of blows makes up for the feebleness of the individual blows.

These experiments show that a vibrating body tends to make other bodies near it vibrate, even if their periods of vibrations be different. Vibrations of this kind, such, for example, as those of B, C, and E, in Experiment 5, and those generated in the sounding-board of pianos, violins, etc., are called *forced vibrations*. But if the period of the incident waves of air be the same as that of the body which they cause to vibrate, the amplitude and the intensity of the vibrations become very great, like that of the pendulum D, and those of the piano strings which gave forth the loud sounds. Such are called *sympathetic vibrations*.

SECTION VII.

PITCH OF MUSICAL SOUNDS.

192. On what Pitch Depends.

Experiment 1. Draw the finger-nail or a card across the teeth of a comb, first slowly and then rapidly. The two sounds produced are commonly described as *low* or *grave*, and *high* or *acute*. The hight of a musical sound is its *pitch*.

Experiment 2. Cause the circular sheet-iron disk A (Fig. 145) to rotate, and hold a corner of a visiting card so that at each hole an audible tap shall be made. Notice that when the separate taps or noises cease to be distinguishable, the sound becomes musical; also, that the pitch of the musical sound depends upon the rapidity of the rotation, *i.e.* upon the frequency of the taps.

Experiment 3. Hold the orifice of a tube, B, so as to blow through the holes as they pass. When the rotation is slow, separate puffs, from which it hardly seems possible to construct a musical sound, are heard. When, however, the ear is no longer able to detect the separate puffs, the sound becomes quite musical, and the pitch rises and falls with the speed.

Pitch depends upon the number of sound-waves striking the ear per second, or upon wave-length; i.e. the greater the number of vibrations per second, or the shorter the wave-length, the higher the pitch.

193. Distinction between Noise and Musical Sound. If the body that strikes the air deal it but a single blow, like the discharge of a firecracker, the ear receives but a single shock, and the result is called a *noise*. If several shocks be slowly received by the ear in succession, the ear distinguishes them as so many separate noises. If, however, the body that strikes the air be in vibration, and deal it a great number of little blows in a second, or if a large number of firecrackers be discharged one after another very rapidly, so that the ear is unable to distinguish the individual shocks, the effect produced is that of one continuous sound, which may be pleasing to the ear; and, if so, it is called a *musical sound*. Continuity alone does not necessarily render a sound musical. There must exist also *regularity* both in the periodicity and the intensity of the impulses. The distinction between music and noise is a distinction between the agreeable and the disagreeable, between *regularity* and *confusion*.

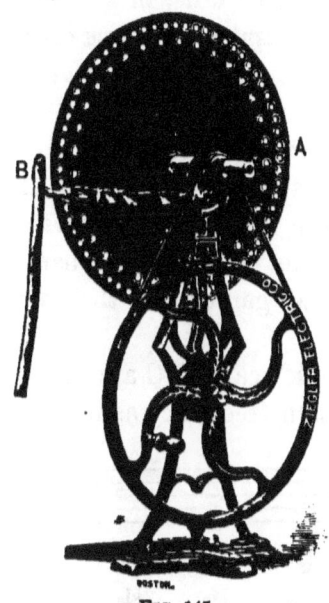

FIG. 145.

194. Musical Scale. Suppose a body, *e.g.* a tuning fork, to make 261 vibrations per second, the sound produced is recognized by our musical sense as the note which corresponds with the so-called middle C (c', or French ut₃) of a piano tuned to the national standard pitch.[1]

The pitch of a sound produced by twice as many vibrations as that of another sound is called the *octave* of the latter. Between two such sounds the voice rises or falls, in a manner very pleasing to the ear, by a definite number of steps called *musical intervals*. This gives rise to the so-called *diatonic scale*, or *gamut*.

The number of vibrations which shall constitute a given note is purely arbitrary, and differs slightly in different countries; but the ratios between the vibration numbers of the several notes of the gamut and the vibration number of the first or fundamental note of the gamut are the same among all enlightened nations.

The successive tones of the diatonic scale of C are related to one another with respect to vibration frequency as follows:

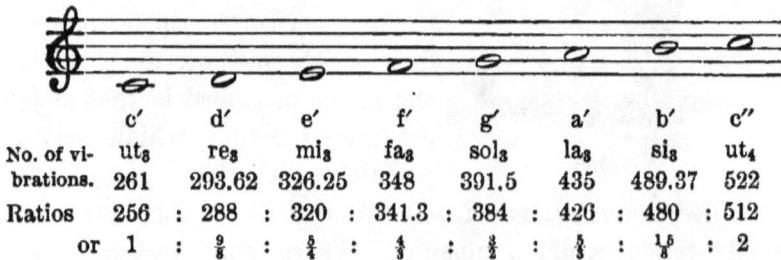

	c'	d'	e'	f'	g'	a'	b'	c''
	ut₃	re₃	mi₃	fa₃	sol₃	la₃	si₃	ut₄
No. of vibrations.	261	293.62	326.25	348	391.5	435	489.37	522
Ratios	256 :	288 :	320 :	341.3 :	384 :	426 :	480 :	512
or	1 :	$\frac{9}{8}$:	$\frac{5}{4}$:	$\frac{4}{3}$:	$\frac{3}{2}$:	$\frac{5}{3}$:	$\frac{15}{8}$:	2

The ear is wholly incapable of determining the number of vibrations corresponding to a given tone, but it is capable of determining with wondrous precision the *ratio* of the vibration numbers of two notes; hence all music must depend upon the recognition of such ratios, and for this reason the vibration ratios given above are of the utmost importance. An octave below c' is c; two octaves below, c_1, and so on. In a similar manner the octaves below any other tone are indicated.

---

[1] In a convention of piano manufacturers held in New York, it was decided that the national pitch, to go into effect July 1, 1892, should be the standard French, Austrian, and Italian pitch of 435 (A_3) double vibrations in a second at 68° F.

EXERCISES.

1. Find the vibration number for each note of the scale of which c'' is the first note.
2. What is the vibration number of c an octave below c'?
3. Find the wave-length corresponding to each note of the scale of which c' is the first, when the temperature of the air is $16°$ C.
4. Find the length of a resonance tube (disregarding its diameter) closed at one end, which will respond to c'' when the temperature is $16°$ C.
5. The same singer may not be able to sing twice alike, *i.e.* in the same key. How is it possible that the singing in both instances may be equally correct?
6. Why does the same bell always give a sound of nearly the same pitch?
7. (a) What is the effect of striking a bell with different degrees of force? (b) What change in the vibrations is produced? (c) What property of the sound remains the same?
8. (a) Strike a key of a piano and hold it down. What is the only change you observe in the sound produced, while it remains audible? (b) What is the cause of this change?

SECTION VIII.

COMPOSITION OF SONOROUS VIBRATIONS AND THEIR RESULTANT WAVE-FORMS.

195. Coexistence and Superposition of Waves. — Interference. When two or more currents of waves traverse the same medium at the same time and in the same or opposite directions, so that one set of waves is, as it were, superposed upon another, all the vibratory motions peculiar to the several waves are imparted to every particle of the medium simultaneously. When two or more systems of waves act on a particle at the same time, they are said to *interfere*. The resultant motion of any particle at a given instant may be found on the principle of parallelogram of motions; or, in case the several motions are parallel and occur at the same time, the resultant is the algebraic sum of the several motions.

This will be best understood by means of graphical representations. In A (Fig. 146) the wave-lines of two coexisting currents of waves having the same wave-length and phase, while the amplitude of one is greater than that of the other, are represented by dotted lines. For example, the amplitudes of the vibrations for the particle a are, respectively, a c and a e. Their algebraic sum is a d. In like manner the displacement of any particle of the medium traversed by the several wave-currents at any instant is determined. The heavy line represents the form of the joint wave resulting from the combination of the two. It will be seen that the only change is one of amplitude or intensity.

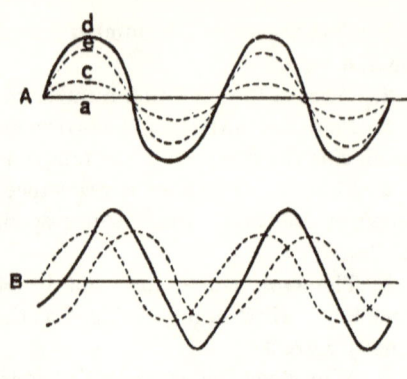

FIG. 146.

In B are two wave-currents whose waves are of the same length and amplitude, but have a difference of phase of ¼ of a period; *i.e.* one is a

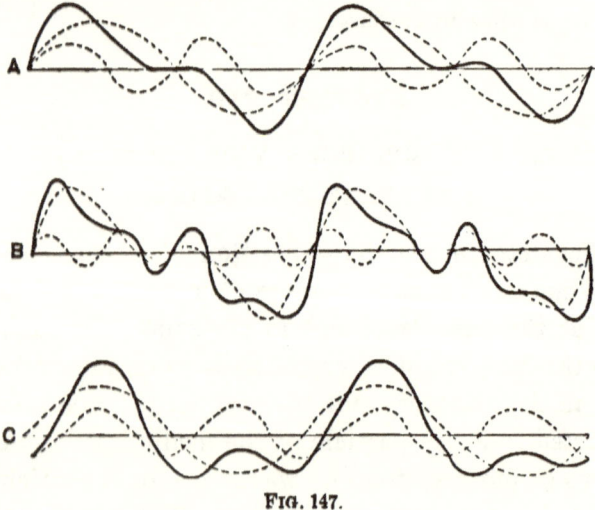

FIG. 147.

quarter of a wave-length behind the other. The result is a wave of the same length but of different phase and amplitude.

In A (Fig. 147) are given two wave-currents whose wave-lengths are as 1 : ¼ and whose phases in the beginning agree. The resultant of this combination with still another of ¼ the wave-length of the longest is shown in B. In C is the same combination as in A, but the phases differ by ¼ of a period of the shorter wave.

In the diagrams given above only transverse vibrations are represented, but the results there depicted apply equally well to longitudinal vibrations and to waves of condensations and rarefactions. In Fig. 148 the heavy

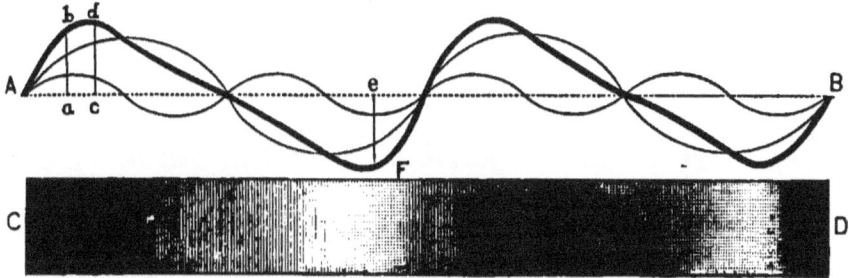

FIG. 148.

line A B is a *typical* representation of the resultant of two currents of aerial sound-waves an octave apart, while the rectangular diagram C D is intended to represent a portion of a transverse section of a body of air traversed by the joint wave corresponding to the heavy wave-line above. The depth of shading in different parts indicates the degree of condensation or rarefaction at those parts.

SECTION IX.

VIBRATION OF STRINGS.

196. Sonometer. This instrument consists of two or more piano wires of different thicknesses stretched lengthwise over

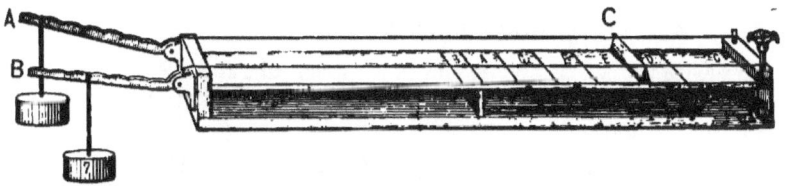

FIG. 149.

a resonance box. One end of each wire is attached to the shorter arm of a bent lever, A or B (Fig. 149), and the tension of the wire is regulated both by the lengths of the longer arms employed and by the magnitude of the weights suspended therefrom. The length of the vibrating portion of the strings is regulated by the sliding bridge C.

Experiment 1. Remove the bridge C, pluck one of the strings with the fingers at the middle point, causing it to vibrate as a whole, and note the pitch of the sound. Place the bridge under the same wire, and move it gradually toward one end of the sonometer, thereby shortening the vibrating portion; the pitch rises as the vibrating portion is shortened. Vary the position of C until a pitch is obtained an octave above the pitch given at first when the entire wire was vibrating. It will be found that the length of the wire which gives the higher note is just half the original length; *i.e. by halving the wire its vibration-number is doubled.* At two thirds its original length, it gives a note at an interval of a fifth above that given by its original length; and generally *the reciprocals of the fractions* (§ 194) *representing the relative vibration-numbers of the several notes of a scale represent the relative lengths of the wires that produce these notes.*

Now increase the tension of the wire; the pitch rises. Increase the tension until the pitch has risen an octave; it will be found that the tension has been increased fourfold.

Next try two wires whose lengths and tension are the same, but whose diameters are (say) as 1 : 2, and whose masses per unit length are consequently as 1 : 4; the pitch given by the wire of greater mass is an octave lower than the pitch given by the other wire.

These conclusions may be summarized thus: *The vibration-numbers of strings of the same material vary inversely as their lengths and the square roots of their masses per unit length, and directly as the square roots of their tensions.*

197. Stationary Vibrations, Nodes, etc.

Experiment 2. Hold one end of a rubber tube about 2 m. long, while the other is fixed, and send along it a regular succession of equal pulses from the vibrating hand. By varying the tension a little, it will be easy to obtain a succession of gauzy

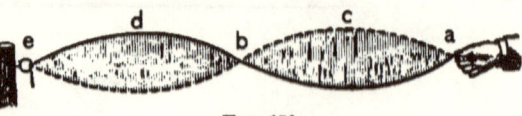

Fig. 150.

spindles (Fig. 150) separated by points that are nearly or quite at rest. Unlike the earlier experiments, the waves here do not appear to travel along the tube; yet in reality they do traverse it. The deception is caused by stationary points being produced by the interference of the advancing and retreating waves.

This interference of direct and reflected waves gives rise to an important class of phenomena called *stationary vibrations*. The points of least motion, as a, b, and e (Fig. 150), are called *nodes* (from fancied resemblance to knots); the points of greatest amplitude, as d and c, are called *antinodes;* and the portions between the nodes are called *venters*.

In a similar manner a string may be made to vibrate in 3, 4, etc., parts, as shown in C, D, and E (Fig. 151). The pitch

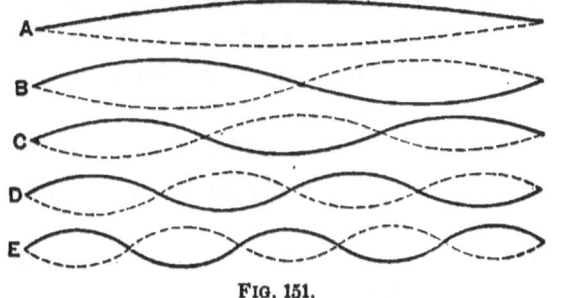

FIG. 151.

of the tone produced by a string when it vibrates as a whole, as in A, is called the *fundamental pitch of the string*. The vibration frequency when the string divides into halves, as in B, is twice as great as before, and consequently the pitch of the tone produced is an octave above that of the fundamental. Generally the vibration frequency varies as the number of venters into which the string divides.

Tones produced by a string or other body when it vibrates in parts are called *overtones* or *partial tones*. If the overtones harmonize (§ 201) with the fundamental of the vibrating body, they are called *harmonics*.

198. Complex Vibrations.

Experiment 3. Press down the C'-key of a piano gently, so that it will not sound, and while holding it down, strike the C-key strongly. In a few seconds release the latter key, so that its damper will stop the vibrations of the string that was struck, and you will hear a sound which you will recognize by its pitch as coming from the C′-wire. Place your finger lightly on the C′-wire, and you will find that it is indeed vibrating. Press down the right pedal with the foot, so as to lift the dampers from all the wires, strike the C-key, and touch with the finger the C′-wire; it vibrates. Touch the wires next to C′, viz. B and D′; they have only a slight forced vibration. Touch G′; it vibrates.

It is evident that the vibrations of the C′ and G′ wires are sympathetic. Now, a C-wire vibrating as a whole cannot cause sympathetic vibrations in a C′-wire; but if it vibrates in halves, it may. Hence, we conclude that when the C-wire was struck it vibrated, not only as a whole, giving a sound of its own pitch, but also in halves; and the result of this latter set of vibrations was that an additional sound was produced by this wire, just an octave higher than the first-mentioned sound.

Again, the G′-wire makes 391.5 vibrations in a second, or three times as many (130.5) as are made by the C-wire; hence, the C-wire, in addition to its vibrations as a whole and in halves, must have vibrated in thirds, inasmuch as it caused the G′-wire to vibrate. It thus appears that a string may vibrate at the same time as a whole, in halves, thirds, etc., and the result is that *a sensation is produced that is compounded of the sensations of several sounds of different pitch.* A sound so simple that it cannot be resolved is called a *tone*.

199. Tones and Notes. A sound composed of many tones is called a *note*.

Not only do stringed instruments produce *notes*, but no ordinary musical instrument is capable of producing a *simple tone*, i.e. a sound generated by vibrations of a single period. In other words, *when any note of any musical instrument is sounded, there is produced, in addition to the primary tone, a number of other tones in a progressive series, each tone of the series being usually of less intensity than the preceding.* The primary or lowest tone of a note is usually sufficiently intense to be the most prominent, and hence is called *the fundamental tone*.

Strings when struck produce many overtones, which vary according to the *place* where they are struck, the *nature of the stroke*, and the *density, rigidity*, and *elasticity* of the string.

200. Beats.

Experiment 4. Strike simultaneously the lowest note of a piano and its sharp (black key next above), and listen to the resulting sound.

You hear a peculiar wavy or throbbing sound, caused by an alternate rising and sinking in loudness. Each recurrence of the maximum intensity is called a *beat*.

Let the continuous curved line A C (Fig. 152) represent a series of waves caused by striking the lower key, and the

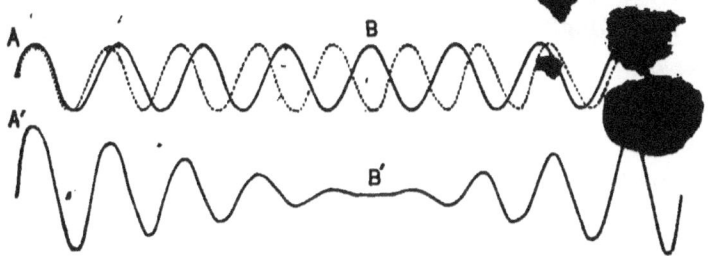

FIG. 152.

dotted line a series of waves proceeding from the upper key. Now, the waves from both keys may start together at A, but as the waves from the lower key are given less frequently, so are they correspondingly longer, and at certain intervals, as at B, condensations will correspond with rarefactions, producing by their interference momentary silence, too short, however, to be perceived ; but the sound as perceived by the ear is correctly represented in its varying loudness by the curved line A′ B′ C′.

It will be apparent from the study of Fig. 152 that exactly one beat will occur in each interval of time during which the acuter of two simple tones performs *one* more vibration than the graver tone.

Hence, *the number of beats per second due to two simple tones is equal to the difference of their respective vibration-numbers.*

201. Origin of Discord and Harmony. Discord produced by two sounds is explained by the fact that the sounds produce *beats,* which do not coalesce because the interval between them is too long.

As the frequency of the beats increases, a point is finally reached where they cease to be recognized as distinct sounds, and where they blend into a more or less pure tone. Beats may thus coalesce to produce beat-tones that are musical.

It must not, however, be inferred that dissonance disappears immediately upon the intermittences becoming too rapid for individual recognition. If two tones form a narrower interval than a minor third, the combined sound is harsh and grating on the ear.

Two tones must be in unison to produce absolutely perfect harmony. The intervals that most nearly approach perfect harmony are, first, that of the *octave*, and secondly, that of the *fifth*.

That two notes sounded together may harmonize, it is essential not only that the pitch of their fundamental tones be so widely different that they cannot produce audible beats, but that no audible beats shall be formed by their overtones, or by an overtone and a fundamental.

Observe that the relation between the vibration-numbers of the fundamentals of C and C', C and G, C and F, and C of any diatonic scale and any note in the same scale, can be expressed in terms of small numbers, *e.g.* 1 : 2, 2 : 3, 3 : 4, etc. (see § 194). Generally, *those notes and only those harmonize whose fundamental tones bear to one another ratios expressed by small numbers; and the smaller the numbers which express the ratios of the rates of vibration, the more perfect is the harmony of two sounds.*

Not only may two notes whose relative vibration frequency is expressible by a simple ratio harmonize, but three or four may concur with the same result. A sound produced by the coexistence of three or more notes is called in music a *chord.* A consonant chord is a *concord;* a dissonant chord is a *discord.*

SECTION X.

QUALITY OF SOUND.

202. Complex Sound-waves. Simple sound-waves can differ only in length and amplitude; consequently the sounds which they produce can differ only in pitch and loudness. Complex sound-waves may differ, as we have seen, in *form,* and this gives rise to a property of sound called *quality* (by musicians, *timbre*). Quality is that property of sound, not due to pitch or intensity, that enables us to distinguish one sound from another.

Although the variety of sounds one hears appears well-nigh infinite, yet no two sounds can differ from each other in any other respect than pitch, loudness, or quality. The length, amplitude, and form of the wave completely determine the wave, and these three elements of a wave are mutually independent, *i.e.* any one may be changed without altering the other two. Loudness depends on amplitude of vibrations, pitch on vibration frequency, and quality on *complexity of the motion of the vibrating particles.*

Let the same note be sounded with the same intensity, successively, on a variety of musical instruments, *e.g.* a violin, cornet, clarinet, accordion, jew's-harp, etc.; each instrument will send to your ear the same number of waves, and the waves from each will strike the ear with the same force, yet the ear is able to distinguish a decided difference between the sounds — a difference that enables us instantly to identify the instruments from which they come. Sounds from instruments of the same kind, but by different makers, usually exhibit decided differences of character. For instance, of two pianos, the sound of one will be described as richer and fuller, or more ringing, or more "wiry," etc., than the other. No two human voices sound exactly alike.

SECTION XI.

ANALYSIS OF SOUND-WAVES.

203. Analysis of Musical Sounds. Every enclosed body of air may act as a resonator to a sound of suitable wave-length. A seashell held to the ear "roars" continually in response to the outside air, which is ever disturbed with waves.

By means of a set of resonators corresponding to the various tones used in music, it is possible to detect the component tones of sounds which are far too complex to be analyzed by the unaided ear. Such a resonator (Fig. 153) is composed of two brass cylinders, one telescoping into the other, the latter being drawn out conically to a small open tip, A, which fits the ear. The length is thus adjustable, and the resonator will respond, according to its length, to any single tone.

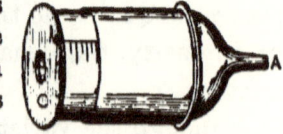

FIG. 153.

When any musical sound is produced near the orifice C of one of these resonators, and suitable adjustments are made, the ear placed at the tip A is able to single out, from the total number of tones composing the note, those overtones to which alone this resonator is capable of responding. By applying one resonator after another to the ear a sound is analyzed into its components. It is thus found, for instance, that the notes of a clarinet are composed only of the odd harmonics, or of tones whose vibration-numbers are in the ratios of $1 : 3 : 5 : 7$.

204. The Phonautograph or Phonograph. Sound-waves, however complex, may be caused permanently to record the succession and variation of their impulses, and thus, as it were, to inscribe their own autograph. Fig. 154 represents the original Edison phonograph.

A metallic cylinder, A, is rotated by means of a crank. On the surface of the cylinder is cut a shallow helical groove running around the cylinder from end to end, like the thread of a screw. A small metallic point, or style, projecting from the under side of a thin metallic disk, D (Fig. 155), which closes one orifice of the mouthpiece B, stands directly over the thread. By a simple device the cylinder, when the crank is turned, is made to advance just rapidly enough to allow the groove to keep constantly under the style. The cylinder is covered with tin foil. The cone F is usually applied to the mouthpiece to concentrate the sound-waves upon the disk D.

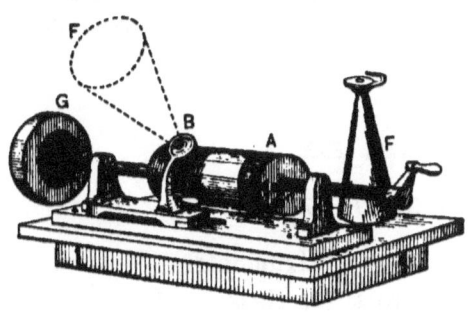

FIG. 154.

Now, when a person directs his voice toward the mouthpiece, the aërial waves cause the disk D to participate in every motion made by the particles of air as they beat against it, and the motion of the disk is communicated by the style to the tin foil, producing thereon impressions or indentations as it passes on the rotating cylinder. The result is that there is left upon the foil an exact representation of every movement made by the style. Some of the indentations are quite perceptible to the naked eye, while others are visible only with the aid of a microscope of high power. Fig. 156 represents a piece of the foil as it would appear inverted after the indentations (here greatly exaggerated) have been imprinted upon it.

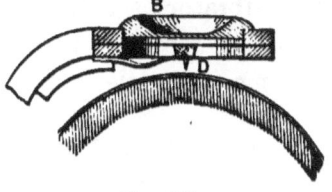

FIG. 155.

The words addressed to the phonograph having been thus impressed upon the foil, the mouthpiece and style are temporarily removed, while the cylinder is brought back to the position it had when the talking began, and then the mouthpiece is replaced. Now, evidently, if the crank be turned in the same direction as before, the style, resting upon the foil beneath, will be made to play up and down as it passes over ridges and sinks into depressions; this will cause the disk D to reproduce the same vibratory movements that caused the ridges and depressions in the foil. The vibrations of the disk are communicated to the air, and through the air to the ear; thus the words spoken to the apparatus may be, as it were, shaken out into the air again at any subsequent time, even centuries after, accompanied by the exact accents, intonations, and quality of sound of the original.

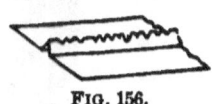

FIG. 156.

Subsequently Edison improved this instrument by replacing the metallic foil by a cylinder of hard wax composition, rotating it by an electric motor, and providing an improved form of style, which engraves upon the wax the most delicate variations of vibratory motions, and thus, as it were, reproduces speech and musical notes with all their delicate shades of expression and modulation.

SECTION XII.

MUSICAL INSTRUMENTS.

205. Classification of Musical Instruments. Musical instruments may be grouped into three classes: (1) stringed instruments; (2) wind instruments, in which the sound is due to the vibration of columns of air confined in tubes; (3) instruments in which the vibrator is a membrane or plate. The first class has received its share of attention; the other two merit a little further consideration.

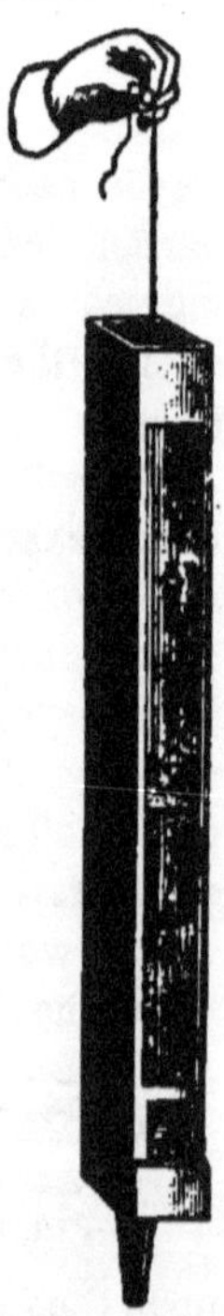

206. Wind Instruments.

The pitch of vibrating air-columns, as well as of strings, varies with the length, and (1) *in both stopped*[1] *and open*[1] *pipes the number of vibrations is inversely proportional to the length of the pipe.* (2) *An open pipe gives a note an octave higher than a closed pipe of the same length.*

Fig. 157 represents an open organ-pipe provided with a glass window, A, in one of its sides. A wire hoop, B, has stretched over it a membrane, and the whole is suspended by a thread within the pipe. If the membrane be placed near the upper end, a buzzing sound proceeds from the membrane when the fundamental tone of the pipe is sounded; and sand placed on the membrane dances up and down in a lively manner. If the membrane be lowered, the buzzing sound becomes fainter, till, at the

FIG. 157.

---

[1] The terms "stopped" and "open" apply to only one end of the pipe; the other, in both kinds, is always open.

middle of the tube, it ceases entirely, and the sand becomes quiet. If the membrane be lowered still further, the sound and dancing recommence, and increase as the lower end is approached.

(3) *When the fundamental tone of an open pipe is produced, its air-column divides into two equal vibrating sections, with the anti-nodes at the extremities of the tube, and a node in the middle.*

If the pipe be stopped, there is a node at the stopped end; if it be open, there is an anti-node at the open end; and in both cases there is an anti-node at the end where the wind enters, which is always, to a certain extent, open.

A, B, and C of Fig. 158 show, respectively, the positions of the nodes and anti-nodes for the fundamental tone and the first and second overtones of a closed pipe; and A′, B′, and C′ show the positions of the same

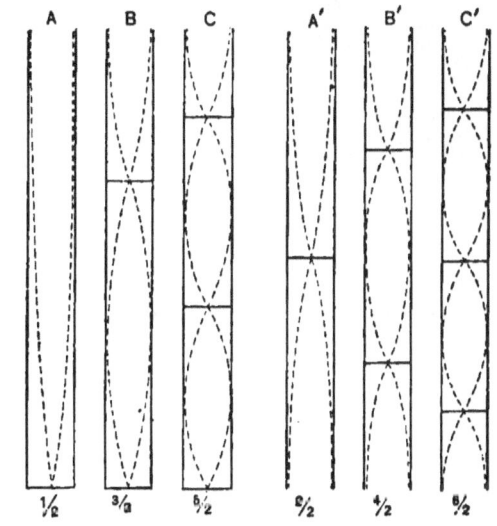

Fig. 158.

in an open pipe of the same length. The distance between the dotted lines shows the relative amplitudes of the vibrations of the air particles at various points along the tube. Now, the distance between a node and the nearest anti-node is a quarter of a wave-length. Comparing, then, A and A′, it will be seen that the wave-length of the fundamental of the closed pipe must be twice the wave-length of the fundamental of the open pipe; hence, the vibration-period of the latter is half that of the former; consequently, the fundamental of the open pipe must be an octave higher than that of the closed pipe.

207. Sounding Plates, etc.

Experiment. Fasten with a screw the elastic brass plate A (Fig. 159) on the upright support. Strew fine sand over the plate, draw a rosined bass bow steadily and firmly over one of its edges near a corner, and

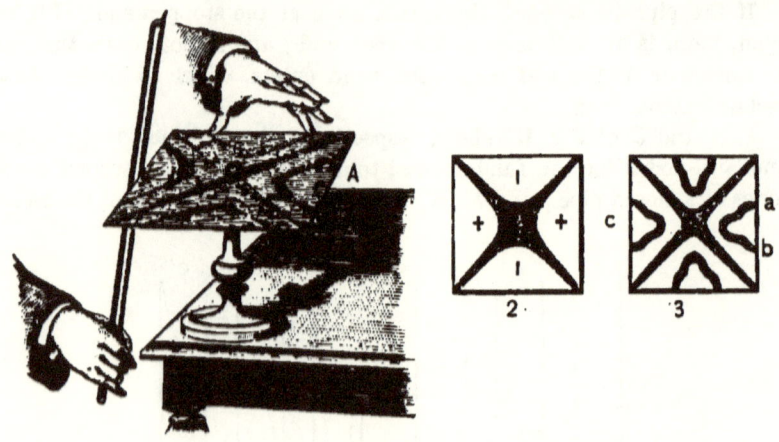

FIG. 159.

at the same time touch the middle of one of its edges with the tip of the finger; a musical sound will be produced, and the sand will dance up and down, and quickly collect in two rows, extending across the plate at right angles to each other. Draw the bow across the middle of an edge, and touch with a finger one of its corners; the sand will arrange itself (2) in two diagonal rows across the plate. Touch, with the nails of the thumb and a finger, two points on one edge of the plate as shown in the figure, and draw the bow across the middle of one of the other edges, and you will obtain additional rows and a shriller note.

By varying the position of the points touched and bowed, a great variety of patterns can be obtained. It will be seen that the effect of touching any point with a finger is to prevent vibration at that point, and consequently a node is there produced.

The whole plate then divides itself up into segments with nodal division lines in conformity with the node thus formed. The sand rolls away from those parts which are alternately thrown into crests and troughs, to the parts that are at rest.

FIG. 60.

208. Bells. A bell or goblet is subject to the same laws of vibration as a plate. If a goblet partly filled with water be bowed at some point, the surface of the water will become rippled with wavelets (Fig. 160) radiating from four points 90° apart, corresponding to the centers of four venters into which the goblet becomes divided, and sprays of water will be thrown from the four quadrants.

SECTION XIII.

VOCAL ORGANS. THE EAR.

209. Vocal Organs. The organ of the voice is a reed instrument situated at the top of the windpipe or trachea. A pair of elastic bands, a a (Fig. 161), called the *vocal chords*, is stretched across the top of the windpipe. The air-passage b between these chords is freely open while a person is breathing; but when he speaks or sings, they are brought nearly together so as to leave only a narrow slit-like opening, thus making a sort of double reed, which vibrates when air is forced from the lungs through the narrow passage, somewhat like the little tongue of a toy trumpet. The sounds are grave or high according to the tension of the chords, which is regulated by muscular action.

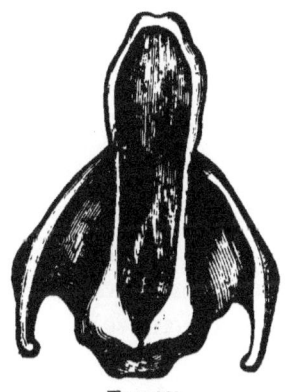

FIG. 161.

The cavities of the mouth and the nasal passages form a compound resonance tube. This tube adapts itself, by its varying width and length, to the pitch of the notes produced by the vocal chords. The different qualities of the different vowel sounds are produced by the varying forms of the resonating mouth-cavity, the pitch of the fundamental tones given by the vocal chords remaining the same. This constitutes *articulation*.

210. The Ear. At the inner end of the outer ear-passage C is the thin membrane D, known as the "drum of the ear," which serves as a partition between the outer ear-passage and

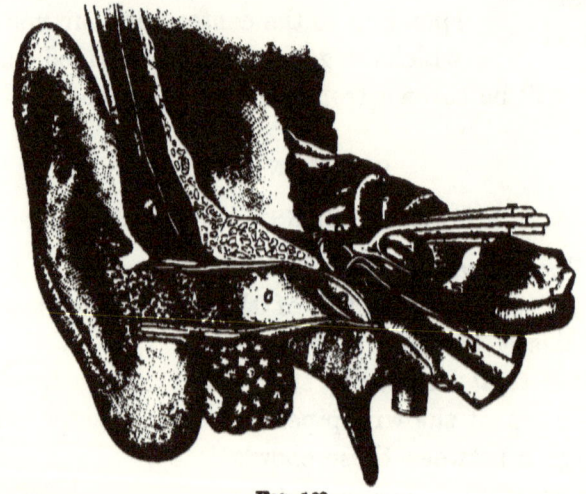

FIG. 162.

the cavity K of the middle ear. A chain of tiny bones stretches across this cavity from the drum to another membranous partition which closes the orifice of the inner ear. N is the Eustachian tube connecting the middle ear-cavity with the back part of the throat; T T are auditory nerves. The middle ear contains air, and the Eustachian tube forms a means of ingress and egress for air through the throat. The inner ear is filled with a transparent liquid.

Now, how does the ear hear? and how is it able to distinguish between the infinite variety of aërial sound-waves so as to interpret correctly the corresponding quality, pitch, and loudness of sound? Sound-waves enter the external ear-passage C as ocean waves enter the bays of the seacoast, are reflected inward, and strike the drum. The air particles, beating against this drum, impress upon it the precise wave-form that is transmitted to it through the air from the sounding body. The motion received by the drum is transmitted by the air in the middle ear-cavity and by the chain of bones to the membranous wall of the inner ear. From the walls of the inner ear project into its liquid contents thousands of fine elastic threads or fibers, called "rods of Corti." These vibratile fibers vary in length and size, and are therefore suited to respond sympathetically to a great variety of vibration-periods. The auditory nerve at this extremity is divided into a large number of filaments, like a cord unraveled at its end, and one of these filaments is attached to each fiber. Now, as the sound-waves reach the membranous wall of the inner ear they set it, and by means of it the liquid contents of the inner ear, into *forced vibration*, and so through the liquid the immersed fibers receive impulses. Those fibers whose vibration-periods correspond to the constituents of the compound wave are thrown into sympathetic vibration. The fibers stir the nerve-filaments, and these transmit the impression to the brain, where in some mysterious manner these disturbances are interpreted as sound of *definite pitch, quality, and intensity*.

EXERCISES.

1. Your ear can easily distinguish a note sounded by a violin from the same note produced by a flute. Explain the cause of this difference.

2. What conditions must be fulfilled that a vibrating body may produce (*a*) sound, (*b*) a musical note?

3. Explain (*a*) why, when a tuning fork is struck and the stem is pressed against a table, it sounds much louder than when held in the

hand; (b) why the sound in the former case does not continue so long as in the latter.

4. State whether or not the velocity of sound in air is affected (a) by the hight of the barometric column, (b) by the temperature of the air; give reasons for your answer in each case.

5. The disturbance produced in the surrounding air by a sounding body has been likened to that caused in a pool by a stone thrown into the water. Point out in what particulars the motions of the air and of the water in the two cases are really similar, and in what dissimilar.

6. The tone given out by an open pipe 3 inches long is found to be caused by 2220 vibrations per second. Calculate from these data the velocity of sound in air.

7. When a sounding body and the ear approach each other, or recede from each other, the pitch of the sound appears to change. Explain.

8. How is an ear trumpet related to a speaking trumpet?

9. Represent by diagrams (a) two waves which have the same wave-length, but of which one has twice the amplitude of the other; also (b) two waves which have the same amplitude, but of which one has double the wave-length of the other.

10. Which is the better conductor of sound, sawdust or solid wood? Why?

11. (a) When a bottle of soda-water is opened a loud sound is heard. Explain. (b) Is it a musical sound? Explain.

12. At one end of a very long tube a pistol is fired. Explain how it is that an observer at the other end of the tube hears the report twice.

13. If a person set his watch by the striking of a clock half a mile distant, when the temperature of the air is 20° C., what will be the magnitude of the error?

14. A rifle is discharged and the echo produced by a cliff is heard in 8 seconds. The temperature of the air is 16° C. What is the distance of the cliff?

15. The waves produced by a man's voice in common conversation are from 8 to 12 feet long. Find the corresponding numbers of vibrations, assuming the velocity of sound to be 1128 feet per second.

16. The maximum resonance of a certain tuning fork is produced when it is placed over a jar 15 inches high. How many vibrations does this fork make in a second?

17. The length of the fundamental wave of a closed pipe is how many times the length of the pipe?

18. Upon what does the pitch of an organ pipe depend?

19. What change occurs in the pitch of the sound made by a circular saw on entering a plank? Explain.

CHAPTER VI.

ENERGY OF ETHER-STRAIN. RADIANT ENERGY. LIGHT.

SECTION I.

RADIANT ENERGY.

211. The Ether. We know matter by its properties; in other words, the existence of any form of matter is to us only an inference from the phenomena to which it gives rise. By evidence of a similar nature we are led to believe in the existence of a medium called *the ether*, pervading all space and penetrating between the molecules of matter, which are imbedded in it and surrounded by it as the earth is surrounded by its atmosphere. We cannot see, hear, feel, taste, smell, exhaust, weigh, or measure it; yet all this, paradoxical as it may seem, furnishes absolutely no evidence that it does not exist.

Phenomena occur just *as* they would occur *if* all space were filled with a medium capable of transmitting motion and energy, and *we can account for all these phenomena on no other hypothesis;* hence our belief in the existence of the medium. *The ether is a medium for the transmission of energy in the form of vibrations.*[1]

212. Radiation. Radiant Energy. The ether is set in vibration by the motion of the molecules of matter. This local disturbance creates ether-waves, and by these waves energy is transferred from body to body by the process

---

[1] The following is Lord Salisbury's witty definition: "Ether is the nominative case of the verb ' to vibrate.' "

called *radiation*. Energy so transmitted is called *radiant energy*, and the body thus emitting energy is called a *radiator*. Radiant energy can be transformed into any other form of energy, and therefore offers no exception to the doctrine of correlation of energy.

213. Effects of Radiant Energy. When radiant energy is received upon the surfaces of our bodies, warmth is felt; when received upon the bulb of a thermometer, rise of temperature is indicated; when received by the eye, the sense of sight may be affected; if it is received upon sensitive photographic plates, upon the leaves of plants, or upon various chemical mixtures, chemical changes may be promoted. Thus it seems that when ether-waves impinge upon objects their energy is transformed, producing effects of different kinds, which are determined by the nature of the body upon which they fall. The effect which most concerns us is that produced when the radiations strike the *eye* and become the means, through this organ, of creating the sensation of *sight*. The eye is an ether sense-organ, much as the ear may be regarded as an air sense-organ.

SECTION II.

LIGHT.

214. Light Defined. Hypotheses. Two widely different hypotheses regarding the nature of light have been propounded. One, the so-called *emission or corpuscular hypothesis*, was supported by Newton (1672), and by most physicists up to the early part of the present century. It assumes that a luminous body (*e.g.* the sun) emits minute material particles (corpuscles) which travel through space in all directions with immense velocity, and that these particles by their impact upon the eye produce the sensation of sight. As a rose emits minute particles which, reaching the nostrils, enable us to smell the rose, so a star is supposed to emit corpuscular matter which, on reaching the eye, enables us to see the star. This hypothesis is now discarded by scientists. The theory which obtains at the present time, called the *undulatory* or *wave-theory*, is based upon the hypothesis that energy is transmitted from body to body, *e.g.* from the sun to the earth, in the form of vibrations or wave-action in the all-pervading ether. According to the latter theory, *light is that vibration of the ether which may be appreciated by the organ of sight.*[1]

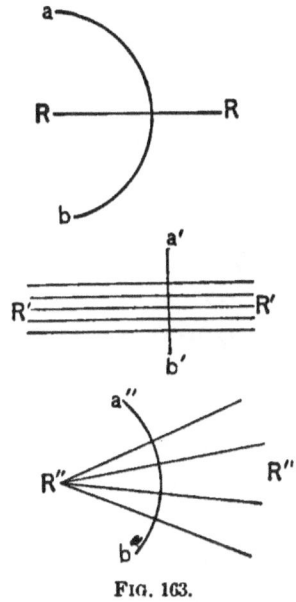

Fig. 163.

215. Luminous and Illuminated Objects. Some bodies are seen by means of light-waves which they generate, *e.g.* the sun, a candle flame, and a "live coal"; they are called *luminous bodies*. Others are seen only by means of light-waves which they receive from luminous bodies and reflect to the eye, and, when thus rendered visible, are said to be *illuminated*; *e.g.* the moon, a man, a cloud, and a "dead" coal.

216. Light-waves Travel in Straight Lines. The paths of light-waves admitted into a darkened room through a small aperture, as indicated by the illuminated dust, are perfectly straight. *An object is seen by means of light-waves which it*

---

[1] It will be shown further on (§ 252) that not all ether-waves are capable of affecting the sight; hence, for the purpose of distinction we apply the term light-waves to those ether-waves only which are capable of producing vision. It is strongly recommended that the student in beginning this branch of science make use of the term *light-waves* instead of light, except when such usage would lead to an inconvenient circumlocution, in order that he may have strongly impressed upon his mind the fact that when he is dealing with light he is dealing with *waves*.

sends to the eye. A small object placed in a straight line between the eye and a luminous point may intercept the light-waves in that path, so that the point becomes invisible. Hence we cannot see around a corner or through a bent tube.

217. Ray, Beam, Pencil. Any line, R R (Fig. 163), which pierces the surface of an ether-wave front, a b, perpendicularly, is called a *ray.* The term *"ray" is but an expression for the direction in which motion is propagated, and along which the successive effects of ether-waves occur.*[1] If the wave-surface a' b' be a plane, the rays R' R' are parallel, and a collection of such rays is called a *beam.* If the wave-surface a" b" be spherical, the rays R" R" have a common point at the center of curvature; and a collection of such rays is called a *pencil.*

218. Transparent, Translucent, and Opaque Substances. Substances are *transparent, translucent,* or *opaque* according to the manner in which they act upon the light-waves which are incident upon them. Generally speaking, those substances are *transparent* that allow objects to be seen through them distinctly, *e.g.* air, glass, and water. Those substances are *translucent* that allow light-waves to pass, but in such a scattered condition that objects are not seen distinctly through them, *e.g.* fog, ground glass, and oiled paper. Those substances are *opaque* that apparently cut off all the light-waves and prevent objects from being seen through them. When bodies intercept light, they are said to cast shadows.

219. Every Point of a Luminous Body an Independent Source of Light-waves. Place a candle flame in the center of a darkened room; each wall and every point of each wall becomes illuminated. Place yourself in any part of the room,

---

[1] In dealing with certain phenomena (*e.g.* reflection of light) we may, to facilitate our study, consider the light as propagated in straight lines or rays; but we must bear in mind that a ray has no material or physical existence, for it is a wave that is propagated, not a ray.

210   ETHER DYNAMICS.

i.e. in any direction from the flame; you are able to see not only the flame, but every point of the flame; hence every point of the flame must emit light-waves in every direction. *Every point of a luminous body is an independent source of light-waves, and emits them in every direction.* Such a point is called a *luminous point.* In Fig. 164 there are represented a few of the infinite number of pencils of light emitted by three luminous points.

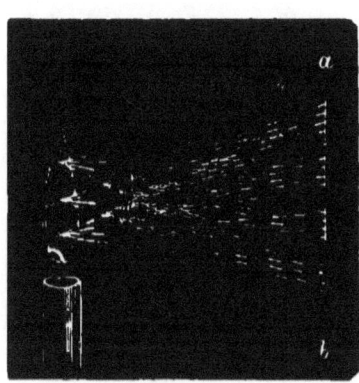

FIG. 164.

220. Images Formed through Small Apertures.

Experiment 1. Cut a hole about 8 cm. square in one side of a box; cover the hole with tin foil, and prick a hole in the foil with a pin. Place the box in a darkened room, and a candle flame in the box near the pin hole. Hold an oiled-paper screen before the hole in the foil; an inverted image of the candle flame will appear upon the translucent paper. An *image* is a kind of picture of an object.

If light-waves from objects illuminated by the sun — *e.g.* trees, houses, clouds, or even from an entire landscape — be allowed to pass through a small aperture in a window shutter and strike a white screen (or a white wall) in a dark room, inverted images of the objects in their true colors will appear upon the screen. The cause of these phenomena is easily understood. When no screen

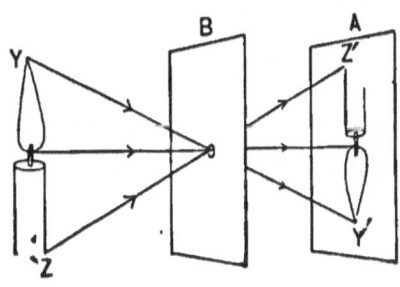

FIG. 165.

intervenes between the candle and the screen A (Fig. 165), every point of the screen receives light from every point of the candle; consequently, at every point on A, images of all the points of the candle are formed. The result of the confusion of images is that no image is distinguishable. But let the screen B, containing a small hole, be interposed; then, since light travels only in straight lines, the point Y' can receive an image only of the point Y, the point Z' only of the point Z, and so for intermediate points; hence a distinct image of the object must be formed on the screen A. *That an image may be distinct, the images of different points of the object must not mix, and therefore all rays from each point on the object must be carried to the corresponding point on the image.*

221. Shadows.

Experiment 2. Procure a piece of tin or cardboard 18 cm. square; place it between a white wall and a candle flame in a darkened room. The opaque tin intercepts the light that strikes it, and thereby excludes light from a space behind it.

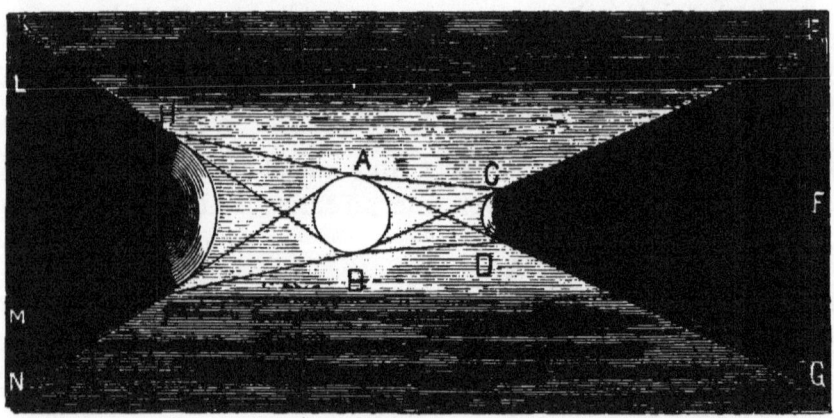

Fig. 166.

This space is called a *shadow*. That portion of the surface of the wall that is darkened is a *section of the shadow*, and represents in form a cross section of the body that intercepts the light. A section of a shadow is frequently for convenience called a shadow. Notice that the shadow is made up of two distinct parts — a dark center, bordered on all sides by a much lighter fringe. The dark center is called the *umbra*, and the lighter envelope is called the *penumbra*.

The umbra is the part of a shadow that gets no light from the luminous body, while the penumbra is the part that gets light from some portion of the body, but not from the whole.

Let A B (Fig. 166) represent a luminous body, and C D an opaque body. The pencil from the luminous point A will be intercepted between the lines C F and D G, and the pencil from B will be intercepted between the lines C E and D F. Hence the light will be wholly excluded only from the space between the lines C F and D F, which enclose the umbra. The enveloping penumbra, a section of which is included between the lines C E and C F, and between D F and D G, receives light from certain points of the luminous body, but not from all.

222. Speed of Light. That light travels with finite speed was first established in 1676 by the Danish astronomer Roemer, then engaged in Paris in observing the eclipses of Jupiter's moons. He made observations on that one of the five of Jupiter's satellites which is nearest to the planet, and which revolves round this planet as the moon does round the earth. At regular intervals the satellite passes behind the planet and is eclipsed within its shadow. The observed intervals, however, were found to be shorter than the mean value when the earth and Jupiter were approaching each other, and longer when they were receding from each other. It was evident that this difference was due to the time consumed by the light in crossing the intervening spaces. From the results of these observations it was calculated that light required 16 minutes and 36 seconds to traverse the diameter of the earth's orbit, approximately 185,000,000 miles. The speed of light thus determined is 192,500 miles per second. It has been determined by later experiments and more reliable methods that this value is too great. The result obtained by Michelson by experimental methods is about 186,380 miles per second. This may be accepted as probably the nearest approximation yet made to the true speed of light in a vacuum. At this rate, light would encircle our earth between seven and eight times in a second.

EXERCISES.

1. Why are images formed through apertures inverted?
2. Why is the size of the image dependent on the distance of the screen from the aperture?
3. Why does an image become dimmer as it becomes larger?
4. Why do we not imprint an image of our person on every object in front of which we stand?
5. Upon what fact does a gunner rely in taking sight?
6. Explain the umbra and penumbra cast by the opaque body H I, Fig. 166.
7. When will a transverse section of the umbra of an opaque body be larger than the object itself?
8. When has an umbra a limited length?
9. What is the shape of the umbra cast by the sphere C D, Fig. 166?
10. If C D should become the luminous body, and A B a non-luminous opaque body, what changes would occur in the umbra and the shadow cast?
11. Why is it difficult to determine the exact point on the ground where the umbra of a church steeple terminates?
12. What is the shape of a section of the shadow cast by a circular disk placed obliquely between a luminous body and a screen? What is its shape when the disk is placed edgewise?
13. The section of the earth's umbra on the moon in an eclipse always has a circular outline. What does this show respecting the shape of the earth?
14. (a) The sensation of sound is how produced? (b) How is the sensation of sight produced? (c) How are sound-waves produced? (d) How are light-waves produced? (e) Which, sound-waves or ether-waves, originate in molecular vibrations? (f) Sound-waves travel in what mediums? (g) Light-waves travel only in what medium?
15. (a) What is radiant energy? (b) Do all bodies emit radiant energy? (c) Do all ether-waves affect the sight? (d) Do all bodies generate light-waves? (e) Is a "dead coal" seen by ether-waves which it generates?

SECTION III.

INTENSITY OF ILLUMINATION.

223. Application of the Law of Inverse Squares to Light. Light diminishes in intensity, and hence in its power to illuminate objects which it strikes, as it recedes from its source. *The intensity of light diminishes as the square of the distance from its source increases.* Calling the quantity of light falling upon a visiting card at a distance of 2 feet from a lamp flame 1, the quantity falling upon the same card at a distance of 4 feet is $\frac{1}{4}$, at a distance of 6 feet it is $\frac{1}{9}$, and so on. This is the meaning of the law of inverse squares, as applied to light.

This law may be illustrated thus: A square card placed (say) 1 foot from a certain point in a candle flame, as at A (Fig. 167), receives from

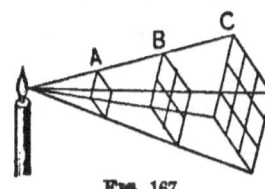

Fig. 167.

this point a certain quantity of light. The same light if not intercepted would go on to B, at a distance of 2 feet, and would there illuminate four squares, each of the size of the card, and being spread over four times the area can illuminate each square with only one fourth the intensity. If allowed to proceed to C, 3 feet distant, it illuminates nine such squares, and has but one ninth its intensity at A. The law is strictly true only when distance from individual points is considered.

224. Unit of Measurement. The unit generally employed in the measurement of the illuminating power of the light emitted by a luminous body is the British *candle-power*. It is the illuminating power of a sperm candle $\frac{7}{8}$ in. in diameter, burning 120 grains to the hour.

225. Photometry. The law just established enables us to compare the illuminating power of one light with that of another, and to express by numbers their relative illuminating powers. The process is called *photometry* (light-measuring), and the instrument employed, a *photometer*.

The Bunsen photometer (Fig. 168) has a screen of paper, S, mounted in a box, B, open in front and at the two ends. The box slides on a graduated bar. The screen has a circular central spot saturated with paraffine, which renders the spot more translucent than other portions of the screen. One side of the screen is illuminated by the light L, whose intensity is

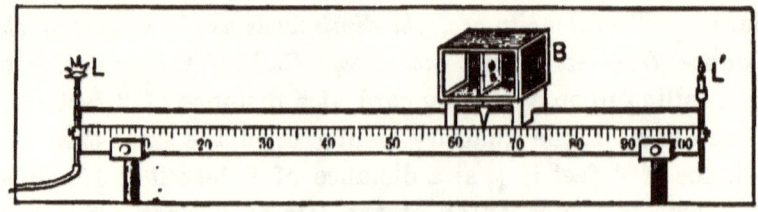

FIG. 168.

to be measured, and the other side by a standard candle, L'. When the screen is so placed that the two sides are equally illuminated by the two lights, the paraffined spot becomes nearly invisible. When one side is more strongly illuminated than the other, the spot appears dark on that side and light on the other. *The candle-power of the two lights is directly proportional to the square of their respective distances from the screen when it is equally illuminated on both sides.*

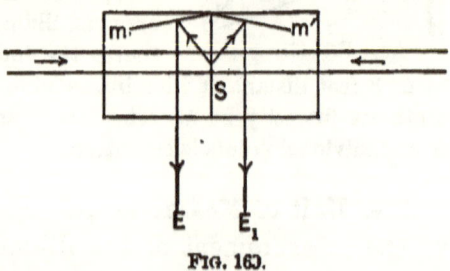

FIG. 169.

In order to render both sides of the disk simultaneously visible, two mirrors, m and m' (Fig. 169), are placed in the box in a vertical position so as to reflect images of the circular spot in the screen S to the eyes at E E$_1$.

QUESTIONS.

1. Suppose that a lighted candle is placed in the center of each of three cubical rooms, respectively 10, 20, and 30 feet on a side, would a single wall of the first room receive more light than a single wall of either of the other rooms, or less?

2. Would one square foot of a wall of the third room receive as much light as would be received by one square foot of a wall of the first room? If not, what difference would there be, and why the difference?

3. If a board 10 cm. square be placed 25 cm. from a candle flame, the area of the shadow of the board cast on a screen 75 cm. distant from the candle will be how many times the area of the board? Then the light intercepted by the board will illuminate how much of the surface of the screen if the board be withdrawn?

4. Give a reason for the law of inverse squares.

5. To what besides light has this law been found applicable?

6. The two sides of a paper disk are illuminated equally by a candle flame 50 cm. distant on one side and a gas flame 200 cm. distant on the other side. (*a*) Compare the intensities of the two lights at equal distances from their sources. (*b*) If the candle be a standard candle, what is the intensity of the gas flame?

SECTION IV.

APPARENT SIZE OF AN OBJECT.

226. Visual Angle. We see an object by means of its image formed on the retina of the eye; and its apparent magnitude is determined by the extent of the retina covered

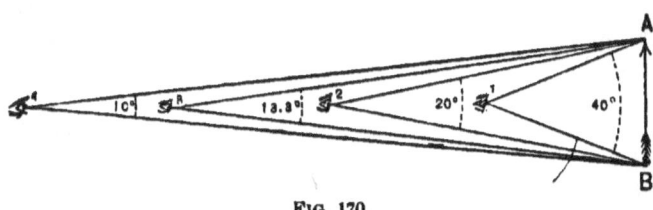

FIG. 170.

by its image. Rays proceeding from opposite extremities of an object, as A B (Fig. 170), meet and cross each other within the eye. Now, as the distance between the points of the blades of a pair of scissors depends upon the angle that the handles form with each other, so the size of the image formed on the retina depends upon the size of the angle, called the *visual angle*, formed by these rays as they enter the eye. But the size of the visual angle diminishes approximately as the distance of the object from the eye increases, as shown in the diagram; *e.g.* at twice the distance the angle is about one half as great; at three times the distance the angle is one third as great; and so on. Hence *distance affects the apparent size of an object.* Our judgment of the size of objects is, however, influenced by other things besides the visual angle which they subtend.

SECTION V.

REFLECTION OF LIGHT.

227. Mirrors. Images. Objects having polished surfaces which reflect light regularly (*i.e.* do not scatter the light), and show images of objects presented to them, are called *mirrors*. The mirror itself, if clean and smooth, is scarcely visible. According to their shape, mirrors are called *plane, concave, convex, spherical, parabolic,* etc.

Experiment 1. (*a*) Look at the mirror M through the hole marked O in the metal band (Fig. 171). You see in the mirror an image of the hole through which you look, but you do not see the image of any of the other holes. Rays that pass through this hole strike the mirror perpendicularly and are said to be *normal* to the mirror. Rays falling upon an object are

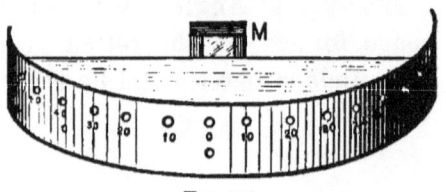

FIG. 171.

called *incident rays.* The point where a ray strikes is called the *point of incidence.* The *reflected rays* in this case are thrown back in the same lines and through the same hole that the incident rays travel. *Rays normal to a mirror after reflection simply retrace their own course.* (*b*) Next hold a candle flame at one of the other holes, *e.g.* at the hole marked 10. You can see the image of the candle flame only through the hole of the same number and at an equal distance on the other side. The angle which an incident ray makes with a line normal at the point of incidence is called the *angle of incidence,* and the angle made with the normal by a reflected ray is called the *angle of reflection.*

218 ETHER DYNAMICS.

Law of Reflection. *The angles of incidence and reflection are in the same plane, and are equal.*

228. Diffused Light.

Experiment 2. Introduce a small beam of light into a darkened room, and place in its path a mirror. The light is reflected in a definite direction. If the eye be placed so as to receive the reflected light, it will see, not the mirror, but the image of the sun, and the light will be painfully intense. Substitute for the mirror a piece of unglazed paper. The light is not reflected by the paper in any definite direction, but is scattered in every direction, illuminating objects in the vicinity and rendering them visible. Looking at the paper, you see, not an image of the sun, but the paper itself, and you may see it equally well from all directions.

The surface of the paper receives light from a definite direction, but reflects it in every direction; in other words, it scatters, or *diffuses*, the light. The difference in the phenomena in the two cases is caused by the difference in the smoothness of the two reflecting surfaces. A B (Fig. 172)

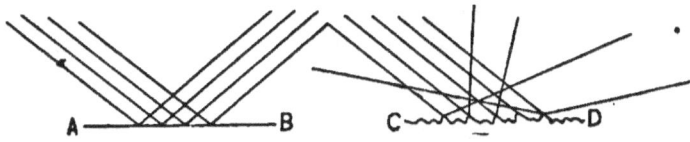

Fig. 172.

represents a smooth surface, like that of glass, which reflects nearly all the rays of light in the same direction, because nearly all the points of reflection are in the same plane. C D represents a surface of paper having the roughness of its surface greatly exaggerated. The various points of reflection are turned in every possible direction; consequently, light is reflected in every direction. Thus, the dull surfaces of various objects around us reflect light in all directions, and are consequently visible from every side. Objects rendered visible by reflected light are said to be *illuminated*.

229. Reflection from Plane Mirrors; Virtual Images. M M (Fig. 173) represents a plane mirror, and A B a pencil of divergent rays proceeding from the point A of an object, A H. By erecting perpendiculars at the points of incidence, or the points where these rays strike the mirror, and making the angles of reflection equal to the angles of incidence, the paths B C and E C of the reflected rays are found.

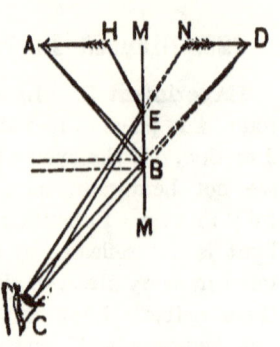

FIG. 173.

Every visible point of an object sends a cone of rays to the eye. The point always appears at the place whence these rays seem to emerge, *i.e.* at the apex of the cone. If the direction of these rays be changed by reflection, or in any other manner, the point will appear to be in the direction of the rays as they enter the eye; thus, the point A appears to lie in the direction C D, and the point H in the direction C N. The exact location of these points may be found by continuing each pencil of rays behind the mirror until it comes to a point, C B at D, C E at N. Thus, the pencils appear to emanate from these points, and the whole body of light-waves received by the eye seems to come from an *apparent object*, N D, behind the mirror. This apparent object is called an *image*. An image is a point or a series of points from which diverging pencils of rays come or appear to come. As of course no real image can be formed back of a mirror, such an image is called a *virtual* or an *imaginary* image. It will be seen, by construction, that *an image in a plane mirror appears as far behind the mirror as the object is in front of it, and is of the same size and shape as the object.*

230. Reflection from Concave Mirrors. Let M M' (Fig. 174) represent a section of a concave spherical mirror, which may be regarded as a small part of a hollow spherical shell having a polished interior surface. The distance in a straight line from M to M' is called the *diameter of the mirror*. C is the center of the sphere, and is called the *center of curvature*. G is the *vertex* of the mirror. A straight line, D G, drawn through the center of the curvature and the vertex is called the *principal axis* of the mirror. A concave mirror may be considered as made up of an infinite number of small plane surfaces.

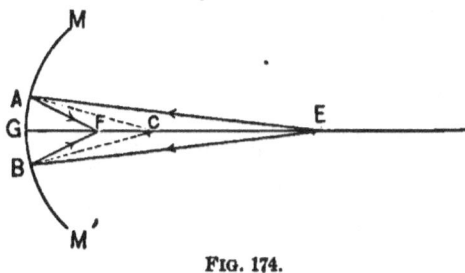

Fig. 174.

All radii of the mirror, as C A, C G, and C B, are perpendicular to the small planes which they strike. If C be a luminous point, it is evident that all light-waves emanating from this point, and striking the mirror, will be reflected to their source at C.

Let E be any luminous point in front of a concave mirror. To find the direction that rays emanating from this point take after reflection, draw any two lines from this point, as E A and E B, representing two of the infinite number of rays composing the divergent pencil that strike the mirror. Next, draw radii to the points of incidence A and B, and draw the lines A F and B F, making the angles of reflection equal to the angles of incidence. Place arrow heads on the lines representing rays to indicate the direction of the motion. The lines A F and B F represent the direction of the rays after reflection.

It will be seen that the rays after reflection are convergent, and meet at the point F, called the *focus*. This point is the focus of reflected rays that emanate from the point E. It is obvious that if F were the luminous point, the lines A E and B E would represent the reflected rays, and E would be the focus of these rays. Since the relation between the two points is such that light-waves emanating from either one are brought by reflection to a focus at the other, these points are called *conjugate foci*. *Conjugate foci are two points so related that the image of either is formed at the other.* The rays E A and E B, emanating from E, are less divergent than rays F A and F B, emanating from a point, F, less distant from the mirror, and striking the same points. Rays emanating from D, and striking the same points A and B, will be still less divergent; and if the point D were removed to a distance of many miles, the rays incident at these points would be very nearly parallel. Hence, rays may be regarded as practically parallel when their source is at a very great distance, *e.g.* the sun's rays. If a sunbeam, consisting of a bundle of parallel rays, as E A, D G, and H B (Fig. 175), strike a concave mirror in a direction parallel to its principal axis, these rays become convergent by reflection, and meet at a point (F) in the principal axis. This point, called the *principal focus*,[1] *is about halfway between the center of curvature and the vertex of the mirror.*

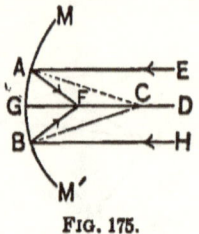

FIG. 175.

On the other hand, it is obvious that *divergent rays emanating from the principal focus of a concave mirror become parallel by reflection.*

The general effect of a concave mirror is to increase the convergence or to decrease the divergence of incident rays.

The following is a *formula for concave mirrors*:

$$\frac{1}{p} + \frac{1}{p'} = \frac{1}{f},$$

in which f represents the distance of the principal focus from the mirror, and p and p' represent the respective distances of any two conjugate foci from the mirror. Evidently, if any

---

[1] The statement that parallel rays, after reflection from a concave mirror, meet at the principal focus, is only approximately true. The smaller the angle subtended at the center by the mirror, the more nearly true is the statement. It is strictly true only of parabolic mirrors. Such are used in the headlights of locomotives.

two of the three quantities involved be given, the third may be calculated. If p and p' have unlike signs, we are to understand that the object and the image are on opposite sides of the mirror; in other words, the image is virtual.

231. Formation of Images. To determine the position and kind of images formed in concave mirrors, of objects placed in front of them, proceed as follows: Locate the object, as D E (Fig. 176). Draw lines, E A and D B, from the extremities of the object through the center of curvature of the mirror, to meet the mirror. These lines are called *secondary axes*. Incident rays along these lines will return by the same paths after reflection. Draw another line from D to any point in the mirror, *e.g.* to F, to represent another of the infinite number of rays emanating from D. Make the angle of reflection C F D′ equal to the angle of incidence C F D, and the reflected ray will intersect the secondary axis D B at the point D′. This point is the conjugate focus of all rays proceeding from D. Consequently, an image of the point D is formed at D′. This image is called a *real image*, because rays actually meet at this point. In a similar manner, find the point E′, the conjugate focus of the point E. The images of intermediate points between D and E lie between the points D′ and E′; and, consequently, the image of the object lies between those points as extremities.

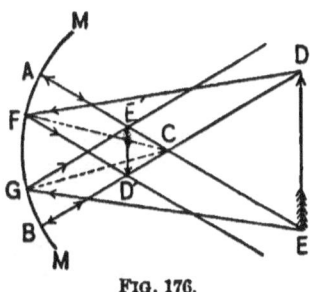

FIG. 176.

It thus appears that *an image of an object placed beyond the center of curvature of a concave mirror is real, inverted, smaller than the object, and located between the center of curvature and the principal focus of the mirror.* A person standing in front of such a mirror, at a distance greater than its radius of curvature, will see an inverted image of himself suspended, as it were, in mid-air.

FORMATION OF IMAGES.

Experiment 3. Hold some object, *e.g.* a rose, as a b (Fig. 177), a few feet in front of a concave mirror. Looking in the direction of the axis of the mirror, you see a small inverted image, A B, of the object, between the center of curvature, C, of the mirror and its principal focus, F.

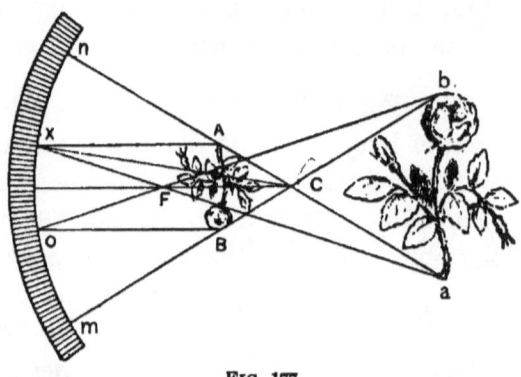

FIG. 177.

Figure 178 shows the path of rays thus reflected as they enter the eye. The observer sees the image of the point A at A'.

Evidently, if A B (Fig. 177) represent an object placed between the principal focus and the center of curvature, then a b will represent the *image* of the object.

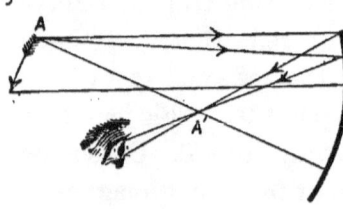

FIG. 178.

Hence, *the image of an object placed between the principal focus and the center of curvature is also real and inverted, but larger than the object, and located beyond the center of curvature.* The image in this case may be projected upon a screen, but it will not be so bright as in the former case, because the light is spread over a larger surface.

Construct an image of an object placed between the principal focus and the mirror, as in Fig. 179. It will be seen in this case that a pencil of rays proceeding from any point of an object, *e.g.* A, has no actual focus, but appears to proceed from a *virtual* focus A', back of the mirror, and so with other points, as B. *The image of an object placed between the prin-*

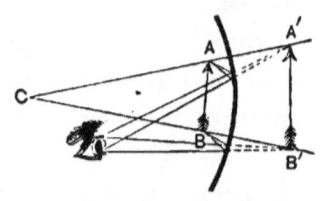

FIG. 179.

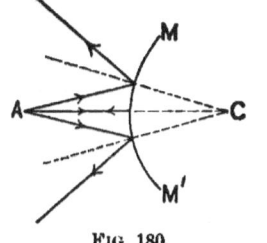

FIG. 180.

cipal focus and the mirror is virtual, erect, larger than the object, and back of the mirror.

The diagram in Fig. 180 suggests the method of finding the disposition of a pencil of rays emanating from any point (*e.g.* A) after reflection from a convex mirror. Construct an image of an object placed in front of a convex mirror.

EXERCISES.

1. With a radius of 8 cm. draw arcs of circles to represent concave mirrors, draw their principal axes, and locate thereon by the letters C and F, respectively, the centers of curvature and principal foci. Construct images of arrows located as follows: (*a*) between the mirror and the principal focus; (*b*) between the principal focus and the center of curvature; (*c*) beyond the center of curvature.

2. An object is 10 feet from a concave mirror; a distinct image of the object is formed 2 feet from the mirror. (*a*) What is the focal length of the mirror? (*b*) Describe the image.

3. The focal length of a concave mirror is 16 inches. At what distance from the mirror will the image of an object which is 18 inches from the mirror appear?

4. Locate and describe the image, if the object in Exercise 3 be placed 12 inches from the mirror.

SECTION VI.

REFRACTION.

232. Introductory Experiments.

Experiment 1. Into a darkened room admit a sunbeam so that its rays may fall obliquely on the bottom of the basin (Fig. 181), and note the place on the bottom where the edge of the shadow D E cast by the side of the basin D C meets the bottom at E. Then, without moving the basin, fill it evenly full with water slightly clouded with milk or with a few drops of a solution of mastic in alcohol. It will be found that the edge of the shadow has moved from D E to D F, and meets the bottom at F. Beat a blackboard eraser, and create a cloud of dust in the path of the beam in the air, and you will discover that the rays G D that graze the edge of the basin at D become bent at the point where they enter the water, and now move in the bent line G D F, instead of, as formerly, in the straight line G D E. The path of the line in the water is now nearer to the vertical side D C ; in other words, this part of the beam *is more nearly vertical than before.*

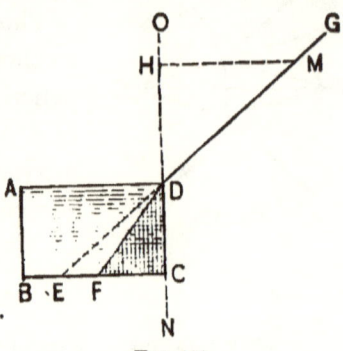

FIG. 181.

Experiment 2. Place a coin, A (Fig. 182), on the bottom of an empty basin, so that, as you look over the edge of the vessel through a small hole in a card, B C, the coin is just out of sight. Then, without moving the card or basin, fill the latter with water. Now, on looking through the aperture in the card, you find that the coin is visible. The beam A E, which

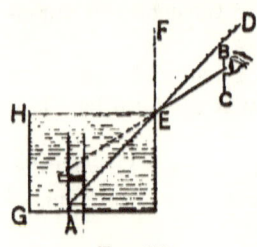

FIG. 182.

formerly moved in the straight line A D, is now bent at E, where it leaves the water, and, passing through the aperture in the card, enters the eye. Observe that as the beam passes from the water into the air it is turned farther from a vertical line, E F ; in other words, *the beam is farther from the vertical than before.*

Experiment 3. Thrust a pencil obliquely into water ; it will appear shortened, and bent at the surface of the water, and the immersed portion will appear elevated.

Experiment 4. Place a piece of wire (Fig. 183) vertically in front of the eye, and hold a narrow strip of thick plate glass horizontally across the wire, so that the light-waves from the wire may pass obliquely through the glass to the eye. The wire will appear to be broken at the two edges of the glass, and the intervening section will appear to the right or the left according to the inclination of the glass ; but if the glass be not inclined to the one side or the other, the wire does not appear broken.

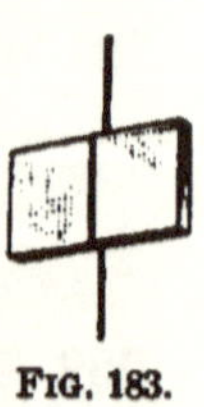

FIG. 183.

When a ray of light passes from one medium into another of different optical density, it is bent or *refracted* at the interface between the two mediums, unless it meet this plane perpendicularly. In the latter case there is no refraction. If it pass into a denser medium, it is refracted toward the perpendicular to this plane; if into a rarer medium, it is refracted from the perpendicular. It is not universally true that the denser mediums are the more highly refracting. The refractive power of water is less than that of alcohol or of oil of turpentine. A substance which has a higher refractive power than another is said to be *optically denser*.

The angle G D O (Fig. 181) is called the *angle of incidence;* F D N, the *angle of refraction;* and E D F, the *angle of deviation.*

233. Cause of Refraction. Foucault and others have proved by careful experiments that the speed of light is less in water than in air. It is less in glass than in water, and much less in diamond than in glass. Every transparent substance has its own rate of transmission. It would seem that there is an interaction between the ether and the molecules of matter such that in different mediums the ether-waves are unequally retarded.

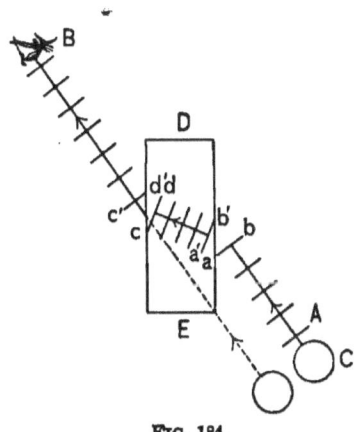

FIG. 184.

Let the series of parallel lines A B (Fig. 184) represent a series of wave-fronts leaving an object, C, and passing through a rectangular piece of glass, D E, and constituting a beam. Every point in a wave-front moves with equal velocity as long as it traverses the same medium; but the point a of a given wave-front, a b, enters the glass first, and its velocity is impeded, while the point b retains its original velocity; so that, while the point a moves to a', b moves to b', and the result is that the wave-front assumes a new direction (very much in the same manner as a line of soldiers executes a wheel), and a ray or a line drawn perpendicularly through the series of waves is turned out of its original direction on entering the glass. Again, the extremity c of a given wave-front, c d, first emerges from the glass, when its velocity is immediately quickened; so that while d advances to d', c advances to c', and the direction of the ray is again changed. The direction of the ray after emerging from the glass is parallel to its direction before entering it, but it has suffered a lateral displacement.

It is evident that if the ray enter the new medium in a direction perpendicular to its surface, *i.e.* with its wave-front parallel to this surface, all parts of the wave-front will be retarded simultaneously and no refraction will take place.

234. Index of Refraction. The deviation of light-waves in passing from one medium into another depends upon the optical density of the mediums and the angle of incidence. It diminishes as the angle of incidence diminishes, and is zero when the incident ray is normal. It is highly important, when the angle of incidence is known, to be able to determine the direction which a ray will take on entering a new medium. Describe a circle around the point of incidence A (Fig. 185) as a center; through the

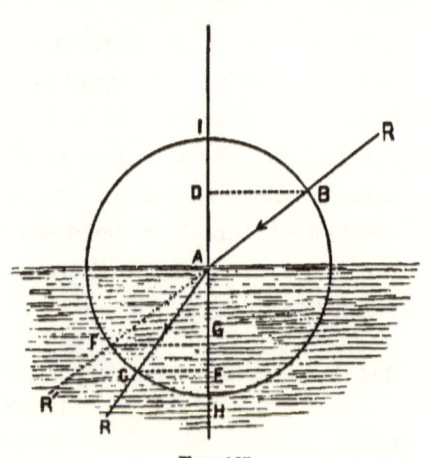

FIG. 185.

same point draw I H perpendicular to the surfaces of the two mediums, and to this line drop perpendiculars B D and C E from the points where the circle cuts the ray in the two mediums. Then suppose that the perpendicular B D is $\frac{8}{10}$ of the radius A B.; now, this fraction $\frac{8}{10}$ is called the *sine* of the angle D A B. Hence, $\frac{8}{10}$ is the *sine of the angle of incidence.* Again, if we suppose that the perpendicular C E is $\frac{6}{10}$ of the radius, then the fraction $\frac{6}{10}$ is the *sine of the angle of refraction.* The sines of the two angles are to each other as $\frac{8}{10} : \frac{6}{10}$, or as 4 : 3. The quotient (in this case $\frac{4}{3} = 1.33 +$) obtained by dividing the sine of the angle of incidence by the sine of the angle of refraction (generally expressed decimally) is called the *index of refraction*. The incident ray may be more or less oblique, yet the quotient (*i.e.* the index of refraction) remains the same.

235. Indices of Refraction. The index of refraction for light-waves in passing from air into water is approximately $\frac{4}{3}$, and from air into glass, $\frac{3}{2}$; of course, if the order be reversed, the reciprocal of these fractions must be taken as the indices; *e.g.* from water into air the index is $\frac{3}{4}$; from glass into air, $\frac{2}{3}$. When a ray passes from a vacuum into a medium, the refractive index is greater than unity, and is called the *absolute index of refraction*. The *relative index of refraction, from any medium, A, into another, B, is found by dividing the absolute index of B by the absolute index of A*.

The refractive index varies with wave-length. The following table is intended to represent *mean indices* for light-waves:

TABLE OF ABSOLUTE INDICES.

Lead chromate	2.97	Spirits of turpentine	1.48
Diamond (about)	2.5	Alcohol	1.37
Carbon disulphide	1.64	Humors of the eye (about)	1.35
Flint glass (about)	1.61	Pure water	1.33
Agate	1.54	Air at 0° C. and 760 mm.	
Canada balsam	1.53	pressure	1.000294
Crown glass (about)	1.53		

236. Critical Angle; Total Reflection. Let S S' (Fig. 186) represent the boundary surface between two mediums, and A O and B O incident rays in the more refractive medium (*e.g.* glass); then O D and O E may represent the same rays, respectively, after they enter the less refractive medium (*e.g.* air). It will be seen that, as the angle of incidence is increased, the refracted ray rapidly approaches the surface O S. Now, there must be an angle of incidence (*e.g.* C O M) such that the angle of refraction will be 90°; in this case the

incident ray C O, after refraction, will just graze the surface
O S. This angle (C O M), which must not be exceeded if the ray
is to pass out into the air, is called *the critical* or *limiting angle*.
Any incident ray, making a larger angle with the normal than
the critical angle, as L O, cannot emerge from the medium,
and consequently is not refracted. Experiment shows that

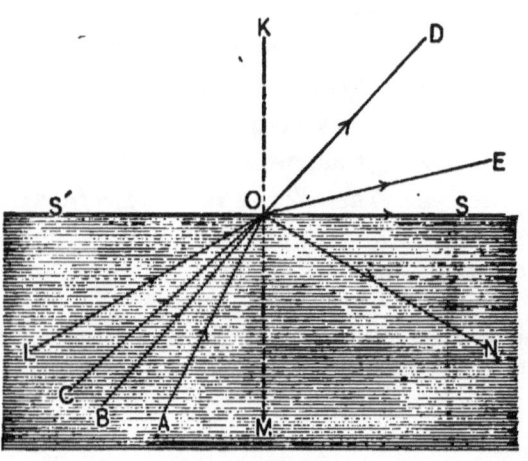

FIG. 186.

all such rays undergo internal reflection ; *e.g.* the ray L O is
reflected in the direction O N. Reflection in this case is so
nearly perfect that it has received the special name *total
reflection*. *Total reflection occurs when rays in the more
refractive medium are incident at an angle greater than the
critical angle.*

Surfaces of transparent mediums, under these circumstances, constitute the best mirrors possible. The critical angle diminishes as the
refractive index increases. For water it is about $48\frac{1}{2}°$; for flint glass,
38° 41'; and for the diamond, 23° 41'. Light-waves cannot, therefore,
pass out of water into air with a greater angle of incidence than $48\frac{1}{2}°$.
The brilliancy of gems, particularly of the diamond, is due in part to
their extraordinary power of reflection, arising from their large indices
of refraction or the smallness of their critical angles.

237. Illustrations of Refraction and Total Reflection.

Experiment 5. Observe the image of a candle flame reflected by the surface of water in a glass beaker, as in Fig. 187.

Experiment 6. Thrust the closed end of a glass test-tube (Fig. 188) into water and incline the tube. Look down upon the immersed part

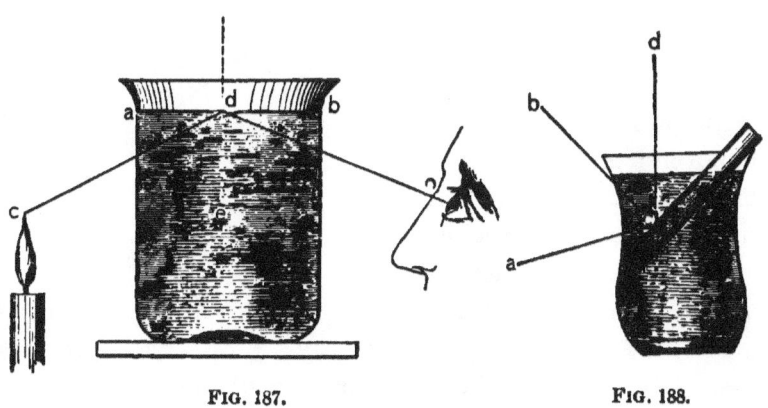

FIG. 187.   FIG. 188.

of the tube, and its upper surface will look like burnished silver, or as if the tube contained mercury. Fill the test-tube with water, and immerse as before; the total reflection which before occurred at the surface of the air in the submerged tube now disappears. Explain.

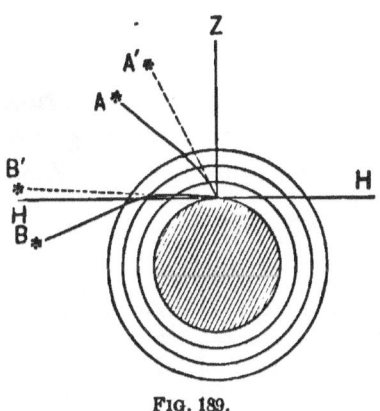

FIG. 189.

A ray of light from a heavenly body, A (Fig. 189), undergoes a series of refractions as it reaches successive strata of the atmosphere of constantly increasing density, and to an eye at the earth's surface appears to come from a point, A', in the heavens. The general effect of the atmosphere on the path of light that traverses it is such as to increase the apparent altitude of the heavenly bodies. It enables us to see a body, B, which is actually below the horizon H H, and prolongs the apparent stay of the sun, moon, and other heavenly bodies, above the horizon. Twilight is due to both refraction and reflection of light by the atmosphere.

SECTION VII.

PRISMS AND LENSES.

238. Optical Prisms. An optical prism is a portion of a transparent medium bounded by two plane surfaces inclined to each other. Fig. 190 represents a transverse section of a common form of prism. Let A B be a ray of light incident upon one of its surfaces. On entering the prism it is refracted *toward* the normal, and takes the direction B C. On emerging from the prism it is again refracted, but now *from* the normal in the direction C D.

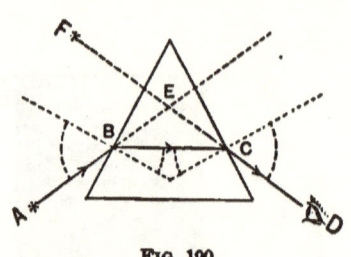

FIG. 190.

The object that emits the ray will appear in the direction D E F. Observe that the ray A B, at both refractions, is bent toward the thicker part, or base, of the prism.

239. Lenses. Any transparent medium bounded by surfaces of which at least one is curved, is a *lens*.

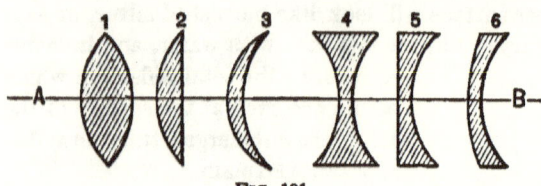

FIG. 191.

Lenses are of two classes, converging and diverging, according as they collect rays or cause them to diverge. Each class comprises three kinds (Fig. 191):

CLASS I.		CLASS II.	
1. Bi-convex	Converging, or convex lenses, thicker in the middle than at the edges.	4. Bi-concave	Diverging, or concave lenses, thinner in the middle than at the edges.
2. Plano-convex		5. Plano-concave	
3. Concavo-convex (or meniscus)		6. Convexo-concave	

A straight line normal to both surfaces of a lens and passing through their centers of curvature, as A B, is called its *principal axis*. There is a point in the principal axis of every lens, either at or near its center of volume, called its *optical center*, so placed that rays of light which pass through this point and the lens *suffer no change of direction*, though there may be a slight lateral displacement. In lenses 1 and 4 it is halfway between their respective curved surfaces.

240. Effect of Lenses. *The general effect of all convex lenses is to cause transmitted rays to converge; that of concave lenses, to cause them to diverge.* In-

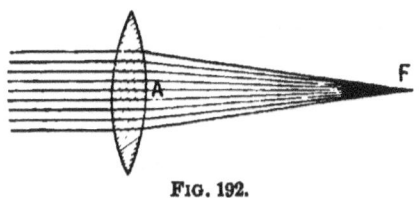

FIG. 192.

cident rays parallel to the principal axis of a convex lens are brought to a focus, F (Fig. 192), at a point in the principal axis. This point is called the *principal focus,* i.e. it is the focus of incident rays parallel to the principal axis. It may be found by holding the lens so that the rays of the sun may fall perpendicularly upon it, and then moving a sheet of paper back and forth behind it until the image of the sun formed on the paper is brightest and smallest. The focal length is the distance from the optical center of the lens to the center of the image on the paper. The

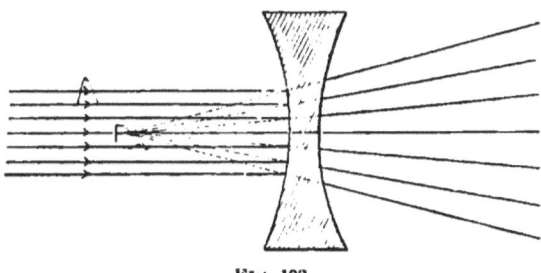

FIG. 193.

shorter the focal length the more powerful is the lens; that is, the more quickly are the parallel rays that traverse different parts of the lens brought to cross one another.

A pencil of rays, emitted from the principal focus F (Fig. 192) as a luminous point, becomes parallel on emerging from a convex lens. If the rays emanate from a point nearer the lens, they diverge after egress, but the divergence is less than before; if from a point beyond the principal focus, the rays are rendered convergent. A concave lens causes parallel incident rays to diverge as if they came from a point, as F (Fig. 193). This point is therefore its principal focus. It is, of course, a *virtual focus.*

Every lens has a *principal focus;* it is the point to which parallel rays are caused to converge, or from which, after deflection, they appear to diverge, as the case may be.

241. Conjugate Foci. When a luminous point, S, beyond the principal focus (Fig. 194) sends rays to a convex lens, the emergent rays converge to another point, S'; rays sent from S' to the lens would converge to S. Two points thus related are called *conjugate foci.* The fact that rays which emanate from one point are caused by convex lenses to collect at one point, gives rise to real images, as in the case of concave mirrors.

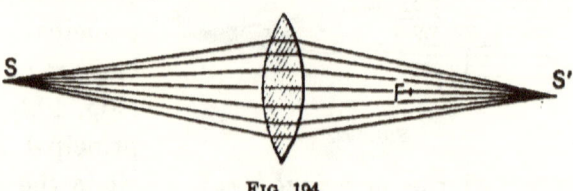

FIG. 194.

242. Law of Converging Lenses.

Lenses, like mirrors, have conjugate foci at distances p and p' from the optical centers. In converging lenses the principal focal distance and the distance of their conjugate foci (or distance of object and image) are related according to the formula

$$\frac{1}{p} + \frac{1}{p'} = \frac{1}{f}.$$

Hence, the law of converging lenses : *The reciprocal of the principal focal length is equal to the sum of the reciprocals of any two conjugate focal lengths.*

When a pencil of light comes from an infinite distance (*i.e.* when its rays are parallel), $p = \infty$; then $p' = f$, and the rays converge at the principal focus. Conversely, if a pencil come from the principal focus, $p = f$; hence $p' = \infty$; that is, no image is formed.

234 ETHER DYNAMICS.

If the object (*i.e.* the source of light) be at a distance less than infinity, but greater than $2f$, the image is real, and is on the other side of the lens at a distance greater than f and less than $2f$. Conversely, if the object be at a distance greater than f, but less than $2f$, the image is at a distance greater than $2f$.

243. Images Formed. Fairly distinct images of objects may be formed through *very small* apertures (§ 220); but owing to the small quantity of light that passes through the aperture, the images are very deficient in brilliancy. If the aperture be enlarged, brilliancy is increased at the expense of distinctness. *A convex lens enables us to obtain both brilliancy and distinctness at the same time.*

Experiment. By means of a porte-lumière, A (Fig. 195), introduce a horizontal beam of light into a darkened room. In its path place some object, as B, painted in transparent colors or photographed on glass. (Transparent pictures are cheaply prepared by photographers for sunlight and lime-light projections.) Beyond the object place a convex lens, L, and beyond the lens a screen, S S. The object being illuminated by the beam of light, all the rays diverging from any point, a, are bent by the

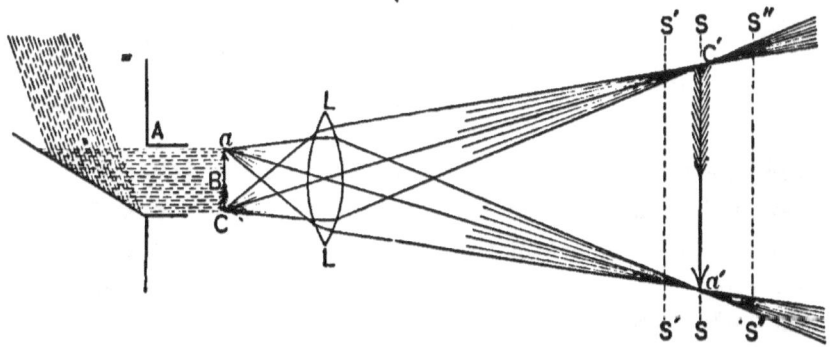

FIG. 195.

lens so as to come together at the point a'. In like manner, all the rays proceeding from C are brought to the same point C'; and so also for all intermediate points. Thus, out of the numberless rays emanating from each of the points on the object, those that reach the lens are guided by it, each to its own appropriate point in the image. It is evident that there must result an image both bright and distinct, provided the screen be suitably placed, *i.e.* at the place where the rays meet. But if the screen be placed at S' S' or S'' S'', it is evident that a blurred image will be formed. Instead of moving the screen back and forth, in order to "focus" the rays properly, it is customary to move the lens.

Figure 196 shows more accurately the form of the image produced by the ordinary convex lens. It is apparent that if the center of the image A' B' be properly focused upon a screen the extremities of the image will be a little out of focus, and *vice versa*.

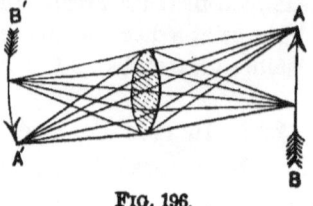

FIG. 196.

244. Construction of Images Formed by Convex Lenses. Given the lens L (Fig. 197), whose principal focus is at F, and object A B in front of it; any two of the many rays from A will determine where its image a is formed. Two rays that can be traced easily are, one along the secondary axis A O a,

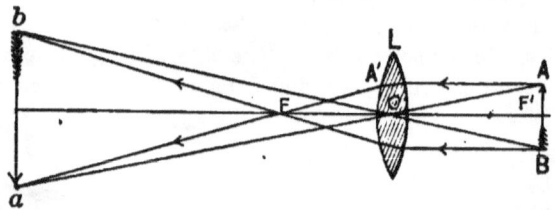

FIG. 197.

and one A A' parallel to the principal axis; the latter will be deviated so as to pass through the principal focus F, and will afterward intersect the secondary axis at some point, a; therefore this is the conjugate focus of A. Rays can be similarly traced for B and all intermediate points along the arrow. Thus, a *real inverted image* is formed at a b.

The linear dimensions of an object and of its image formed by a convex lens are proportional to their respective distances from the center of the lens.

245. Virtual Images. Since rays that emanate from a point nearer the lens than the principal focus diverge after egress, it is evident that their focus must be virtual and on the same side of the lens as the object. Hence, *the image of an object placed nearer the lens than the principal focus is virtual, magnified, and erect*, as shown in Fig. 198. A convex lens used in this manner is called a *simple microscope*.

246. Simple Microscope. As its name implies, the microscope is an instrument for viewing minute objects. The simple microscope consists of a single converging lens so placed that the object is between the principal focus and the lens. It magnifies by increasing the visual angle.

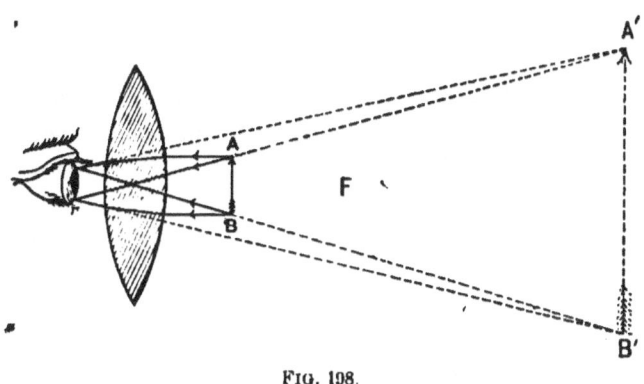

Fig. 198.

The *magnifying power* of the lens is simply the ratio between the apparent linear dimension of the image and the corresponding dimension of the object, *e.g.* A′ B′ : A B (Fig. 198), or it is the ratio between the visual angles under which the eye would see image and object, if both were placed at the distance of distinct vision.[1] If the lens be of short focus, as is usually the case, the magnifying power is approximately the ratio of the distance of distinct vision to the focal length. Thus a lens of $\frac{1}{2}$ inch focal length would magnify 20 to 24 times.

---

[1] For normal eyes, an object to be seen most distinctly must be placed at a distance of 10 to 12 inches; hence this is regarded as the distance of distinct vision.

247. Diverging Lenses. Since the effect of concave lenses is to render transmitted rays divergent, pencils of rays emitted from A and B (Fig. 199) diverge after refraction, as if they came from A' and B', and the image appears to be at A'B'. Hence, *images formed by concave lenses are virtual, erect, and smaller than the object.*

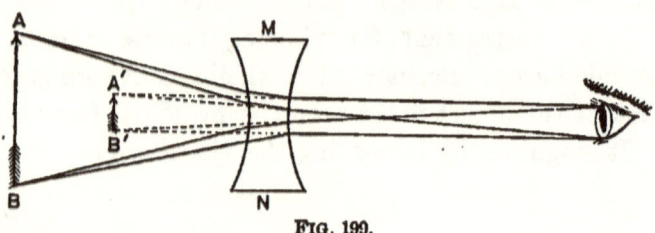

FIG. 199.

248. Spherical Aberration. In all ordinary convex lenses the curved surfaces are spherical, and the angles which incident rays make with the little plane surfaces, of which we may imagine the spherical surface to be made up, increase rapidly toward the edge of the lens. Thus, while those rays

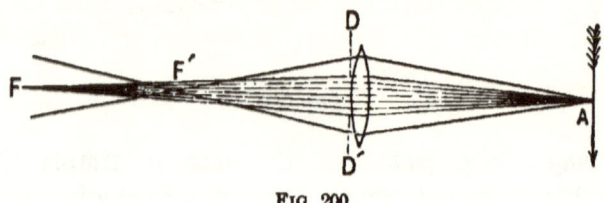

FIG. 200.

from a given point of an object which pass through the central portion (Fig. 200) meet approximately at the same point F, those which pass through the marginal portion are deflected so much that they cross the axis at nearer points, *e.g.* at F'; so a blurred image results. This wandering of the rays from a single focus is called *spherical aberration.*

No lens with spherical surfaces can bring all the rays to the same focus. The evil may be in a measure corrected by interposing a diaphragm, D D', provided with a central aperture smaller than the lens, so as to cut off those rays that pass through the marginal part of the lens. But it can be wholly corrected only by properly modifying the curvature of the surfaces of the lens. A lens having surfaces thus modified is said to be *aplanatic*.

EXERCISES.

1. What must be the position of an object with reference to a converging lens, that its image may be real and magnified?

2. A photographic transparency is placed between a porte-lumière and a bi-convex lens, 16 inches from the latter. How many diameters is a distinct image of the transparency multiplied on a screen 20 feet distant?

3. A transparency whose dimensions are 3 × 4 inches is placed 16 inches from the lens. At what distance from the lens must the screen be that it may receive a distinct image of the transparency that shall cover a surface 3 × 4 feet?

4. What is the focal length of the lens used in the last case?

5. With a converging lens the image of a candle is thrown on a screen 6 feet from the candle, and the focal length of the lens is 16 inches. Find the distances of the candle and of the screen from the lens.

6. A luminous point is 3 inches from a convex lens having a focal length of 5 inches. Find the position of the image.

7. If the candle and the screen be 3 feet apart, and the lens be midway between them, what is the focal length?

8. Find the focal length of a lens which throws the image of an object 5 feet distant on a screen 3 feet distant.

9. About what is the focal length of a simple microscope that magnifies 30 times?

10. About how many times does a lens of 2 inches focal length magnify?

11. (*a*) What is meant by the "power" of a simple microscope? (*b*) What is necessary that it may have great power?

12. If an object be at twice the focal distance of a convex lens, how will the length of the image compare with the length of the object?

13. To an eye whose distance of distinct vision is 25 cm., how many diameters will a lens of 1 cm. focus magnify?

14. Show that a concave air lens in water has the same effect on incident light as a convex water lens in air.

SECTION VIII.

PRISMATIC ANALYSIS OF LIGHT. SPECTRUMS.

249. Analysis of Sunlight.

Experiment 1. Place a disk with an adjustable slit in the aperture of a porte-lumière so as to exclude from a darkened room all light-waves except those which pass through the slit. Near the slit interpose a double-convex lens of (say) 10-inch focus. A narrow sheet of light will traverse the room and produce an image, A B (Fig. 201), of the slit on a white screen placed in its path. Now place a glass prism, C, in the path

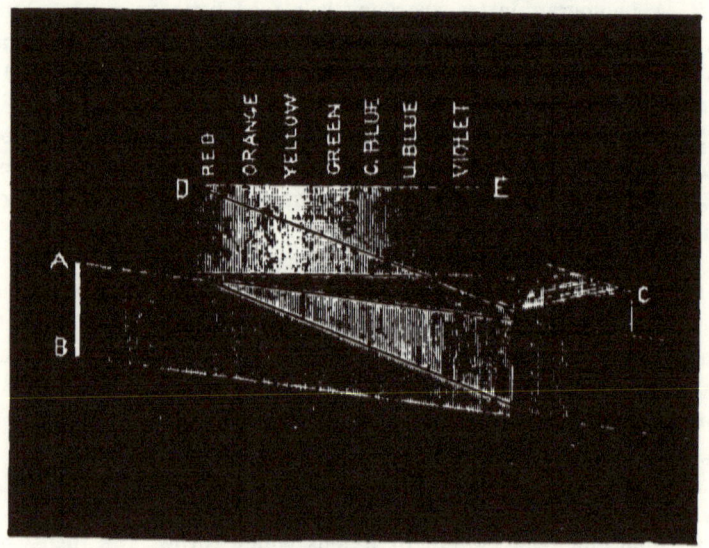

FIG. 201.

of the narrow sheet of light and near to the lens, with its edge vertical. (1) Not only is the light now turned from its former path, but that which before was a narrow sheet is, after emerging from the prism, spread out fan-like into a wedge-shaped body, with its thickest part resting on the screen. (2) The image, before only a narrow, vertical band, A B, is now drawn out into a long horizontal ribbon, D E. (3) The image, before white, now presents all the colors of the rainbow, from red at one end to violet at the other; it passes gradually through all the gradations of orange, yellow, green, blue, and violet. (The difference in deviation between the red and the violet is purposely much exaggerated in the figure.)

From this experiment we learn (1) that *white light is not simple* in its composition, *but the result of a mixture of colors.*[1] (2) *The colors of which white light is composed may be separated by refraction.* (3) *The separation is due to the different degrees of deviation which colors undergo by refraction.* Red, which is always least turned aside from a straight path, is the least refrangible color. Then follow orange, yellow, green, blue, and violet, in the order of their refrangibility. The many-colored ribbon of light D E is called the *solar spectrum.*[2] This separation of white light into its constituents is called *dispersion.* The number of colors of which white light is composed is really infinite, but we have names for only seven of them; viz. *red, orange, yellow, green, cyan-blue, ultramarine-blue,* and *violet;* and these are called the *prismatic colors.* The names of the blues are derived from the names of the pigments which most closely resemble them.

The *rainbow* is a solar spectrum on a grand scale. It is the result of refraction, total reflection, and dispersion of sunlight by falling raindrops.

250. Synthesis of White Light. The composition of white light has been ascertained by the process of analysis; it can be verified by *synthesis;* i.e. the colors after dispersion may be reunited, and the result of the reunion is white light.

Experiment 2. Place a second prism (2) in such a position $\triangle \triangledown$ that light which has passed through one prism (1), and been refracted and decomposed, may be refracted back, and the colors will be reblended, and a white image of the slit will be restored on the screen.

251. Chromatic Aberration. There is in ordinary convex lenses a serious defect, to which we have not before referred, called *chromatic aberration,* the correction of which has

---

[1] Newton (1666) was the first to recognize the true import of this phenomenon, *i.e.* to refer the colors to the heterogeneity of white light.

[2] A succession of colors in the order of their refrangibility, obtained from any source of light, is called a *spectrum.*

demanded the highest skill. The convex lens both *refracts* and *disperses* the light-waves that pass through it. The tendency, of course, is to bring to a focus the more refrangible rays, as the violet, much sooner than the less refrangible rays, such as the red. The result is a disagreeable coloration of the images that are formed by the lens, especially by those portions of the light-waves that pass through the lens near its edges. This evil may be overcome very effectually by combining with the convex lens a plano-concave lens. Now, if a crown-glass convex lens be taken, a flint-glass concave lens may be prepared that will correct the dispersion of the former without neutralizing all its refraction.[1] A compound lens composed of these two lenses cemented together (Fig. 202) constitutes what is called an *achromatic lens*.

FIG. 202.

252. Cause of Color and Dispersion. The color of light is determined by *vibration-frequency*, or, in other words, by the corresponding *wave-length*. The light-waves diminish in length from the red to the violet. As pitch depends on the frequency with which aërial waves strike the ear, so color depends upon the frequency with which ether-waves strike the eye. The difference between violet and red is a difference analogous to the difference between a high note and a low note on a piano.

The speed of propagation in a vacuum appears to be the same for all wave-lengths. But in a refracting medium the short waves are more retarded than the longer ones, hence they are more refracted. This is the cause of dispersion. Each wave-length has its own refractive index, or, since vibration-frequency corresponds to color, every simple color has its special refractive index. Light composed of waves all of the same (or nearly the same) length is called *homogeneous*

---

[1] The refractive and the dispersive powers of the two lenses are not proportional.

or *monochromatic* light. The yellow light emitted by the flame of a Bunsen burner or alcohol lamp when common salt is sifted upon it is approximately monochromatic. Ordinary white light is a mixture of long and short ether-waves.

From well-established data, determined by a variety of methods, physicists have calculated the number of waves that succeed one another for each of the several prismatic colors, and the corresponding wave-lengths; the following table contains the results. The letters A, C, D, etc., refer to Fraunhofer's lines (see § 258).

		Length of waves in millimeters.	No. of waves per second.
Dark red	A	.000760	395,000,000,000,000
Orange	C	.000656	458,000,000,000,000
Yellow	D	.000589	510,000,000,000,000
Green	E	.000527	570,000,000,000,000
C. Blue	F	.000486	618,000,000,000,000
U. Blue	G	.000431	697,000,000,000,000
Violet	H	.000397	760,000,000,000,000

There is a limit to the sensibility of the eye as well as of the ear. The range of vibrations appreciable by the eye lies approximately between the highest and the lowest numbers indicated in the above table; *i.e.* if the succession of waves be much more or much less rapid than is indicated by these numbers, the sensation of sight is not produced.

It is evident that *the frequency of the waves emitted by a luminous body, and consequently the color of the light emitted, must depend on the rapidity of the vibratory motions of the molecules of that body, i.e. upon its temperature.*

This has been shown in a convincing manner as follows: The temperature of a platinum wire is slowly raised by passing a gradually increasing current of electricity through it. At a temperature of about 540° C. it begins to emit light; and if the light be analyzed by a prism, it is shown that only red light is emitted. As the temperature rises, there will be added to the red of the spectrum, first yellow, then green, blue, and violet, successively. When it reaches a white heat, it emits all the prismatic colors. It is significant that a white-hot body emits more red light than a red-hot body, and likewise more light of every color than at any lower temperature. The conclusion is that *a body which emits white light sends forth simultaneously waves of a variety of lengths.*

253. Continuous Spectrums. The spectrum produced by the platinum is continuous; that is, the band of light is unbroken. If the spectrum be not complete, as when the temperature is too low, it will begin with red, and be continuous as far as it goes. *All luminous solids and liquids give continuous spectrums.*

A gas, kerosene, or candle flame does not give the spectrum of a vapor, but gives that of the solid particles of carbon in a state of incandescence; hence the continuous spectrums which these flames afford.

254. Spectroscopes. Instruments for the observation of spectrums are called *spectroscopes*. The essential part of the apparatus is the "dispersion piece," which is usually a glass prism. Instead of looking at the spectrum with the naked eye, it is usually better to view it through a small telescope, which serves to magnify it. Fig. 203 represents the simplest form of the Kirchhoff and Bunsen spectroscope. A flint-glass prism, A, receives light through an adjustable slit at the end of a tube, B, called the *collimator*. At the opposite end of this tube is a converging lens, and the slit is located at its principal focus, so that rays diverging from the slit are rendered parallel by the lens.

It is often necessary to have some means of determining the positions of certain lines (to be described hereafter) observed in the spectrum. The usual method is to have a second tube, somewhat like the collimating tube, so placed that the rays from a light (*e.g.* a candle flame, C, in Fig. 203), after passing through a transparent plate (inside the tube), on which a fine scale is engraved, and through a lens, by which they are made parallel, are reflected from the nearest face of the prism, and pass into the telescope along with the beam of light under analysis. Thus the eye while viewing the spectrum through the telescope D sees also a magnified image of the scale coinciding with the spectrum.

255. Bright-line Spectrums. If a platinum wire be dipped in a solution of common salt and placed in the almost colorless flame of a Bunsen burner (Fig. 203), the flame will become colored a deep yellow. Examining the flame with the spectroscope, you find instead of a continuous spectrum, such as is described in § 253, only a bright narrow line of yellow in the yellow part of the spectrum. Your spectrum consists essentially of a single [1] bright yellow line on a comparatively dark ground. (See Sodium, Plate I, frontispiece.)

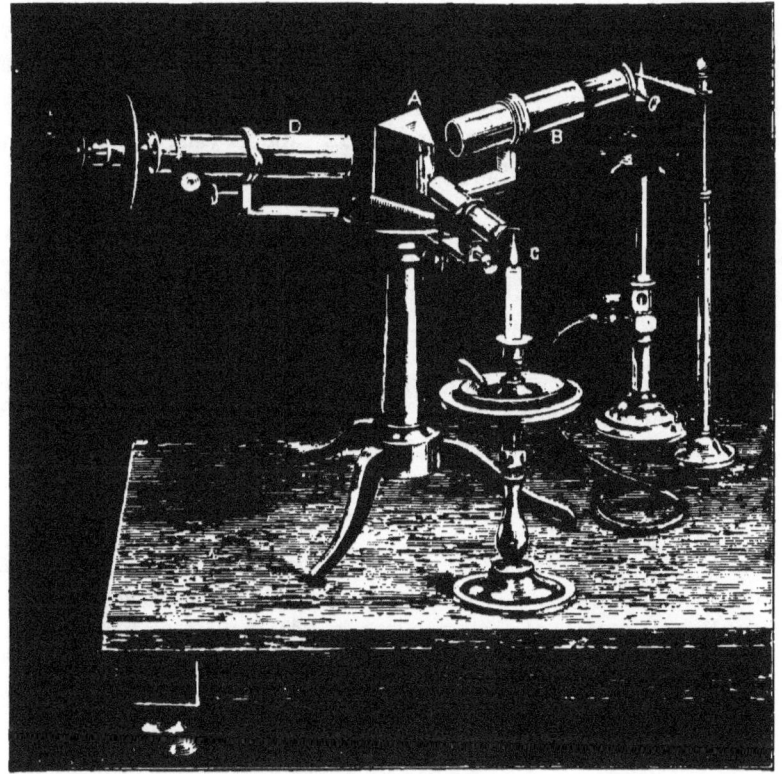

FIG. 203.

Plate I also exhibits the spectrums obtained when salts of lithium, strontium, and potassium are, respectively, introduced into the flame. Each of these salts contains a different metal; *e.g.* common salt contains the metal sodium; the other substances used successively contain respec-

[1] Spectroscopes of higher dispersive power show that the sodium line is really a double line divided by a narrow interval.

tively the metals lithium, potassium, and strontium. These metals, when introduced into the flame, are vaporized, and we get their spectrums when in a gaseous state. *All incandescent gases, unless under great pressure, give discontinuous, or bright-line, spectrums, and no two gases give the same spectrum.*

256. Spectrum Analysis. Molecules of different substances, *e.g.* sodium, lithium, etc., have their own peculiar rates of vibration, and each emits ether-waves whose lengths correspond to the rates of vibration, and hence each produces its own distinctive bright-line spectrum. Hence has arisen a new chemical analysis, wherein substances are detected by observing the bright lines of their spectrums, a branch of physical chemistry called *spectrum analysis*.

It is only in the gaseous state, however, that the molecule is free to exhibit its special rate of vibration; when they are packed closely together in a solid or liquid, their motions are cramped, their periodicity is lost, and all manner of vibrations are induced. Hence spectrums of solids and liquids are continuous, *i.e.* the rates of vibrations are so many in number as to leave no gaps in their spectrums.

Many chemical compounds are decomposed into their elements, and the elements are rendered gaseous, at a temperature that is at, or below, the temperature necessary for incandescence. In that case the spectrum given is the combined spectrums of the elements.

257. Reversed or Dark-line Spectrum. This type of spectrum may be studied as follows: Arrange apparatus in a dark room, as in Fig. 204. N is the nozzle of a stereopticon (p. 262) containing only the condensing lens; T and S are two tin plates, in the latter of which a slit is cut. Allow a beam of calcium light to pass through the slit in S, and thence through the converging lens L and the prism P, and form a spectrum on a screen, H. Hold in the flame of a Bunsen burner, B, a pellet of sodium; it burns vividly, and the calcium light has to pass through the intensely yellow flame. We should naturally expect that the yellow of the spectrum would now be more intensely illuminated, but, instead, a *dark* band in the yellow now appears. It is not really black, but *comparatively* dark.

Next hold the plate T between the burner and the condensers so that the calcium light may be cut off from the upper portion of the slit, leaving the light from the sodium flame alone to pass through this part of the slit. The spectrum R formed by this part consists of a bright yellow line on a dark ground, being the radiation spectrum of sodium. (It should be borne in mind that the image of the slit is inverted.) The other half, A, shows a dark line on the continuous spectrum. We thus have, contiguous to each other, the bright-line spectrum of sodium and its *reversed, dark-line,* or *absorption* spectrum. If you use salts of lithium, potassium, strontium, etc., in a similar manner, in every case you will find your spectrum crossed by dark lines where you would expect to find bright lines.

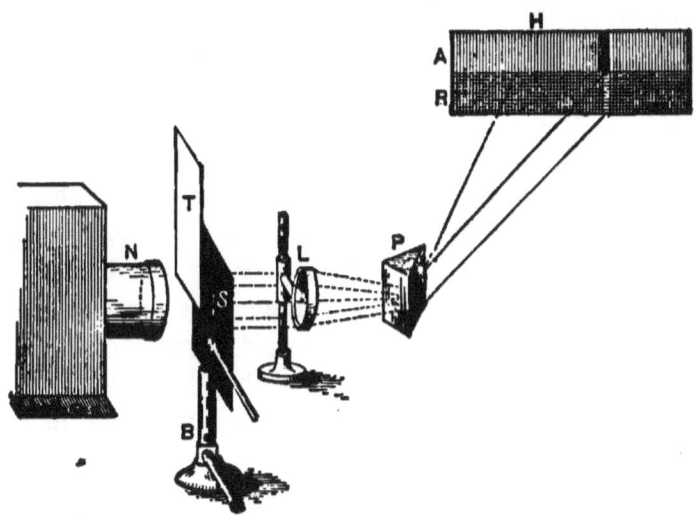

FIG. 204.

It thus appears that *the vapors of different substances absorb or quench the very same rays that they are capable of emitting when made self-luminous,* very much, it would seem, as a given tuning fork selects from various sounds only those of a definite wave-length corresponding to its own vibration-period. The dark places of the spectrum receive light in full force from the salted flame; but the light is so feeble, in comparison with those places illuminated by the calcium light, that the former appear dark by contrast. Light transmitted through certain liquids (as sulphate of quinine and blood) and certain solids (as some colored glasses) produces band spectrums. These spectrums are obtained only when light passes through mediums capable of absorbing rays of a certain wave-length; hence, they are commonly called *absorption spectrums*. Since a given vapor causes dark lines precisely where it would cause bright lines if it were itself the only radiator of light, dark-line spectrums are frequently called *reversed spectrums*. There are, then, three kinds of spectrums: *continuous spectrums*, produced by luminous solids, liquids, or, as has been found in a few instances, gases under great pressure; *bright-line spectrums*, produced by luminous vapors; and *absorption spectrums*, produced by light that has been sifted by certain mediums.

258. Fraunhofer's Lines. The spectrum of sunlight, when the apparatus employed in Experiment 1, § 249, is properly adjusted, is observed to contain a large number of dark lines transverse to its length. These were first mapped by Fraunhofer (1814), who distinguished several of the more prominent ones by letters of the alphabet; hence the dark lines of the solar spectrum have received the name of *Fraunhofer's lines*.

So far as has been discovered, no two substances have a spectrum consisting of the same combination of lines; and, in general, different substances very rarely possess lines appearing to be common to both. Hence, when we have once observed and mapped the spectrum of any substance, we may ever after be able to recognize the presence of that substance when emitting light, whether it is in our laboratory or in a distant heavenly body.[1]

---

[1] By examination of the reversed spectrum of the sun, we are able to determine with certainty the existence there of sodium, calcium, copper, zinc, magnesium, hydrogen, and many other known substances. Again, from our knowledge of the way in which a reversed spectrum can be produced, we may conclude that the sun consists of a luminous solid, a liquid, or an intensely heated and greatly condensed gas (called a *photosphere*), and that this nucleus is surrounded by an atmosphere of cooler vapor, in which exist at least all the substances just named. The moon, and planets that are visible only by reflected sunlight, give the same spectrums as the sun, while those that are self-luminous give spectrums which differ from the solar spectrum.

259. Infra-red and Ultra-violet Rays. The energy of ether-waves is capable, as has been before observed, of producing calorific, luminous, or chemical effects, according to the nature of the bodies upon which it falls. When a sensitive thermoscope is passed along the spectrum, heat effects are observed throughout the visible spectrum, and for considerable distances beyond at each extremity. All ether-waves are capable of producing heating effects.

It thus appears that the solar spectrum is not limited to the visible spectrum, but extends beyond at each extremity, and spectroscopic analysis, besides sifting the waves of one color from those of another, is able to sift out rays which do not produce the sensation of light from those which do. Those rays that lie beyond the red are called the *infra-red* rays, while those that lie beyond the violet are called the *ultra-violet* rays. The infra-red rays are of longer vibration-period, and the ultra-violet of shorter period, than the luminous waves.

260. Only One Kind of Radiation. The fact that radiant energy produces three distinct effects — *viz.* luminous, heating, and chemical — has given rise to a prevalent idea that there are three distinct kinds of radiation. There is, however, absolutely no proof that these different effects are produced by different kinds of radiation. Science recognizes in radiations no distinctions but periods, wave-lengths, and wave-forms. *The same radiation that produces vision can generate heat and chemical action.*

The fact that the infra-red and ultra-violet rays do not affect the eye does not argue that they are of a different nature from those that do, but it does show that there is a limit to the susceptibility of the eye to receive impressions from radiation. Just as there are sound-waves of too long, and others of too short period to affect the ear, so there are ether-waves, some of too long, and others of too short period to affect the eye.

SECTION IX.

COLOR.

261. Color by Absorption. Color is a sensation; it has no material existence. The term "yellow light" means, primarily, a particular sensation; secondarily, it means the physical cause of this sensation, *i.e.* a train of ether-waves of a particular frequency.

Experiment 1. By means of a porte-lumière introduce a beam of sunlight into a dark room. With the slit and prism form a solar spectrum. Between the slit and the prism introduce a ruby-colored glass; all the colors of the spectrum except the red are much reduced in intensity.

It thus appears that the color of a colored transparent object, as seen by transmitted light, arises from the unequal absorption of the different colors of white light incident upon it. A red glass absorbs less red light than light of other colors. The color produced by absorption is rarely very pure, the particular hue of the transmitted light being due merely to a predominance of certain colors, and not to the absence of all others. As the absorbing layer is thicker, the resulting color is purer but less intense.

Experiment 2. We have found that common salt introduced into a Bunsen flame renders it luminous, and that the light when analyzed with a prism is found to contain only yellow. Expose papers or fabrics of various colors to this light in a darkened room. *No one of them except yellow exhibits its natural color.*

Experiment 3. Hold a narrow strip of red paper or ribbon[1] in the red portion of the solar spectrum; it appears red. Slowly move it toward the other end of the spectrum; on leaving the red it becomes darker, and when it reaches the green it is quite black or colorless, and remains so as it passes the other colors of the spectrum. Repeat the experiment, using other colors, and notice that only in light of its own color does each strip of paper appear of its natural color, while in all other colors it is dark.

---

[1] Care must be exercised to select only pure colors.

These experiments show that the color of a body seen by light reflected from it depends both upon the color of the light incident upon it and upon the nature of the body.

If a piece of colored glass, A B (Fig. 205), be held near a window so as to receive rays of sunlight obliquely, a portion

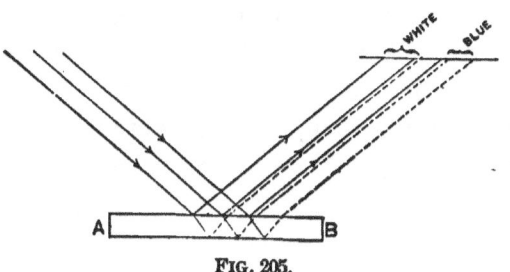

FIG. 205.

of the light will be reflected by the anterior surface of the glass, and, falling upon the white ceiling, will illuminate it with white light. Another portion of the light will enter the glass and be reflected from the posterior surface; this light, having entered the glass and traveled in it a distance a little greater than twice its thickness, will suffer an unequal absorption of its rays, and after emerging from the glass will, if the glass be blue, illuminate a neighboring portion of the ceiling with blue light.

This illustrates the method by which pigments afford color. Thus, the anterior surface of a water-color drawing reflects the white daylight. Most of the light reflected to the eye has, however, passed through the pigment to the white paper beneath, and, being reflected from this, again passes through the layer of pigment before reaching the eye. Certain of the colors which compose white light are extinguished while passing through the pigment, and the color by which the pigment is recognized is the resultant of the unextinguished colors. This is technically called *selective absorption*. Different pigments quench different colors; the unquenched colors determine the color of the pigment.

262. Opalescence. Sky Colors.

Experiment 4. Dissolve a little white castile soap in a tumbler of water; or, better, stir into the water a few drops of an alcoholic solution of mastic, enough to render the water slightly turbid. Place a black screen behind the tumbler, and examine the liquid by reflected sunlight, — the liquid appears to be blue; examine the liquid by transmitted sunshine, — it now appears yellowish red.

Experiment 5. Pour some of the turbid liquid into a small test-tube, and examine it and the tumbler of liquid by transmitted light; the former appears almost colorless, while the latter is deeply colored.

When a medium holds in suspension fine particles of matter, the shorter light-waves are most abundantly reflected, giving a blue color. The blue is purer as the particles are smaller. Objects seen *through* such mediums appear of the complementary hue (see § 267). This phenomenon is called *opalescence*. It accounts for the blue of watery milk, opalescent glass, smoke, and the sky.

Skylight is reflected light. The minute particles (of water, probably) that pervade the atmosphere, like the fine particles of mastic suspended in the water, reflect blue light; while beyond the atmosphere is a black background of darkness. But we must not, from this, conclude that the atmosphere is blue; for, unlike blue glass, but like the turbid liquid, it transmits yellow and red rays freely, so that seen by reflected light it is blue, but seen by transmitted light it is yellowish red.

When the sun is near the horizon, its rays travel a greater distance in the air to reach the earth than when it is in the zenith (see Fig. 189); consequently, there is a greater loss by absorption and reflection in the former case than in the latter. But the yellow and red rays suffer less destruction, proportionally, than the other colors; consequently, these colors predominate in the morning and evening.

263. Mixing Colors. A mixture of all the prismatic colors in the proportion found in sunlight produces white. Can white be produced in any other way?

Fig. 206.

Experiment 6. On a black surface, A (Fig. 206), lay two small rectangular pieces of paper, one yellow and the other blue, about 2 inches apart. In a vertical position between these papers, and from 3 inches to 6 inches above them, hold a slip of plate glass, C. Looking obliquely down through the glass you may see the blue paper by transmitted light.

waves and the yellow paper by reflection. That is, you see the object itself in the former case, and the image of the object in the latter case. By a little manipulation the image and the object may be made to overlap each other, when both colors will apparently disappear, and in their place the color which is the result of the mixture will appear. In this case it will be white, or rather *gray*, which *is white of a low degree of luminosity*. If the color be yellowish, lower the glass; if bluish, raise it.

Experiment 7. With the rotating apparatus, rotate the disk (Fig. 207) which contains only yellow and blue. The colors (*i.e.* the sensations) so blend in the eye as to produce the sensation of gray.

Fig. 208 represents "Newton's disk," which contains the seven prismatic colors arranged in a proper proportion to produce gray when rotated.

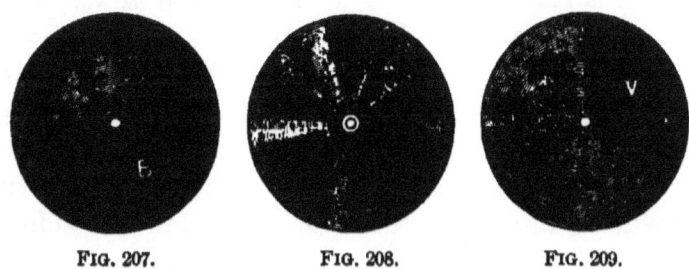

FIG. 207.   FIG. 208.   FIG. 209.

In like manner, you may produce white by mixing purple and green; or, if any color on the circumference of the circle (see Complementary Colors, Plate I) be mixed with the color exactly opposite, the resulting color will be white. Again, the three colors, red, green, and violet, arranged as in Fig. 209, with rather less surface of the green exposed than of the other colors, will give gray. Green mixed with red, in varying proportions, will produce any of the colors in a straight line between these two colors in the diagram (Plate I); green mixed with violet will produce any of the colors between them; and violet mixed with red gives purple.

All colors are represented in the spectrum, except the purple hues. The latter form the connecting link between the two ends of the spectrum. Our color chart (Plate I) is intended to represent the sum total of all the sensations of color. By means of this chart we may determine the result of the (optical) mixture of any two colors, as follows: Find the places occupied upon the chart by the two colors which are to be mixed, and unite the two points by a straight line. The color produced by the mixture will invariably be found at the center of this line.

264. Mixing Pigments.

Experiment 8. Mix a little of the two pigments chrome yellow and ultramarine blue, and you obtain a green pigment.

The last three experiments show that mixing certain colors, and mixing pigments of the same name, may produce very different results. In the first experiments you mixed colors; in the last experiment you did not mix colors, and we must seek an explanation of the result obtained. If a glass vessel with parallel sides, containing a blue solution of sulphate of copper, be interposed in the path of the light-waves which form a solar spectrum, it will be found that the red, orange, and yellow waves are cut out of the spectrum, *i.e.* the liquid absorbs these waves. And if a yellow solution of bichromate of potash or picric acid be interposed, the blue and violet waves will be absorbed. It is evident that, if both solutions be interposed, all the colors will be destroyed except the green, which alone will be transmitted; thus:

      Canceled by the blue solution,   R̶ O̶ Y̶ G B V.
      Canceled by the yellow solution,   R O Y G B̶ V̶.
      Canceled by both solutions,   R̶ O̶ Y̶ G B̶ V̶.

In a similar manner, when white light strikes a mixture of yellow and blue pigments on the palette, it penetrates to some depth into the mixture; and, during its passage in and out, all the colors except the green are destroyed; so the mixed pigments necessarily appear green. But when a mixture of yellow and blue waves enters the eye, we get, as the resultant of the *combined* sensations produced by the two colors, the sensation of white; hence, a mixture of yellow and blue gives white.

The color square 3 (Plate I) represents the result of the mixture of pigments 1 and 2; while 4 represents the result of the optical mixture of the same colors.

265. Theory of Color-vision.

The generally accepted theory of color-vision is that of Dr. Young (1801-2), verified by Maxwell and Helmholtz. It supposes the existence of three color sensations, red, green, and violet. These excited simultaneously, and with proper intensities, produce the sensation of white light. Combined in twos, they produce the intermediate color sensations. Thus, red and green sensations combined give yellow or orange; green and violet give blue, etc. The longer light-waves excite the sensation of red; together with those somewhat shorter, they excite both red and green, thus giving yellow, and so on. Strictly speaking, light-waves of any length excite all three sensations; but usually either one or two of them greatly predominate.

266. Color Blindness. In this defect in vision, one of the three color sensations is either wanting or deficient, usually that of red; so that the colors perceived are reduced to those furnished by the remaining two sensations, *viz.* green and violet. This causes the red-blind person to confound reds, greens, and grays. In some rare cases the sensation of green or violet is the one deficient.

267. Complementary Colors.

Experiment 9. On a piece of gray paper lay a circular piece of blue paper 15 mm. in diameter. Attach one end of a piece of thread to the colored paper, and hold the other end in the hand. Place the eyes within about 15 cm. of the colored paper, and look steadily at the center of the paper for about fifteen seconds; then, without moving the eyes, suddenly pull the colored paper away, and instantly there will appear on the gray paper an image of the colored paper, but the image will appear to be yellow. This is usually called an *after-image*. If yellow paper be used, the color of the after-image will be blue; and if any other color given in the diagram (Plate I), the color of its after-image will be the color that stands opposite to it.

This phenomenon is explained as follows: When we look steadily at blue for a time, the eyes become fatigued by this color, and less susceptible to its influence, while they are fully susceptible to the influence of other colors; so that when they are suddenly brought to look at white, which may be regarded as a compound of yellow and blue, they receive a vivid impression from the former, and a feeble impression from the latter; hence, the predominant sensation is yellow. Any two colors which together produce white are said to be *complementary* to each other. The complement of green is purple — a compound color not existing in the spectrum. The opposite colors in the diagram (Plate I) are complementary to one another.

268. Effect of Contrast. When different colors are seen at the same time, their appearance differs more or less from that observed when they are seen separately. Thus, a red object (*e.g.* a red rose) appears more brilliant if a green object be seen in juxtaposition with it. Such effects are said to be due to *contrast*.

When any two colors given in the circle (Plate I) are brought into contrast, as when they are placed next each other, the effect is to move them farther apart in the color scale. For example, if red and orange be brought in contrast, the orange assumes more of a yellowish hue, and the red more of a purplish hue. Colors that are already as far apart as possible, *e.g.* yellow and blue, do not change their hue, but merely cause each other to appear more brilliant.

269. Colors Produced by Interference. We recall that two sets of sound-waves may so combine as to neutralize each other and produce silence. For example, the phenomenon of "beats," or the alternate increase or diminution of intensity of sound, is due to the interference of two sets of sound-waves in the same and opposite phases respectively. If radiation be wave-motion, similar phenomena ought to occur under proper conditions.

Experiment 10. Press firmly together with an iron clamp two polished pieces of thick plate glass. Bands of colors will be seen arranged around the point of pressure in curves more or less regular.

Newton's method of studying these colors was very simple and effective, and the phenomena exhibited are known as "Newton's rings." By this method a convex lens of very small curvature is gently pressed upon a piece of plate glass (Fig. 210), and beautiful circular interference color bands

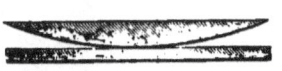

FIG. 210.   FIG. 211.

encircle the point of contact (Fig. 211). It will be seen that the film of air between the lens and the plate increases in thickness from the point of contact radially. Now if light-waves be incident on the lens, a portion will be reflected from its curved surface and another portion from the surface of the plate-glass on which the lens rests. Since the latter portion has farther to travel than the former by twice the thickness of the air-film between the two surfaces, and since the film gradually increases in thickness from the point of contact outward, it is apparent that the two sets of reflected waves will meet at certain places in like phase and at other places in opposite phase, causing intensification of illumination in the former instance and an extinction of light in the latter. If the incident light be red, a series of concentric red rings will alternate with dark rings, each shading off into the other. If violet light be used, the color rings will be closer together, since the wave-lengths are shorter. If white light be used, at every point some one color will be destroyed, leaving its complementary at that point. Thus, the point of contact between the lens and plate is surrounded by rainbow-like color bands.

Examples of color produced by interference : Colors of the soap-bubble, of a film of oil on water, of oxides formed on certain metals when heated, and of striated surfaces like mother-of-pearl.

SECTION X.

SOME OPTICAL INSTRUMENTS.

270. Compound Microscope. When it is desired to magnify an object more than can be done conveniently and with distinctness by a single lens, two convex lenses are used, — one, O (Fig. 212), called the *objective*, to form a magnified real image, a' b', of the object a b; and the other, E, called the *eye-piece*, to magnify this image so that the image a' b' appears of the size a" b". Instead of looking at the object as when we use a simple lens, we look at the real inverted image (a' b') of the object.

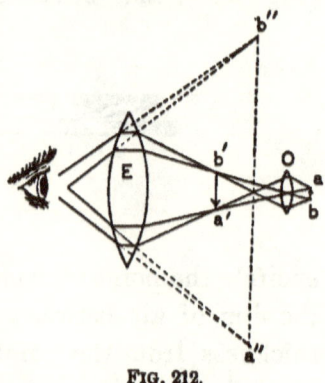

FIG. 212.

This represents the simplest possible form of the *compound microscope*. In practice, however, the construction is more complicated.

Fig. 213 represents a perspective and a sectional view of a simple form of a modern compound microscope. The body of the instrument consists of a series of brass tubes movable one within another. In the upper end H is the *ocular* or *eye-piece*. It consists of two plano-convex lenses, c and n, the former called the *eye-lens*, the latter called the *field-lens*.[1]

Microscopes should have an achromatic objective. This consists of two to four achromatic lenses (the achromatic triplet, the most common form, is represented on an enlarged scale at L in Fig. 213), combined so as to act as a single lens of short focus. By the use of several lenses the aberrations can be better corrected than with a single lens.

---

[1] The advantages derived from the use of two lenses in the eye-piece are as follows:

1. *The combination diminishes spherical aberration and thereby increases the flatness of the field.* The images a' b' and a" b" (Fig. 213) are in reality curved, in consequence of the spherical aberration caused by the objective. The effect of the field-lens is to correct this curvature in a measure.

2. *The combination increases the field of view*, so that a larger area of the object is made visible at the same view.

3. *The combination diminishes chromatic aberration.*

258   ETHER DYNAMICS.

The object to be examined is placed on a stage, S, and, if the object be transparent, it is strongly illuminated by focusing light upon it by means of a concave mirror, M. If the object be opaque, it is illuminated by light directed upon it obliquely from above by the converging lens N.

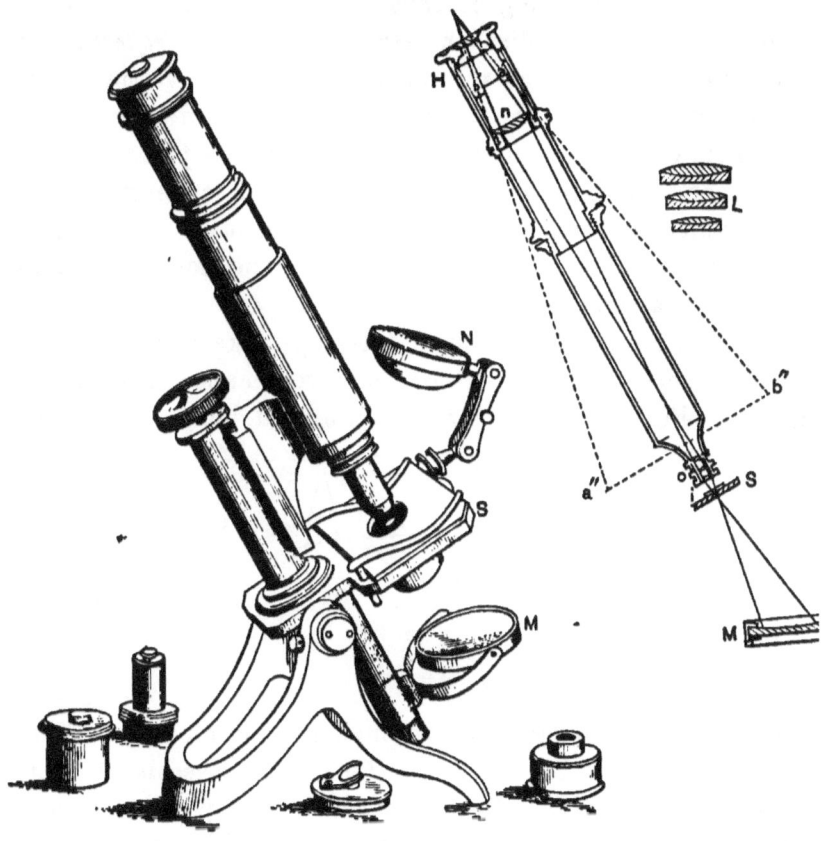

FIG. 213.

271. Magnifying Power. The magnifying power of a compound microscope is the product of the respective magnifying powers of the object-glass and the eye-piece; that is, if the first magnify 20 times and the other 10 times, the total magnifying power is 200. The magnifying power is determined experimentally by means of a micrometer scale, for a description of which the student is referred to technical works on microscopy.

272. Telescopes. Telescopes are used to view (scope) objects afar off (tele). They are classified as *astronomical* or *terrestrial*, according as they are designed to be used in viewing heavenly bodies or terrestrial objects; *reflecting* or *refracting*, according as the objective is a concave mirror or a converging lens. The terrestrial telescope differs from the astronomical in producing images in their true position without inversion. This is effected by means of an extra object lens, which corrects the inversion of the main object lens. The matter of inversion is of little or no consequence in viewing heavenly bodies.

The refracting astronomical telescope, like the compound microscope, consists essentially of two lenses. The object-glass O (Fig. 214) forms a real diminished image, a b, of the

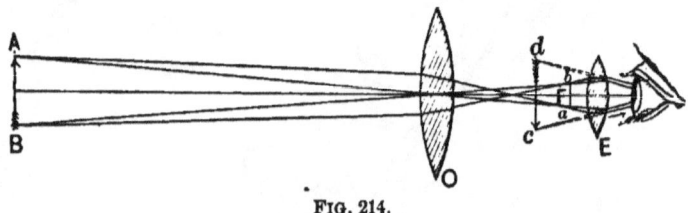

FIG. 214.

object A B; this image, seen through the eye-glass E, appears magnified and of the size c d. The object-glass is of large diameter, in order to collect as much light as possible from a distant object for a better illumination of the image.

> This telescope is analogous to the microscope, but the two instruments differ in this respect: in the microscope, the object being very near the object-glass, the image is formed much beyond the principal focus, and is greatly magnified, so that both the object-glass and the eye-piece magnify; while in the telescope, the heavenly body being at a great distance, the incident rays are practically parallel, and the image formed

by the object-glass is much smaller than the object. The only magnification which can occur is produced by the eye-piece, which ought therefore to be of high power. The magnifying power of this instrument equals approximately *the focal length of the object-glass divided by the focal length of the eye-piece.*

273. The Human Eye. Fig. 215 represents a horizontal section of this wonderful organ. Covering the front of the eye, like a watch-crystal, is a transparent coat, 1, called the *cornea*. A tough membrane, 2, of which the cornea is a continuation, forms the outer wall of the eye, and is called the *sclerotic coat*, or "white of the eye." This coat is lined on the interior with a delicate membrane, 3, called the *choroid coat;* the latter consists of a black pigment, which prevents internal reflection. The inmost coat 4, called the *retina*, is formed by expansion of the optic nerve O. The muscular tissue ii is called the *iris;* its color determines the so-called "color of the eye." In the center of the iris is a circular opening, 5, called the *pupil*, whose function is to regulate, by involuntary enlargement and contraction, the quantity of light-waves admitted to the posterior chamber of the eye. Just back of the iris is a tough, elastic, and transparent body, 6, called the *crystalline lens*. This lens divides the eye into two chambers; the anterior chamber 7 is filled with a limpid liquid called the *aqueous humor;* the posterior chamber 8 is filled with a jelly-like substance called the *vitreous humor*. The lens and the two humors constitute the refracting apparatus.

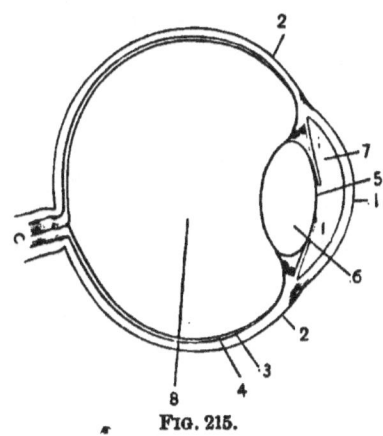
Fig. 215.

The eye may be likened to a photographer's camera, in which the retina takes the place of the sensitized plate. Images of outside objects are projected by means of the

DEFECTS OF VISION. 261

crystalline lens, assisted by the refraction of the humors, upon this screen, and the impressions thereby made on this delicate membrane of nerve filaments are conveyed by the optic nerve to the brain. Fig. 216 illustrates the manner

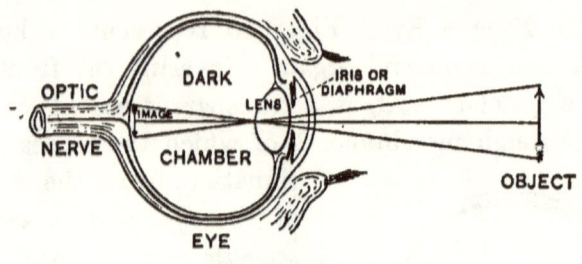

FIG. 216.

in which the image of an object is formed on the retina, except that no attempt is made to represent the several refractions. It will be seen that the image is inverted, but by some *mental* act it is made to appear upright.

With the ordinary camera the distance of the lens from the screen must be regulated to adapt itself to the varying distances of outside objects, in order that the images may be properly focused on the screen. In the eye this is accomplished by changing the convexity of the lens. We can almost instantly and unconsciously change the lens of the eye so as to form on the retina a distinct image of an object miles away, or of one only a few inches distant. The nearest limit at which an object can be placed so as to form a distinct image on the retina is about five inches. On the other hand, the normal eye in the passive state is adjusted for objects at an infinite distance.

274. Defects of Vision. *Myopia* (short-sightedness) is caused by the excessive length of the globe from front to back, so that the images of all but near objects are formed in front of the retina. Remedy: use diverging lenses. *Hypermetropia* (long-sightedness) occurs when the axis of the globe is so short that the image of an object is back of the retina unless the object is held at an inconvenient distance, in which case it

262   ETHER DYNAMICS.

tends to become indistinct. Remedy: use converging lenses. *Presbyopia* is due to loss of accommodation power, so that while vision for distant objects remains clear, that for near objects is indistinct. This defect is incident to advancing years, and is due to progressive loss of elasticity of the crystalline lens. Remedy: converging lenses. *Astigmatism* is caused by an inequality in the curvature of the cornea in different meridians, so that when, for example, a diagram like Fig. 217 is held at a distance, vertical lines will be in focus and horizontal lines will be out of focus and will appear blurred and indistinct, or *vice versa*. Remedy: lenses of cylindrical curvature. But, for this, as well as for all other defects or troubles of the eyes, consult a skilled oculist, and the earlier the better.

FIG. 217.

Advice *to all:* Do not overstrain or overtax the eyes, or use them in insufficient or excessive light, in flickering light such as that of a gas-jet, or in unsteady light such as that in a moving vehicle; and avoid, so far as practicable, sudden changes of light, such as lightning flashes, etc.

275. Stereopticon. This instrument is extensively employed in the lecture room for producing on a screen magnified images of small, transparent pictures on glass, called *slides;*

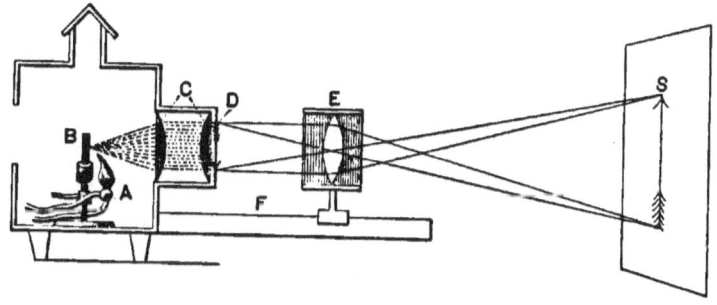

FIG. 218.

also for rendering a certain class of experiments visible to a large audience by projecting them on a screen. The *lime light* is most commonly used, though the electric light is preferred for a certain class of projections. The flame of an oxyhydrogen blowpipe, A (Fig. 218), is directed against a stick of lime, B, and raises it to a white heat. The radiations from the lime are condensed by means of a convex lens, C, called the *condensing lens* (usually two plano-convex lenses are used), so that a larger quantity of radiations may pass through the convex lens E, called the *projecting lens*. The latter lens produces (or projects) a real, inverted, and magnified image of the picture on the screen S. The mounted lens E may slide back and forth on the bar F so as properly to focus the image.

SECTION XI.

THERMAL EFFECTS OF RADIATION.

276. Heat not Transmitted by Radiation. We have learned that heat may travel *through* matter (by conduction) and *with* matter (by convection), and it is sometimes stated that there is a third method by which it travels, *viz.* "radiation." Heat itself is not transferred by radiation at all; heat generates radiation (*i.e.* ether-waves) at one place, and radiation produces heat at another; it is *radiation* which travels, not heat. It does not exist as heat in the intervening space, and therefore does not necessarily heat the substance filling that space. Heat can flow only one way, *viz.* from a given point to a point that is colder; radiation travels in all directions. The sun sends us no heat, but it sends radiations which the earth transforms into heat; but it should be borne in mind that while they are radiations they are not heat, and *vice versa*.

277. Diathermancy and Athermancy. The character of any given body determines largely what becomes of the radiations which strike it. If the nature of the body be such that its molecules can accept the motion of the ether, the ether vibrations are said to be *absorbed* by the body, and the body is thereby heated; *i.e.* the undulations of ether are transformed into molecular energy, or *heat*. Glass, for instance, allows the sun's radiations to pass very freely through it, and very little is transformed into heat. But if the glass be covered with the soot of a candle flame, the soot will absorb the radiations, and the glass become heated. Observe how cold window-glass may remain, while radiations pour through it and heat objects in the room. *Only those radiations that a body absorbs heat it; those that pass through it do not affect its temperature.* Bodies that transmit radiations freely are said to be *diathermanous*, while those that absorb them largely are called *athermanous*.

These terms bear the same relation to the transmission of radiant energy of any and all wave-lengths as do transparency and opacity to the transmission of light or visible radiations. The most diathermanous substance known is rock salt. A solution of iodine in carbon bisulphide absorbs almost completely the rays of the visible spectrum, but transmits almost completely the invisible infra-red rays. A plate of alum acts in the reverse manner, transmitting the visible and absorbing the invisible. Among liquids carbon bisulphide is exceptionally transparent to all forms of radiation; while water, transparent to short waves, absorbs the longer waves, and is thus quite athermanous.

Experiment 1. Bring the bulb of an air thermometer into the focus of a burning-glass exposed to the sun's rays. The radiation concentrated on the enclosed air scarcely affects this delicate instrument.

Experiment 2. Cover the outside of the bulb of the air thermometer with lampblack and repeat the last experiment. The lampblack absorbs the radiant energy, and the heat conducted through the glass to the enclosed air raises its temperature and causes it to expand and rapidly push the liquid out of the stem.

Dry air is almost perfectly diathermanous. All of the sun's radiations that reach the earth pass through the atmosphere, which contains a vast amount of aqueous vapor. This vapor, like water, is comparatively opaque to long waves; hence it modifies very much the character of the radiations which reach the earth.

This fact, together with what we have learned from Experiment 1, enables us to understand the method by which our atmosphere becomes heated. First, that portion of the radiant energy from the sun which comes to us in the form of relatively long waves is stopped by the watery vapor in the air, which is thereby heated. The portion that comes to us in short waves, escaping this absorption, heats the earth in falling upon it. The warmed earth loses its heat — partly by conduction to the air, still more largely by radiation outward. The form of radiation, however, has been greatly changed; for now, coming from a body at a low temperature, it is chiefly in long waves that the energy is transmitted; while, as we have seen, it was largely in the form of short waves that the earth received its heat. But it is exactly these long waves which are most readily absorbed by the atmosphere; hence the atmosphere, or rather the aqueous vapor of the atmosphere, acts as a sort of trap for the energy which comes to us from the sun.

Remove the watery vapor (which serves as a "blanket" to the earth) from our atmosphere, and the chill resulting from the rapid escape of heat by radiation would probably put an end to all animal and vegetable life. Glass does not screen us from the sun's heat, but it can very effectually screen us from the heat radiated from a stove or any other terrestrial object. Glass is diathermanous to the sun's radiations (simply because they have already lost most of the very long waves by atmospheric absorption), but quite athermanous to other radiations. This is well illustrated in the case of hotbeds and greenhouses. The sun's rays pass through the glass of these enclosures almost unobstructed, and heat the earth; but the radiations given out in turn by the earth are such as cannot pass out through the glass, and hence the heat is retained within the enclosures.

278. All Bodies Emit Radiations. Hot bodies *usually* part with their heat much more rapidly by radiation than by all other processes combined. But cold bodies, like ice, emit radiations, even when surrounded by warm bodies. This must be so from the nature of the case, for all bodies that we are acquainted with are at some temperature; their molecules are therefore in a state of motion, and being surrounded by ether they cannot move without imparting some of their motion to the ether. But in order that a body become colder by radiation it must lose more heat by radiation than it receives,

279. Prévost's Theory of Exchanges. Let us suppose that we have two bodies, A and B, at different temperatures — A warmer than B. Radiation takes place, not only from A to B, but from B to A; but, in consequence of A's excess of temperature, more radiation passes from A to B than from B to A, and this continues until both bodies acquire the same temperature. At this point radiation by no means ceases, but each now gives as much as it receives, and thus equilibrium is kept up. This is known as "Prévost's Theory of Exchanges."

280. Good Absorbers; Good Radiators.
Experiment 3. Select two small tin boxes of equal capacity,— one should be bright outside, while the other should be covered thinly with soot from a candle flame. Cut a hole in the cover of each box large enough to admit the bulb of a thermometer. Fill both boxes with hot water, and introduce into each a thermometer. They will register the same temperature at first. Set both in a cool place, and in half an hour you will find that the thermometer in the blackened box registers several degrees lower than the other. Then fill both with cold water, and set them in front of a fire or in the sunshine, and it will be found that the temperature in the blackened box rises the more rapidly.

As bodies differ widely in their absorbing power, so they do in their radiating power, and it is found to be universally true that *good absorbers are good radiators, and bad absorbers are bad radiators.* In both cases much depends upon the character of the surface as well as of the substance. Bright, polished surfaces are poor absorbers and poor radiators; while tarnished, dark, and roughened surfaces absorb and radiate rapidly. Dark clothing absorbs and radiates more rapidly than light clothing.

EXERCISES.

1. What objections can you raise to the term "radiant heat"?
2. Explain why the temperature of a hotbed is above that of the surrounding air.
3. How could you separate the dark radiation of an electric arc lamp from the luminous radiation?

4. How can you demonstrate the existence of ether-waves of greater length than the light-giving waves?

5. Ice appears to radiate cold. Explain the phenomenon by Prévost's theory.

6. What parts of the spectrum are invisible to the eye?

7. How can you prove the existence of invisible solar rays?

GENERAL EXERCISES.

1. What is light?

2. State points of resemblance and points of difference between light-waves and sound-waves. Which can traverse a vacuum (as regards matter)?

3. Two books are held, respectively, 2 feet and 7 feet from the same gas flame. Compare the intensities of the illumination of their respective pages.

4. (a) What is the general effect of a concave mirror on light-waves? (b) What kind of lens produces a similar effect?

5. How can a beam of light be bent?

6. When red and green sensations coexist, what is the resulting sensation?

7. Why do white surfaces appear gray at twilight?

8. How are objects heated by the sun?

9. What evidences can you give that the earth receives energy from the sun?

10. What phenomenon shows that ether-waves do not traverse all substances with equal speed?

11. Why does not winking interfere with vision?

12. What effect has the refractive action of the atmosphere upon the apparent position of the sun and the duration of daylight?

13. A small bright image of the sun is projected on a card held 16 inches from a convex lens. How far must the card be held from the lens to receive a distinct image of a candle flame which is at a distance of 18 inches from the lens?

14. Account for the dazzling whiteness of snow.

15. What is the focal distance of a convex lens when the distances of the image and object are, respectively, 5 and 36 cm. from the lens?

16. A candle flame illuminates a screen 15 inches distant. How many candle flames at a distance of 5 feet would be required to produce an equal illumination?

17. (a) What do we mean by a white body? (b) What by a black body? (c) What is meant by white light?

CHAPTER VII.

ELECTROSTATICS.

SECTION I.

INTRODUCTION.

281. Electrification. Certain bodies, provided the conditions are suitable, acquire by contact and subsequent separation (or more readily by friction) the property of attracting light bodies, such as pieces of tissue paper, etc. For example, glass rubbed with silk, and sealing wax or ebonite rubbed with woolen cloth, manifest this property by picking up scraps of paper, etc. Bodies in this state are said to *be electrified*, or *charged with electricity*. An electric charge is supposed to be due to a strained condition of the ether surrounding the charged body.[1]

Experiment 1. Rub a rubber comb with a woolen cloth, or draw it a few times through your hair (if dry). Hold the comb over a handful of bits of tissue paper; the papers quickly jump to the comb, stick to it for an instant, and then leap energetically from it. The papers are first attracted to the comb, but in a short time acquire some of its electrification, and then are repelled.

282. Two Kinds of Electrification.

Experiment 2. Suspend a ball of elder pith, C (Fig. 219), by a silk thread. Electrify a glass rod, D, with a silk handkerchief, and present it to the ball; attraction at first occurs, followed by repulsion soon after contact. Next excite a stick of sealing wax or a rubber comb with a woolen cloth and present it to the ball, which is repelled by the electrified glass; the ball is attracted by the electrified wax or rubber.[2]

---

[1] The electric charge has its origin in the contact of dissimilar molecules.

[2] "When two substances have different molecular velocities at their common surface of mutual contact, the molecules hamper one another, and energy is lost; this energy takes the form of energy of electrical displacement." — DANIELL.

It is evident (1) that *there are two kinds of electrification;* (2) that *bodies similarly electrified repel one another,* and that *bodies oppositely electrified attract one another.*

Glass rubbed with silk is said to receive a *vitreous charge.* On the other hand, the wax, on being rubbed with woolen cloth, receives a *resinous charge.* The vitreous charges are said to be *positive* (+ E), and the resinous charges *negative* (— E).

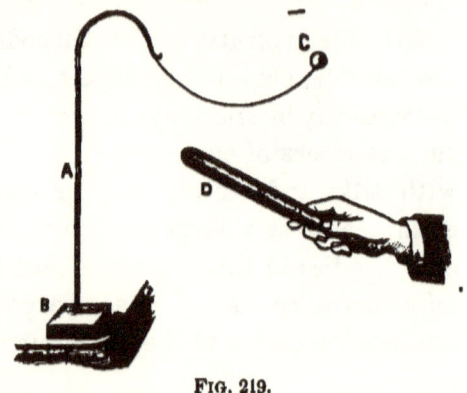

Fig. 219.

Experiment 3. Once more electrify a stick of sealing wax with a woolen cloth, and present it to the pith ball, and after the ball is repelled bring the surface of the flannel which had electrified the rod near the ball; the ball is attracted by it, showing that the rubber is also electrified, and with the opposite kind to that which the sealing wax possesses.

One kind of electrification is never developed alone. When two substances are rubbed together, and one becomes electrified, electrification of the opposite kind is always developed on the other.

283. Electrification a Form of Potential Energy. When small pieces of glass and silk are rubbed together, it is found that after they are pulled apart they attract each other with a definite and measurable force, and that this force varies inversely as the square of the distance between them.

The strained ether between them is thought to operate like strained India-rubber bands, pulling the two bodies together. Work is required in order to separate the excited bodies, and the bodies thus separated possess *potential energy of electrical separation.*

284. What is Electricity? The student naturally asks this never-answered question.[1] Provisionally we shall regard electricity as that which is transferred from one body to another body when the two become oppositely electrified. Electricity is *not* a form of energy.[2] It is quite true that electricity *under pressure* or *in motion* possesses energy; in the same sense water and air under like conditions possess energy, but no one presumes to call them forms of energy.

Our methods of "producing electricity" are, so far as we know, merely methods of disturbing electrical equilibrium.

FIG. 220.

285. The Electroscope. This is an instrument used to detect the presence of electrification in a body, and to determine its kind. It usually consists of two strips of gold foil, A B (Fig. 220), suspended from a brass rod within a glass jar. To the upper end of the rod is fixed a metal disk, C. On the opposite sides of the interior of the jar are two strips of metal foil, D and E, of sufficient hight to be touched by the strips A and B on their extreme divergence.

(1) If an unelectrified body be brought near the disk C, no change takes place in the two strips of foil A and B; but if an electrified body be brought near the disk, the strips diverge,

---

[1] "This is at once an important and a difficult question. Many who ask it never doubt the existence of electricity. To the scientific mind the question presents itself rather in the form — Is there such a thing as electricity? Cannot electrical phenomena be traced back, like all others, to the properties of the ether and of ponderable matter?" — HERTZ.

[2] There is no mechanical equivalent of a quantity of electricity, as there is of a quantity of heat; but there is a mechanical equivalent of *electrical energy*.

thus indicating the existence of a charge of electricity in the body.

(2) If the electroscope be charged by contact with an excited body, the strips will remain in a divergent position. While they are in this condition, if a body similarly charged be brought near the disk, the strips will diverge *more;* but if an unexcited body or a body oppositely electrified be brought near the disk, the strips will collapse.

286. Conduction.

Experiment 4. (*a*) Rub a brass tube, held in the hand, with warm silk. Bring it near the disk of the electroscope; the leaves are unaffected. (*b*) Wrap a piece of sheet rubber around one end of the tube and hold this end in the hand, and rub as before. Bring it near the disk of the electroscope; notice that the leaves diverge. (*c*) Repeat the last operation, but before bringing the tube near the disk touch the tube with a finger; the leaves no longer show signs of electrification.

In the first (*a*) and last (*c*) operations electricity escaped through the hand and body to the earth; in the second (*b*) it was prevented from escaping by the intervening sheet rubber. Substances which allow electricity to spread over them, *i.e.* substances which offer little resistance to the flow of electricity, are called *conductors*. Those which offer great resistance to its passage are called *non-conductors, insulators,* or *dielectrics.*

Some of the best insulating substances are *dry air, ebonite, shellac, resins, paraffine, glass, silks,* and *furs.* On the other hand, metals are exceedingly good conductors. Moisture injures the insulation of bodies; hence, experiments succeed best on dry, cold days of winter, when the moisture of the air is least liable to be condensed on the surfaces of apparatus, *especially if the latter be kept warm.*

Water cannot be retained in a reservoir unless its walls be of sufficient strength; so a body, in order to retain a charge of electricity, must be surrounded by something that will offer sufficient resistance to the escape of electricity.

272   ETHER DYNAMICS.

This entity which corresponds to the walls of the reservoir is termed the *dielectric*. It may be the air or any of the so-called non-conductors of electricity. A body thus surrounded is said to be *insulated*. There is no limit to the quantity of electricity with which a body can be charged, provided the charge be not conducted away.

SECTION II.

INDUCTION.

287. Electricity Acts across a Dielectric.

Experiment 1. Figure 221 represents an empty eggshell covered with tin foil to make it a good conductor. It is suspended from a glass rod by a silk thread. (a) Electrify a glass rod and bring it near the shell. The shell moves toward the rod. (b) Next introduce a glass plate between the rod and shell. The shell approaches the rod as before.

FIG. 221.

The chief lesson we learn from this experiment is that *electricity acts across a dielectric*. In (a) the dielectric was air; in (b), air and glass.

288. To Determine what actually Happens on an Insulated Conductor when an Electrified Body is Brought Near.

Experiment 2. (a) Suspend, as above, two shells so as to touch each other, end to end, as in Fig. 222, thus making practically one conductor. Bring near to one end of the shells a sealing-wax rod, D, excited with — E. While the rod is in this position carry a thin strip of tissue paper, C, along the shells. The paper is attracted to the shells, but most strongly at the ends. In the middle of the conductor, where the shells touch each other, there is little if any electrification.

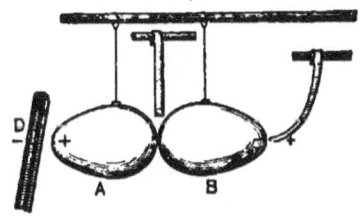

FIG. 222.

(b) While the rod D is still in position, separate B from A, then remove D. Test each shell with the tissue paper; both are found to be excited.

CHARGING BY INDUCTION. 273

(c) Charge an electroscope with + E. Then bring A near it; the leaves diverge, showing that A is charged with + E. Bring B near the electroscope; the leaves collapse, showing that B is charged with − E.

(d) Finally bring the two shells near each other; they attract each other. Allow them to touch each other, and then test each with the tissue paper or the electroscope; it will be found that both have become discharged.

From the above operations we learn that when an electrified body is brought near but not in contact with an insulated conductor, the electrified body acts across the dielectric upon the conductor, repelling electricity of the same kind to the remote side of the conductor, and attracting the opposite kind to the side near to it. Such electrical action is called *induction*. The electrified body which produces the action is called the inducing body; the charge of electricity thus produced is called *induced electricity*.

289. Charging by Induction.

Experiment 3. Take a proof plane, E (Fig. 223) (which consists of an insulating handle of glass or gutta-percha, terminating at one end with a thin metal disk, F, about the size of a 5-cent nickel), and connect it with an electroscope, G, by a fine wire, H. Bring a stick of sealing wax electrified as before with − E near the eggshell conductor. Holding the proof plane by the insulating handle, bring the disk near the end of the conductor charged by in-

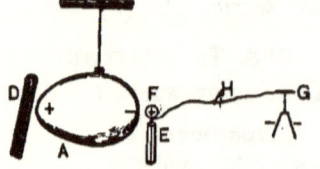

FIG. 223.

duction with − E. The − E will act inductively upon the continuous conductor consisting of disk, wire, and electroscope, charging the end nearest itself (*i.e.* the disk) with +E and the remote end (*i.e.* the leaves) with .− E. The leaves of the electroscope show the presence of a charge by their divergence.

Now, while everything is in the position indicated by the cut, touch with the finger any part of the continuous conductor; the leaves of the electroscope instantly collapse. The − E with which the leaves had been charged, being *free*, is discharged through your body. But the + E concentrated on the disk of the proof plane is *bound* by the attraction of the charge of $-$ E on the end of the shell nearest it, and cannot escape. Remove the finger from the electroscope and the proof plane from the influence of the shell; the leaves again diverge.

The last phenomenon is explained as follows: After $-$ E has been discharged from the continuous conductor, there is left an excess of $+$ E; but this excess is all concentrated in the disk F so long as it remains near the negative charge of the shell. But as soon as F is removed from the influence of the shell, the charge spreads over the entire conductor, and the leaves, which receive a portion of the charge, diverge. The conductor is said to be charged by *induction*.

Experiment 4. To electrify the shell by induction, bring the excited wax near it, touch the shell with a finger, remove the finger, and finally remove the rod. The proof plane being connected with the electroscope and being charged with $-$ E, bring F near to the shell A; the leaves collapse, showing that the shell is charged with $+$ E, which draws the $-$ E away from the leaves.

Observe that when a body becomes charged by induction the charge which it receives is *opposite in kind to that of the inducing body*.

290. Charging by Conduction.

Experiment 5. Disconnect the proof plane from the electroscope. Charge the electroscope with $-$ E and the shell with $+$ E; touch the shell with the disk of the proof plane, then hold the disk near the electroscope; the divergent leaves collapse, showing that the disk bears $+$ E which it received by *conduction* from the shell when they were brought in contact. Of course the charge is the same kind as that of the body which communicated it.

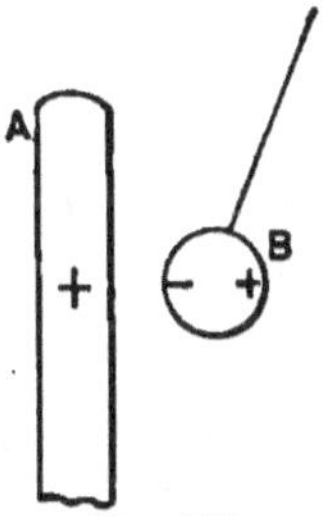

FIG. 224.

291. Induction Precedes Attraction. When a pith ball is brought near an electrified glass rod, the $+$ E on the rod A (Fig. 224) induces $-$ E on the side of the ball B, nearest A, and repels $+$ E to the farther side.

The + E of A and the − E of B therefore attract each other; likewise the + E of A and the + E of B repel each other; but since the former charges are nearer each other than the latter are, the attraction exceeds the repulsion.

SECTION III.

ELECTRICAL POTENTIAL.

292. Electrostatics and Electro-kinetics. Electricity may be at rest, as in a charged body, or it may be in motion, as in the case of a charged body connected by a conductor with the earth, when it is discharged through the conductor to the earth. It will be shown later on that as long as a flow of electricity continues, the conductor along which it flows has properties different from those of a simple electrified body. That branch of electrical science which treats of the properties of simple electrified bodies is called *Electrostatics*, because in these bodies electricity is supposed to be *at rest;* and that branch which treats of electricity in motion is called *Electro-kinetics*.

293. Potential. The fundamental fact of electricity is that *we are able to place bodies in different electrical conditions. A charge of electricity* (which implies an abnormal electrical condition) *is the foundation of all electrical phenomena.* We are now to discuss the meaning and use of the very important term *potential*, as it is employed in electrical science.

a. When a charged conductor is connected with the earth, a transfer of electricity takes place between the body and the earth.

b. If the body be charged with + E, we say arbitrarily that electricity passes from the body to the earth; but if the body be charged with − E, we say that electricity passes from the earth to the body.

c. If two insulated charged conductors be connected with each other, electricity may or may not pass from one to the other. Whether electricity passes from one to the other, and in what direction it passes, if at all, depends upon the so-called *potentials* of the conductors.

d. If two bodies have the same potential, no transfer of electricity takes place between them; but if they have different potentials, there will be a transfer, and the body *from* which the electricity flows is said to be at a *higher potential* than the body *to* which it flows.

294. Definition of Potential. The potential of a conductor may, therefore, be defined provisionally as *the electrical condition of that conductor which determines the direction of the transfer of electricity.*

The term potential is relative. It is important to have a standard of reference whose potential is considered to be zero, just as it is convenient in stating the elevations or depressions of the earth's surface to give the distances above or below sea level, which is taken as the zero of hight. For experimental purposes the earth is usually assumed to be at zero potential. A body charged with $+ E$ is understood to be one that has a higher potential than that of the earth, and a body charged with $- E$ is one that has a lower potential than that of the earth.

295. Analogies. Potential is analogous, in many respects, to (1) temperature and to (2) liquid level.

(1) When we say that the temperature of air is 20° or $- 10°$ C., we mean that its temperature is 20° above or 10° below the standard temperature of reference, *viz.* that of melting ice. If two bodies at different temperatures be placed in thermal communication, heat will pass from the body at a higher temperature to the one at a lower, and will continue to do so until both are at the same temperature.

(2) If two vessels containing water at different levels be put in communication at their bottoms by a pipe, water will flow from the one at a higher level to the one at a lower until the water is at the same level in both vessels.

Temperature is not heat; level is not water; and potential is not electricity, but merely the *state* of the conductor which determines the direction of transfer of electricity. (See definition of temperature, § 126.)

SECTION IV.

ATMOSPHERICAL ELECTRICITY.

296. Lightning. Franklin, by his historic experiment with the kite, in 1752, proved the exact similarity of lightning and thunder to the light and crackling of the electric spark. Certain clouds which have formed very rapidly are highly charged, usually with $+$ E, but sometimes with $-$ E. The surface of the earth and objects thereon immediately beneath the cloud are, of course, charged inductively with the opposite kind of electricity. The opposite charges on the earth and on the cloud hold each other prisoners by their mutual attraction, the air serving as an intervening dielectric.

As condensation progresses in the cloud its potential rises (or sinks). This process continues till the difference of potential between the cloud and the earth becomes great enough to produce a discharge through the air.

It is the accumulation of induced charges on elevated objects, such as buildings, trees, etc., that offers an intensified attraction for the opposite electricity of the cloud in consequence of their greater proximity, and renders such objects especially liable to be struck by lightning.

The clouds gather electricity from the atmosphere. Our knowledge of the method by which the atmosphere becomes charged is very limited,

We see what we call a "flash of lightning." What we really see is merely particles of air heated temporarily to incandescence. Lightning strokes last for a very brief time, — perhaps a millionth of a second, — though the sensation produced on the retina of the eye lasts longer.

Lightning rods may be very effective in protecting buildings from lightning strokes, since electricity is more likely to pass along the rods of metal, which are good conductors, than through the building. But the rods should either run several feet into water or be connected with a large metal plate buried deep in a stratum of earth that is always moist. Lightning rods may be worse than useless unless there is a good electrical connection between them and the earth.

EXERCISES.

1. Our knowledge of the existence of electricity, and the possibility of employing electric energy in the performance of nearly every species of work, are due to what fundamental fact?

2. (a) Is the energy of a charge of electricity potential or kinetic? (b) In what is a charge of electricity supposed to consist? (c) How does the energy of a charge differ from the energy of a discharge?

3. State in full the difference in the operations of charging by conduction and charging by induction.

4. When a discharge takes place, what becomes of the electric energy?

5. State how it may be shown that there are two kinds of electrification.

6. To what is electrical potential analogous?

7. How can you ascertain the kind of electrification a body has?

8. Do electrostatics and electro-dynamics treat of different *kinds* or different *states* of electricity?

9. On what condition will electricity pass from one body to another body; or from a point in a given body to another point in the same body?

10. (a) When glass is electrified by rubbing it with silk, does its potential become higher or lower than that of the earth? (b) Has sealing wax, after being rubbed with woolen cloth, a higher or a lower potential than that of the earth?

BENJAMIN FRANKLIN.

CHAPTER VIII.

ENERGY OF ELECTRIC FLOW. ELECTRO-KINETICS.

SECTION I.

VOLTAIC. CELLS.   ELECTRIC CIRCUITS.

297. Introductory Experiments.

APPARATUS REQUIRED. A tumbler ⅔ full of water, into which have been poured two or three tablespoonfuls of strong sulphuric acid; a strip of sheet-copper, and two pieces of rolled zinc, each about 5 inches long, 1½ inches wide, and at least $\frac{3}{16}$ of an inch thick (a piece of No. 16 copper wire 12 inches long should be soldered to one end of each piece of metal, and the soldering covered with asphaltum paint); 2 yards of silk-insulated No. 18 copper wire;

FIG. 225.

two double connectors (Fig. 225), which serve to join two wires without the inconvenience of twisting them together. One of the zincs should be amalgamated as follows: First dip the zinc, with the exception of ½ inch at the soldered end, into the acidulated water; then pour mercury over the wet surface, and finally rub the surface, now wet with mercury, with a cloth. (To ensure complete amalgamation, it is best to repeat this operation.)

Experiment 1. (*a*) Put the unamalgamated zinc into the tumbler containing acidulated water. Bubbles of hydrogen gas arise from the surface of the zinc.

(*b*) Remove this zinc and introduce the amalgamated zinc. No bubbles (or at least very few) arise from the latter (provided that the zinc be properly amalgamated).

(*c*) Put the copper strip into the liquid, but do not allow the two metals or their wires to touch. No bubbles arise from either metal. Connect the wires of the two metals with a double connector; copious bubbles arise from the copper strip, but very few from the zinc strip. Bubbles escaping from the copper seem to indicate that chemical action is taking place between the metal and the liquid. But experience will teach you that the appearance is deceptive, as you will find that in no case is copper consumed.

(*d*) Substitute the unamalgamated zinc for the amalgamated; bubbles rise abundantly from the surfaces of both the zinc and copper.

Lesson Learned. Unamalgamated zinc is acted on by the liquid under all circumstances; amalgamated zinc is not acted on by the liquid unless the copper strip is also in the liquid, and not then *unless the metals be connected.* If then we would at any time stop the action, we have only to disconnect the metals. It seems also that the wire connecting the two metals serves some important purpose in keeping up this action.

If plates of two dissimilar metals be placed in a liquid of a class which we will term an *electrolyte* (*i.e.* one which is capable of being decomposed by a current of electricity), it may be shown by actual experiment[1] that the free end of the wire connected with one plate is charged with $+$ E, and the free end of the wire connected with the other plate is charged with $-$ E. Hence, in the experiment above we conclude that if the two oppositely charged bodies be brought into contact, *i.e.* if the free end of the wire leading from the copper plate (called the *positive electrode*) be brought to touch the free end of the wire leading from the zinc (called the *negative electrode*), a discharge will occur between the oppositely charged bodies. In this case, however, the discharge is a continuous one; in other words, there is a continuous flow or current of electricity as long as the contact is preserved.

That difference in quality in virtue of which zinc and copper placed in acidulated water can give rise to an electric current is called their *electro-chemical difference,* and

[1] See author's "Principles of Physics," p. 463.

the zinc is said to be *electro-positive* to the copper in the liquid.[1]

298. The Voltaic Cell. Two solids differing electro-chemically (of which zinc is almost invariably one) placed in an electrolytic liquid constitute what is called a *galvanic* or *voltaic*[2] *cell* (or *pair*). One of these plates must be more actively attacked by the liquid than the other; the plate most acted upon is called the *electro-positive element*, and the other the *electro-negative element*.

The greater the disparity between the two elements with reference to the action of the liquid on them, the greater the difference in potential; hence the greater the current.

In the following *electro-motive series* the substances are so arranged that the most electro-positive, or those most affected by dilute sulphuric acid, are at the beginning, while those most electro-negative, or those least affected by the acid, are at the end. The arrow indicates the direction of the current through the liquid.

+ Zinc. Iron. Tin. Lead. Copper. Silver. Platinum. Carbon. −

⟶

It will be seen that zinc and carbon are the two substances best adapted to give a strong current.

When the wires from the two plates are joined, the discharge of the two plates would produce electrical equilibrium were there not some means of maintaining a difference

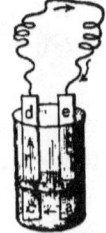

FIG. 226.

---

[1] The nomenclature in use, by which the zinc plate is called the *electro-positive element* and at the same time the *negative pole* of the combination, is at first perplexing to the student. Let him bear in mind that electricity always flows from a point of high potential to a point of relatively low potential. For example, let a current originating in a voltaic cell at point a (Fig. 226) follow the direction indicated by the arrows; then point c must be negative with reference to point a, but positive with reference to point d; again, point d, while negative to point c, is positive to point e.

[2] A single voltaic couple is usually termed a cell; a combination of cells, a battery.

of potential between the two plates. This is accomplished by the chemical action between the liquid and the electro-positive element, and at the expense of the chemical potential energy of the electrolyte. *A voltaic cell is, therefore, a contrivance which converts potential energy of chemical separation into electrical energy.*

299. Electrical Circuit. This term is applied to the entire path along which electricity flows, and it comprises the battery itself and the wire or other conductor connecting the elements.[1] The operations of bringing the two extremities of the wire into contact and separating them are called, respectively, *closing and opening,* or *making and breaking, the circuit.* Opening a circuit at any point and filling the gap with an instrument of any kind, so that the current is obliged to traverse it, is called *introducing the instrument into the circuit.*

FIG. 227.

300. Importance of Amalgamating the Zinc. Commercial zinc contains impurities, such as carbon, iron, etc. Fig. 227 represents a zinc element having on its surface a particle of carbon, A, purposely magnified. If such a plate be immersed in dilute sulphuric acid, the particles of carbon will form with the zinc numerous voltaic circuits. This occasions a great waste of materials, because, when the regular circuit is broken, *local action,* as it is called, still continues. If mercury be rubbed over the surface of the zinc, it dissolves a portion of the zinc, forming with it a semi-liquid amalgam, which covers up the impurities.

301. Polarization of the Negative Element.

Experiment 2. Construct a voltaic cell composed of dilute sulphuric acid and plates of copper and zinc. Introduce into the circuit a galvanoscope (§ 310) and note the deflection of the needle when the circuit

[1] It was an early discovery in telegraphic history that a complete metallic circuit is not necessary, but that, in common parlance, the earth can be used as a " return circuit." This type of circuit is represented by a cell with a wire leading from one element to any convenient point of the earth, and a second wire leading from the other element to any other point of the earth, even though it be many miles distant from the first point.

is first closed. Watch the needle for a time. Little by little this deflection decreases, and meantime bubbles of hydrogen gas collect on the copper plate. This accumulation of gas gives rise to what is called "polarization of the negative element or plate."

We already understand that difference of potential is indispensable to a flow of electricity. Difference of potential gives rise to something analogous to a force, which causes the flow of electricity. The greater the difference of potential, the greater is this agent which puts the electricity in motion. But a deposit of hydrogen on the copper raises, in some measure, the potential of this (negative) element and thereby diminishes the potential difference between the two elements. Hence, the current is (in technical language) "weakened."

The remedy for this is to prevent the deposit of hydrogen upon the negative element. The usual method is to employ in addition to the dilute sulphuric acid (*i.e.* the exciting liquid) some substance which will combine with the hydrogen as soon as it is liberated. A substance used for this purpose is termed a *depolarizer*. A mixture of a solution of crystals of potassium dichromate in water with a suitable quantity of sulphuric acid is used as a depolarizer in the so-called *dichromate cells*.

302. Grenet Cell. This is a potassium dichromate cell in which two carbon plates, C C (Fig. 228), electrically connected, and a zinc plate, Z, suspended between them by a brass rod, A, are immersed in the mixed liquid referred to above.

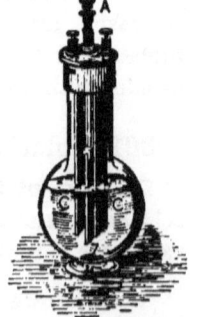

FIG. 228.

This combination furnishes a much more energetic and constant current than would be furnished if only dilute sulphuric acid were used.

303. Bunsen Cell. A plan generally adopted to keep the depolarizing liquid away from the zinc plate, where it is not wanted and only does harm, is to place the carbon plate in an unglazed, porous, earthen cup, and to surround it with the depolarizing substance. This arrangement, called a two-fluid cell, is that adopted by Bunsen (Fig. 229) and others.

304. Leclanché Cell. There is a class of galvanic cells in which the negative element is protected from polarization by means of metallic oxides. Of these the best known is the Leclanché cell (Fig. 230). In this cell the carbon plate C is contained in a porous cup, P, and packed

Fig. 229

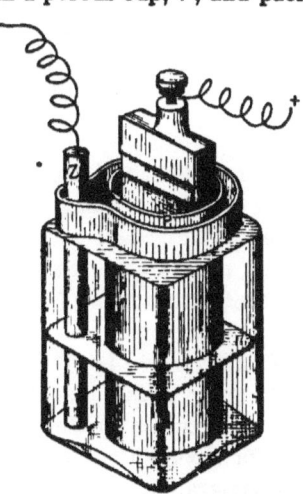
Fig. 230.

round with fragments of gas-retort coke and manganese peroxide. The manganese compound has a strong affinity for the hydrogen. Nevertheless, the elements quickly polarize when in action. They need periodical rest to recover their normal condition. Such are called *open-circuit cells*, since they are suited for work only on lines kept open or disconnected most of the time, as in telephone and bell-ringing circuits. The zinc rod Z is immersed in a solution of ammonium chloride, which is the exciting liquid.

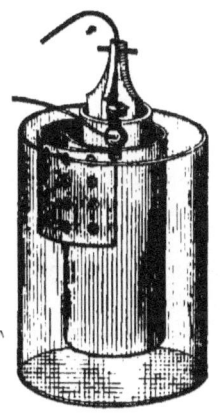
Fig. 231.

305. Daniell Cell.

Leaving the hydrogen-generating batteries, we will examine briefly another form incapable of this species of polarization. The Daniell cell (Fig. 231) uses a solution which, instead of depositing hydrogen, deposits copper upon a copper negative plate, and hence is free from hydrogen polarization. It contains a copper negative and a zinc positive plate. The copper plate is immersed in a solution of copper sulphate, the zinc in a solution of zinc sulphate or dilute sulphuric acid, and a porous cup separates the two liquids. By the electrolytic action the zinc combines with the sulphuric acid (H_2SO_4), forming zinc sulphate ($ZnSO_4$), thereby setting hydrogen free. This hydrogen, while on its way to the negative element or the copper plate, meets the copper sulphate solution ($CuSO_4$), which it decomposes, forming sulphuric acid again (H_2SO_4), and setting free the copper, which is deposited on the copper plate.

EXERCISES.

1. (a) What are electrodes? (b) What are the essential parts of a voltaic cell? (c) What metal is almost invariably used for the positive element? (d) Name several substances that are quite commonly used for the negative element. (e) What happens when the electrodes are brought in contact? (f) What purpose does joining the two elements serve?

2. Why ought not the elements of a voltaic cell to touch each other?

3. What is the function of a voltaic cell?

4. If a current passes points A, B, C, and D in a circuit successively, (a) which point is positive with reference to all the others, and which point is negative with reference to all the others? (b) State the relation of point B to each of the other points.

5. With what propriety is the zinc element of a voltaic cell called the *positive element* and the *negative electrode* of a voltaic system?

6. (a) What do you understand by the "polarization of the negative element"? (b) How is it caused? (c) What harm does it cause? (d) How is it commonly prevented?

7. Which, electricity or electrification, is the result of work done?

8. (a) What is meant by "local action"? (b) Why is it objectionable? (c) How is it prevented in some measure in certain cells?

9. What kind of cells are suitable for only "open circuit" systems?

10. Which of the several cells that have been described will yield a current most nearly uniform or constant? Why?

11. (a) What do you understand by an open-circuit battery? (b) For what purposes only are they suited? (c) Which of the several kinds of cells that have been described is especially suited for closed-circuit work?

12. Consult technical works on electricity and ascertain how the terms "galvanic" and "voltaic" became associated with electro-chemical cells.

SECTION II.

EFFECTS PRODUCIBLE BY AN ELECTRIC CURRENT.

306. Classification of Effects. The several effects producible by an electric current may be classified as *electrolytic, magnetic, thermal,* and *physiological.*

307. Electrolysis.

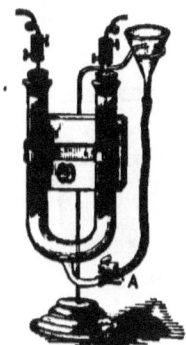

Fig. 232.

Experiment 1. Take a dilute solution of sulphuric acid (1 part by volume to 10), pour some of it into the funnel (Fig. 232), so as to fill the U-shaped tube when the stoppers are removed. Place the stoppers which support the platinum electrodes tightly in the tubes. Connect with these electrodes the battery [1] wires. Instantly bubbles of gas arise from both electrodes, accumulating in the upper part of the tube and forcing the liquid back into the funnel. Introduce a glowing splinter into the gas surrounding the + electrode: it relights and burns vigorously, showing that the gas is oxygen. Invert the glass tube, remove the rubber tube, allow the gas which had accumulated about the — electrode to escape at A, and apply a lighted match to it: the gas burns; it is hydrogen.

The volume of hydrogen is just double that of the oxygen liberated in the same time. The process by which a compound substance is separated into its constituents by an electric current is called *electrolysis,* and a compound that may be thus decomposed is called an *electrolyte.* The electrode by which the current enters the electrolyte is called the *anode,* and that by which the current leaves, the *cathode.* Those constituents that appear at the anodes are called *anions;* those that appear at the cathodes are called *cations.* Anions are electro-negative and cations are electro-positive; hence, they are attracted to electrodes that are oppositely electrified.

---

[1] A battery consisting of not less than two Grenet or Bunsen cells connected in series will be required.

MAGNETIZING EFFECT OF ELECTRIC CURRENT. 287

Thus, the anion *oxygen*, being electro-negative, is attracted to the anode, or positive electrode, and the cation *hydrogen*, being electro-positive, is attracted to the cathode, or negative electrode.

When a chemical salt is electrolyzed, the base appears at the cathode, and the acid at the anode. In general, it will be found that in both the battery and the decomposing cell, hydrogen, bases, and metals appear at the plates *toward* which the current flows.

Experiment 2. Prepare a solution of potassium iodide. Make a paste by boiling pulverized starch in water. Take a small portion of this paste and stir it into the solution. Wet a piece of writing-paper with the liquid thus prepared. Spread the wet paper smoothly on a piece of tin, *e.g.* on the bottom of a tin basin (Fig. 233). Press the negative electrode of the battery (of not less than two cells) against an uncovered part of the tin.

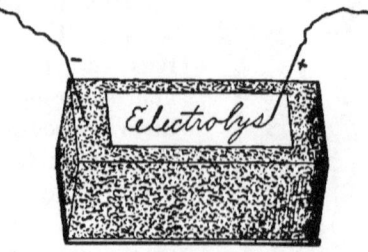

FIG. 233.

Draw the positive electrode over the paper. A mark is produced upon the paper as if the electrode were wet with a purple ink. In this case the potassium iodide is decomposed, and the iodine combining with the starch forms a purplish-blue compound.

308. Magnetizing Effect of an Electric Current. Electro-magnets.

Experiment 3. (*a*) Wind an insulated copper wire in the form of a spiral round a rod of soft iron (Fig. 234). Pass a current of electricity through the spiral, and hold an iron nail near the end of the rod. Observe, from its attraction for the nail, that the rod is magnetized. A magnet may be provisionally defined as a body which attracts iron.

(*b*) Break the circuit; the rod loses its magnetism and the nail drops.

The iron rod is called a *core*, the coil of wire a *helix*, and both together are called an *electro-magnet*. In order to take advantage of the attraction of both ends, or *poles*, of the magnet, the rod is frequently bent into a U-shape (A, Fig. 235). Often two iron rods are used, connected by a rectangular piece of iron, as a in B of Fig. 235. The method of winding is such that if the iron core of the U-magnet were straightened, or the two spools were placed together end to end, one would appear as a continuation of the other. A piece of soft iron, b, placed across the ends and attracted by them, is called an *armature*. The piece of iron a is called a *yoke*.

FIG. 234.

309. Deflection of the Magnetic Needle by a Current.

Experiment 4. (*a*) Place the apparatus (Fig. 236) so that the magnetic needle, which points (nearly) north and south, shall be parallel with the wires W_1 and W_2. Introduce the + electrode of a battery into screw-cup T_2 and the — electrode into screw-cup T_1, and pass a current through the upper wire. At the instant the circuit is closed the needle swings on its axis, and after a few oscillations comes to rest in a position which forms an angle with the wire bearing the current.

 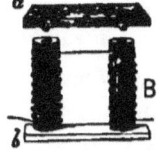

FIG. 235.

(*b*) Break the circuit by removing one of the wires from the screw-cup. The needle, under the influence of the magnetic action of the earth, returns to its original position.

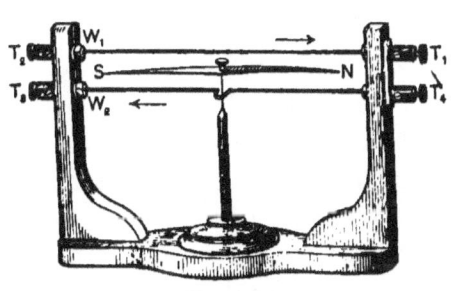

FIG. 236.

(*c*) Reverse the current by inserting the + electrode of the battery into screw-cup T_1 and the — electrode into screw-cup T_2. Again there is a deflection of the needle, but the direction of the deflection is reversed; that is, the north-pointing pole (N-pole), which before turned to the west, is now deflected toward the east.

(*d*) Place your *right hand* above the wire, with the palm *towards* the wire and with the fingers pointing in the same direction as that in which the current is flowing, and extend your thumb at right angles to the direction of the current (Fig. 237). You observe that your thumb points in the *same* direction as the N-pole of the needle *under* the current-bearing wire.

(*e*) Reverse the current again (so that it shall flow northward), place your right hand as before (*viz.* with the palm towards the wire and with the fingers pointing in the same direction as the current); your outstretched thumb still points in the *same* direction as the N-pole of the needle.

(*f*) Introduce the + electrode of the battery into screw-cup T_3 and the — electrode into screw-cup T_4, so that the current shall flow northward *under* the needle. Place the right hand as directed before, except

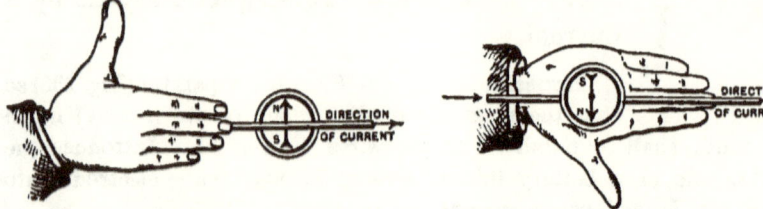

FIG. 237. Right hand above the wire; needle below it.

FIG. 238. Right hand below the wire; needle above it.

that it must be *under* the wire, so that the wire shall be between the hand and the needle; the thumb will point in the same direction as the N-pole (Fig. 238). Reverse the direction of the current in this wire, and apply the same test; the same rule holds.

The rule for determining the direction of the deflection of the N-pole of a needle when the direction of the current is known is this: *Place the outstretched right hand over or under the wire, so that the wire shall be between the hand and the needle, with the palm towards the needle, the fingers pointing in the direction of the current, and the thumb extended laterally at right angles to the direction of the current; then the extended thumb will point in the direction of the deflection of the N-pole.*

It will be observed that a deflection is reversed either by reversing the current or by changing the relative positions of the wire and needle, *e.g.* by carrying the needle from above the wire to a position below it.

The force exerted by the current upon the needle in deflecting it is called an *electro-magnetic force*.

310. Simple Galvanoscope or Current Detector.

Experiment 5. Introduce the + electrode of the battery into screw-cup T_2 (Fig. 236) and the — electrode into screw-cup T_3, so that the current shall pass above the wire in one direction and below it in the opposite direction, as indicated by the arrows. *A larger deflection is obtained than when the current passes the needle only once.*

If the right-hand test be applied, it will be seen that the tendency of the current, both when passing the needle in one direction above and in the opposite direction below, is to produce a deflection in the same direction, and consequently the two parts of the current assist each other in producing a greater deflection.

If a more sensitive instrument, *i.e.* one which will produce considerable deflections with weak currents, be required, then it will be necessary to pass the current through an insulated wire wound many times around the needle. Such an instrument is called a *galvanoscope,* or *current detector,* since one of its important uses is to detect the presence of a current.

311. (3) Thermal and Luminous Effects of the Electric Current.

Experiment 6. Construct a low resistance battery (§ 334) of three or four cells, and introduce into the circuit a platinum wire, No. 30, about ¼ inch long. The wire very quickly becomes white hot, *i.e.* it emits white light, which indicates a temperature of approximately 1900° C.

This experiment illustrates the conversion of the energy of an electric current into heat energy. In this case the energy of the current is consumed in overcoming the *resistance* which the conductor or the circuit offers to its passage. Heat is developed by a current in every part of the circuit, because all substances offer some resistance to a current; in other words, there are no perfect conductors. The small platinum wire offers much greater resistance than an equal length of a larger copper wire; whence the greater quantity

STRENGTH OF CURRENT. 291

of heat generated in this part of the circuit. All of the energy in any electric current that is not consumed in doing other kinds of work is changed into heat.

312. (4) Physiological Effects.

Experiment 7. Place one of the copper electrodes of a single voltaic cell on each side of the tip of the tongue. A slight stinging (not painful) sensation is felt, followed by a peculiar acrid taste.

EXERCISES.

1. (a) Have we any special sense-organ for appreciating electricity? (b) Can we see electricity?

2. What distinction do you make between an electroscope and a galvanoscope?

3. Enumerate the various effects producible by an electrical current, and give in connection with each some industrial applications to which that effect gives rise.

4. Are any of the effects just given producible by electricity when in the static state?

5. Give all the phenomena with which you are familiar that are peculiar to electricity in the statical state.

6. (a) Why are electrical contacts usually made of platinum? (b) Why, in electrolytical experiments, are platinum electrodes usually employed? (c) Why will not platinum electrodes answer for the decomposition of chloride salts? [Carbon electrodes may be used for this purpose.]

SECTION III.

ELECTRICAL QUANTITIES AND UNITS. — OHM'S LAW.

313. Strength of Current. The Ampere and the Coulomb. The magnitude of the effects produced by an electric current depends, among other things, upon the magnitude of the current. Any one of the effects producible by a current may be made the basis of a system of measurement of currents. For example, the quantity of hydrogen gas, or of any metal liberated at the cathode in a given time by electrolysis, is strictly proportional to the magnitude of the current, or, as it is technically termed, the *strength of the current.*

By current strength is meant the number of units of electricity which flows in a given time, or, briefly, the "rate of flow." The *practical* unit of current strength is the *ampere*. It is the current which, passed through a solution of nitrate of silver ("in accordance with standard specifications"), deposits silver at the rate of 0.001118 gram per second. The quantity of electricity transferred by a current of one ampere in one second is called a *coulomb*.[1] The coulomb is the unit of *quantity of electricity*. When the quantity of electricity conveyed by a current in one second is one coulomb, its strength is one ampere. By the strength of a current is meant not its total flow (which would be expressed in coulombs), but a *rate of flow*, and this distinction should be borne in mind.[2]

314. Electro-motive Force. The Volt. Liquid will flow from vessel A to vessel B (Fig. 239), provided the pressure be greater at the extremity M of the pipe C than at the extremity N. The difference in pressure at these two points is proportional to the "head" of water in A, or to the vertical hight D E of the liquid surface in A above the liquid surface in B. We might say that the flow of liquid is due to a *liquid-motive force* arising from the difference of pressure at the points M and N.

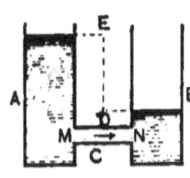

FIG. 239.

---

[1] The definitions of the coulomb and the ampere here given are those of the so-called *international units*, which were adopted as the legal units by act of the United States Congress in July, 1894.

[2] Roughly expressed, the current strength employed in certain practical applications is as follows:

In electric welding, 20 to 50 kilo-amperes.
In arc lighting, 8 to 10 amperes.
In 16 candle-power incandescent lamps, .45 to .75 ampere.
In average telegraphic circuit, 25 to 35 milliamperes.
In alternating current employed in the execution of criminals (New York State), 3 to 8 amperes, and an E.M.F. of about 1900 volts.

Similarly, electricity will flow in a conductor provided there be what may be termed greater *electrical pressure* at one end of the conductor than at the other end. As long as such a difference of pressure is maintained, so long there will exist something that is analogous in many respects to a *current-producing force*. It is for this reason called *electro-motive force* (E. M. F.). *Electro-motive force is that which maintains or tends to maintain a current of electricity through a conductor.* Like a mechanical force, it has a definite direction. It does no work unless it moves electricity.

Difference in electrical pressure we have hitherto assumed to be due to difference of potential. Potential difference may be due to contact of dissimilar substances, as in the voltaic cell, or to the movement of a part of the conductor in a magnetic field, as in the dynamo. In every case it is due to an expenditure of energy of some kind.

The *volt* is the name chosen for the practical unit of E. M. F. and difference of potential. It is the electrical pressure required to maintain a current of one ampere against a resistance of one ohm (§ 315). For purposes where great accuracy is not required, it will answer to consider a volt as the E. M. F. of a Daniell's cell; *i.e.* it is *about* the difference of potential between the zinc and copper elements of this cell, the E. M. F. of a standard Daniell cell being approximately 1.07 volts.

The E. M. F. which causes an electric spark discharge between the brass knobs of a frictional or influence machine when they are separated by an air space of one inch may be as high as 75 kilo-volts.

315. Electrical Resistance. The Ohm. Every substance offers resistance to the passage of a current. Those substances which offer a very powerful barrier are called *insulators*. The unit of resistance is called the *ohm*.

The international ohm is "the resistance offered to an unvarying electric current by a column of mercury at the temperature of melting ice, 14.421 grams in mass, of a constant cross-sectional area, and of the length of 106.3 centimeters"; or about the resistance of 9.3 ft. of No. 30 (American gauge) copper wire (.01 in. diam.).

A million ohms is a *megohm;* a millionth of an ohm is a *microhm.* The resistance of a cube of pure water of 1 cm. edge is 3.75 megohms; *i.e.* pure water is almost a perfect insulator.

The particular resistance [1] of different substances referred to some standard is called *specific resistance* or *resistivity*[1]. The reciprocal of resistivity is called *conductivity.* (See Table of Resistivities in Appendix.)

316. Electrical Power and Electrical Work or Energy. When an electrical current of one ampere flows between two points in a conductor whose difference of potential is one volt, work is done at the expense of electrical energy, and heat or some other form of energy is generated at a rate called one *volt-ampere,* or *watt.* Briefly, *the watt is the rate at which work is done in a circuit where the electro-motive force is one volt and the current is one ampere.*

If a coulomb of electricity flow between two points in a conductor whose difference of potential is one volt, a quantity of work is done, or a quantity of electrical energy is absorbed, that is called a *volt-coulomb,* or a *joule.* Briefly, *a joule is the quantity of work done in one second by a current working at the rate of one watt.*

The watt and the joule are, therefore, units of electrical power and electrical work (or energy), respectively. The volt-coulomb or joule is analogous to the foot-pound, and in the latitude of Washington, D.C., is equivalent to .738 ft.-lb. *Watts = volts × amperes. Joules = volts × coulombs.* Seven hundred and forty-six watts are equivalent to one horse-power.

1 h.p. = 0.746 kilowatt.
1 k.w. = 1.34 h.p.

[1] The termination *-ance* is used for words expressing the *properties* of a body; *e.g.* resistance, conductance, permeance, etc. The termination *-ivity* or *-ility* is used for words expressing the *specific properties* of a substance; *e.g.* resistivity, conductivity, permeability, etc.

317. Résumé. *A unit current is a current maintained by a unit E. M. F. against a unit resistance.*

A unit E. M. F. is the E. M. F. required to maintain a unit current against a unit resistance.

A conductor has a unit resistance when a unit E. M. F. (or a unit difference of potential between its two ends) causes a unit current to pass through it.

A unit of electrical power is the power of a unit current maintained by a unit difference of potential.

A unit of electrical work is the work done by a unit current in a unit time.

318. Ohm's Law. The three factors, current strength (C), electro-motive force (E), and resistance (R), are evidently interdependent. Their relations to one another are stated in the well-known Ohm's law thus: *The current is equal to the electro-motive force divided by the resistance;* or,

$$C = \frac{E}{R}; \text{ whence } E = RC, \text{ and } R = \frac{E}{C}.\text{[1]}$$

Hence, the strength of a current is directly proportional to the E. M. F. and inversely proportional to the resistance.

If E represent the fall of potential, R the resistance, and C the strength of current between *any two points in a circuit*, then any two of the three quantities being given, the third may be calculated.

This famous law is the basis of a large portion of electrical measurements commonly made.

319. Important Laws. The following formulas and laws relating to the electric current will be found convenient for reference: (1) P (watts) = C (amperes) × E (volts). (2) The watt = $\frac{1}{746}$ horse-power. Hence, $\frac{CE}{746}$ = power in horse-power. (3) Substituting in (1) the value of C (Ohm's formula), we have $P = \frac{E^2}{R}$. Or (4), substituting in (1) the

---

[1] Resistance is often defined as *the ratio of E. M. F. to the current strength.*

value of E (Ohm's formula), we have $P = C^2 R$; *i.e. power is proportional to the square of the current strength when* R *is constant, and to the resistance when* C *is constant.*

The number of units of heat developed in a conductor is proportional (1) *to its resistance,* (2) *to the square of the strength of the current, and* (3) *to the time the current is flowing.*

A current of one ampere flowing through a resistance of one ohm develops therein 0.00024 calorie of heat per second. Hence, H (calories) = C^2 (amperes) × R (ohms) × t (seconds) × 0.00024.

The amount of chemical decomposition produced by a current in a given time varies as the strength of the current. On this principle is based the voltameter, which measures the strength of a current by the amount of chemical action it effects in a given time.

The mass in grams of an element deposited by electrolysis is found by multiplying its electro-chemical equivalent (*i.e.* the mass in grams of the element deposited by one ampere in one second) *by the strength of the current in amperes, and this product by the time in seconds during which the current electrolyzes.*

320. Relation of Resistance, Potential Difference, Current Strength, and Energy Expended in an Electric Circuit. Deductions from Ohm's Law.

Let the line A B (Fig. 240) represent the length of a circuit (say 1000 feet), and the line A C the total fall of potential; then, obviously, the slope of the line C B will represent the average rate of fall of potential throughout the circuit. But suppose that the line for equal lengths of the conductor varies in resistance. Thus, assume that one tenth the resistance and consequently one tenth the fall of potential (C d) is included in the first quarter (A a), or 250 feet; then that the next 250 feet (a b, being very fine wire, perchance) represents one half the total resistance; that the next 250 feet (b c) represents one fourth the total resistance; and that the remaining resistance, fifteen hundredths, is in the last section (c B) of 250 feet. Then the lines C a', a' b', b' c', and c' B represent, respectively, the rate of fall of potential in each section of 250 feet.

The above is a deduction from Ohm's law; for, since $C = \dfrac{E}{R}$, and C is the same in every section of the conductor, it follows that in every section E (fall of potential) must be proportional to R.

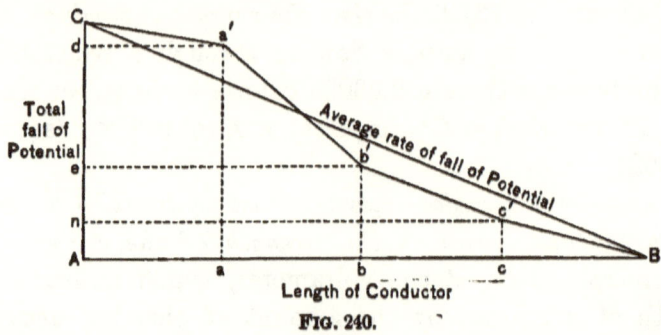

Fig. 240.

Again, since the power consumed or work performed in any circuit is proportional to the resistance when the current strength is constant (formula (4), § 319), and since the current strength is the same in every section of the circuit, it follows that the energy expended in any given section of a circuit is proportional to the resistance of the section. We conclude, therefore, that *the fall of potential along a conductor and the work done* (or energy expended) *are proportional to the resistance encountered in the different parts of the conductor.*

EXERCISES.

1. What E. M. F. is required to maintain a current of 1 ampere against a resistance of 1 ohm?

2. An E. M. F. of 10 volts will maintain a current of 5 amperes against what resistance?

3. What current ought an E. M. F. of 20 volts to maintain against a resistance of 5 ohms?

4. A voltmeter applied each side of an electric lamp shows a difference of potential of 40 volts. What current flows through the lamp, if it have a resistance of 10 ohms?

5. The resistance between two points in a circuit is 10 ohms. An ammeter (an instrument which measures the strength of a current in amperes) shows that there is a current strength in the circuit of 0.5 ampere. What is the difference in potential between the points?

6. (a) If 150 coulombs of electricity be transferred through a circuit in 30 seconds, what is the average current strength? (b) What quantity of electrical work is done?

7. A current of 4 amperes flows 5 minutes. What quantity of electricity is transferred?

8. When the difference of potential between two points in a circuit is 80 volts and the resistance between these points is 40 ohms, what quantity of electricity will pass between the points in 1 minute?

9. (a) Is it proper to speak of the *power* or of the *energy* of an electrical current? (b) Of the *energy* or of the *power* absorbed by an electric lamp? (c) Of the *energy* or of the *power* absorbed in a lamp in 5 minutes?

10. How much heat is generated per minute in a 16 candle-power electric lamp having a resistance of 140 ohms and a fall of potential of 110 volts, when a current of .75 ampere is maintained in it?

11. The electro-chemical equivalent of copper is .000328 g. If in a Daniell's cell the electro-negative element increased .6 g. in 25 minutes, what was the average current strength during the time?

12. A current of 40 amperes is sent over a line of 4 ohms resistance. What is the total fall of potential in the line?

13. If 200 coulombs be transferred in 40 seconds, what is the average current strength?

SECTION IV.

INSTRUMENTS FOR MEASUREMENT OF ELECTRIC CURRENTS.

321. Galvanometer. This is an instrument for measuring current strength by means of the deflection of a magnetic needle when placed in the field of the current. It is so constructed that either the deflection angle itself, or some function of it, is proportional to the current strength.

A very simple form of this instrument is represented in sectional elevation and plan in Fig. 241. It consists of an insulated wire wound many times around a magnetic needle. A card graduated like that of a mariner's compass is placed beneath the needle so that the number of

GALVANOMETERS. 299

degrees of deflection may be read from it. This form of instrument is much used to detect the presence of a current, to locate faults, etc.

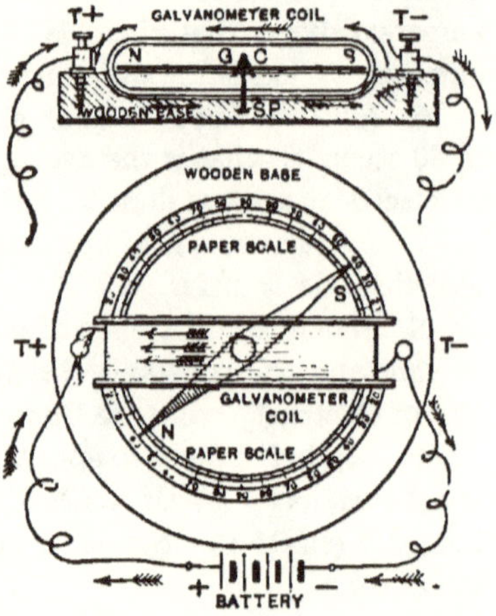

322. Tangent Galvanometer. A tangent galvanometer is one so constructed that the current passing through it is proportional to the tangent of the angle of deflection produced. To this end it is necessary that the needle be very short (not more than $\frac{1}{30}$) in comparison with the diameter of the coil.

It consists of a large vertical coil (C, Fig. 242) in the center of which is either a small compass needle or a needle suspended by a silk fiber. A needle thus placed in the field of a current is acted on by a mechanical couple tending to place it at right angles to the plane of the coil, and is deflected until it is balanced by the opposing couple due to the earth's magnetism.

Fig. 241.

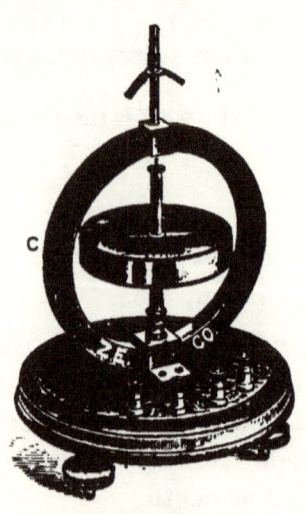

Fig. 242.

When the scale is divided into degrees, the corresponding tangents are found by consulting a table of tangents. (See Appendix). When the strengths of two currents are to be compared, it is only necessary to obtain deflections with each current, and compare the tangents of the angles.

323. Ammeter. If the galvanometer be calibrated so as to read in amperes, we shall have a direct-reading ampere-meter, or *ammeter*, as it is more commonly called. There is a great variety of ammeters in use, for a description of which the student is referred to technical works on the subject.

EXERCISES.

1. (*a*) Compare the strengths of two currents which produce deflections in a tangent galvanometer of 10° and 5°, respectively. (*b*) Of 87° and 88°. (*c*) In which of the two cases are the current strengths more nearly proportional to the angles of deflection?

2. The tendency of an electric current is to place a magnetic needle at right angles to itself. Why does the needle never quite attain this angle?

3. (*a*) What is an ammeter? (*b*) How can a galvanometer be converted into an ammeter? (*c*) Does a galvanometer measure the strength of a current directly?

SECTION V.

RESISTANCE OF CONDUCTORS.

324. External and Internal Resistance. For convenience the resistance of an electric circuit is divided into two parts, the *external* and the *internal*. External resistance includes all the resistance of a circuit except that of the generator, while that of the latter is termed internal resistance.

When the external resistance in a circuit is considered separately from the internal, Ohm's formula must be converted thus (calling the former R, and the latter r):

$$C = \frac{E}{R + r}.$$

If the electrical dimensions of a cell be $E = 1$ volt, and $r = 1$ ohm, and the connecting wire be short and stout, so that R may be disregarded, then the cell yields a current of one ampere. If by any means the internal resistance of this cell can be decreased one half, it will then be capable of yielding a two-ampere current if the other conditions remain the same.

325. External Resistance.

Experiment 1. Introduce into a circuit a galvanometer, and note the number of degrees the needle is deflected. Then introduce into the same circuit the wire on the spool numbered 4 on the platform [1] S (Fig. 243). (The wire on any one of the five spools on this platform can at any time be introduced into a circuit by connecting the battery wires with the binding screws on each side of the spool to be introduced.)

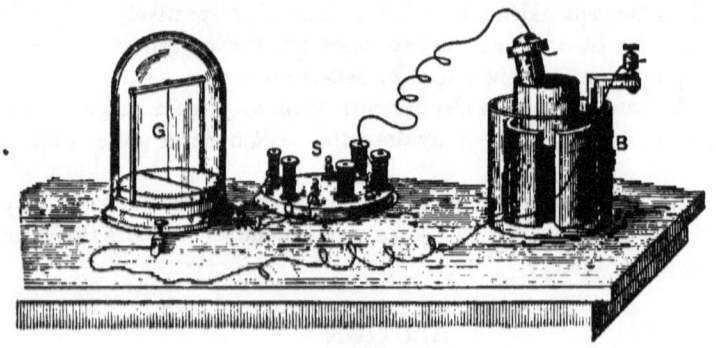

FIG. 243.

The deflection is now less than before. The copper wire on this spool is 16 yards in length; its size is No. 30 of the Brown and Sharpe wire gauge. When this spool is in circuit, the circuit is 16 yards longer than when the spool is out. The effect of lengthening the circuit is to weaken the current, as shown by the diminished deflection.

Experiment 2. Next, substitute Spool 2 for Spool 4. This contains 32 yards of the same kind of wire as that on Spool 4. The deflection is still smaller.

The weakening of the current by introducing these wires is caused by the resistance which the wires offer to the current, much as the friction between water and the interior of a pipe impedes, to some extent, the flow of water through it. The longer the pipe the greater is the resistance to the flow.

If the wire on the spools had been the only resistance in the circuit, then, when Spool 2 was in the circuit, the resistance would have been double what it was when Spool 4 was in the circuit, and the current, with double the resistance, would have been half as strong.

(1) *Other things being equal, the resistance of a conductor varies as its length.*

[1] A platform of spools containing wire of different (known) sizes, lengths, and material, so arranged that any one, two, or more can be introduced into the circuit.

Experiment 3. Next substitute Spool 1 for Spool 2. This spool contains 32 yards of No. 23 copper wire,—a thicker wire than that on Spool 2,—but the length of the wire is the same. The deflection is now greater than it was when Spool 2 was in circuit. This indicates that the larger wire offers less resistance.

Careful experiments show that (2) *the resistance of all conductors varies inversely as the areas of their cross sections. If the conductors be cylindrical, it varies inversely as the squares of their diameters.*

Experiment 4. Substitute Spool 5 for Spool 1, and compare the deflection with that obtained when Spool 4 was in the circuit. The deflection is smaller than when Spool 4 was in circuit. The wire on these two spools is of the same length and size, but the wire of Spool 5 is German silver. It thus appears that German silver offers more resistance than copper.

(3) *In obtaining the resistance of a conductor, the specific resistance of the substance must enter into the calculation.* (See Table of Specific Resistances in the Appendix.)

The resistance of metal conductors increases slowly with a rise of temperature of the conductor. The resistance of German silver is affected less by changes of temperature than that of most metals; hence its general use in standards of resistance. The resistance of carbon diminishes with a rise of temperature. The resistance of carbons in electric lamps is less when hot than when cold.

326. Internal Resistance.

Experiment 5. Connect with the galvanometer the copper and zinc strips used in Experiment 1, Section 1, and introduce the strips into a tumbler nearly full of acidulated water. Note the deflection. Then raise the strips, keeping them the same distance apart, so that less and less of the strips will be submerged. As the strips are raised, the deflection becomes smaller. This is caused by the increase of resistance in the liquid part of the circuit, as the cross section of the liquid lying between the two strips becomes smaller.

(4) *The internal resistance of a circuit, other things being equal, varies inversely as the area of the cross section of the liquid between the two elements.*

THE RESISTANCE BOX.

In a large cell the area of the cross section of the liquid between the elements is larger than in a small cell, and consequently the internal resistance is less. This is the only way in which the size of the cell affects the current.

Obviously, the resistance of the battery would be increased by any increase of the distance between the elements, since this increases the length of the liquid conductor; but as this distance is usually made as small as convenient, and is kept invariable, it demands little of our attention.

EXERCISES.

1. The resistance of 1000 feet of No. 24 copper wire (diameter = .511 mm.) is 26.284 ohms. What length of this wire would have a resistance of .5 ohm?
2. What is the resistance of 100 feet of No. 30 copper wire (diameter = .255 mm.)?
3. What is the resistance of 30 feet of No. 30 German-silver wire (the resistance of copper and German silver being as 1 : 12.8)?

SECTION VI.

MEASUREMENT OF RESISTANCE.

327. Description of the Resistance Box.

Fig. 244 represents a cylindrical box containing a series of coils of German-silver wire, whose resistances range from 0.1 ohm to 50 ohms, so that the total resistance is 160 ohms. These coils consist of insulated and doubled wires, the terminals of each being connected with brass blocks A, B, C, etc. (Fig. 245). When the brass plugs 1, 2, etc., are inserted between these blocks, the coils are short circuited, so that practically the whole current passes through the plugs from block to block; but when a plug is withdrawn the current is obliged to traverse the corresponding coil. Thus, by withdrawing the proper plugs, any desired resistance within the capacity

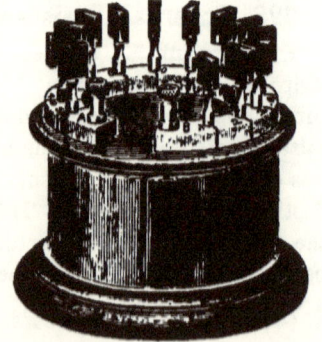

FIG. 244.

of the box may be thrown into the circuit. The resistance box is introduced into the circuit by connecting the battery terminals with the screw-cups A and B (Fig. 244).

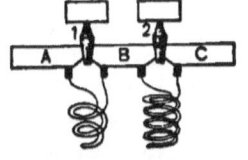

FIG. 245.

328. Wheatstone Bridge.

Fig. 246 represents a perspective view of the bridge (as modified by the author), and Fig. 247 represents a diagram of the essential electrical connections. The battery wires are connected with the bridge at the binding screws B B'. A galvanometer, G", is connected at G G', a resistance box, r, at R R, and the conductor, x, whose resistance is sought, at X X.

When the circuit is closed by means of the key T, the current, we will suppose, enters at B ; on reaching the point A it divides, one part flowing *via* the branch A G B', and the other *via* the branch A D B'. If points D and G in the two branches be at different potentials and a connection be made between them through the galvanometer G', by closing the key S there will be a current through this wire and through the galvanometer, and a deflection of the needle will be produced. But if the points D and G have the same potential, there will be no cross current through the bridge wire and no deflection. In § 320 it was demonstrated that the fall of potential along a circuit is everywhere proportional to the resistance. If, therefore, we have a divided circuit (§ 331), consisting of two branches, A G B' and A D B' (Fig. 247), of any resistance whatever, and if we select points G and D so that the resistance on both sides of them in each branch are in the proportion A G : G B' = A D : D B', then the fall of potential through A G will be the same as that through A D and there will be no difference

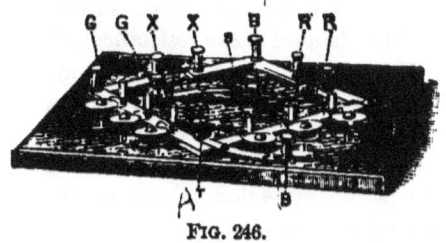

FIG. 246.

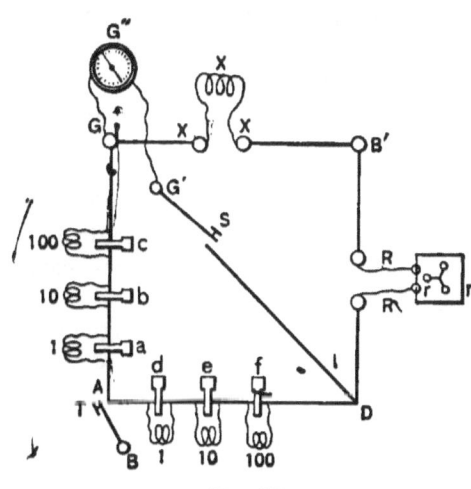

FIG. 247.

of potential between G and D, and there will be no flow of current through the bridge G D and galvanometer G″, however large a current may be flowing through the divided circuit. Between A and D, and A and G, there are three coils of wire having resistances, respectively, of 1, 10, and 100 ohms. One or more of these coils are introduced into the circuit by removing the corresponding plugs a, b, c, d, e, and f. As the other connections between A and D, and A and G, have no appreciable resistance, being for the most part short brass bars, the only practical resistance between these points is that introduced at will through the coils. Similarly, between points D and B′ the only practical resistance is that introduced at will through the resistance box, and between the points G and B′ the resistance is the resistance (x) sought.

It is apparent, then, that in using the bridge after the connections are properly made through the several instruments and certain known resistances are introduced between A and D and A and G, we have simply to regulate the resistance through the resistance box so that there will be no deflection in the galvanometer; then we are sure that the above proportion is true. The first three terms of the proportion being known, the fourth term, which is the resistance sought, is computable.[1]

If the same resistance be introduced between points A and G as between A and D, it is evident that the resistance in the resistance box r must be made equal to the unknown resistance x in order that there may be no deflection in the galvanometer. Consequently, when this result is obtained, the resistance of x may be read from the resistance box.

Experiment. Measure the resistance of each of the several spools of wire used above, electro-magnets, electric lamps, etc., using the bridge. Place the switches of the resistance box on the zero studs. Make connections as in the description above. Then close the circuit at T, and afterwards the bridge at S. There will probably be a deflection in the galvanometer. Regulate the resistance through the resistance box, throwing in or taking out resistance according as one or the other tends to reduce the deflection (the process is much like that of weighing), until there is no deflection. Then compute the resistance sought according to the above proportion.

---

[1] The accuracy of the results obtained largely depends upon so choosing resistances of the bridge as to make the arrangement have maximum sensibility, and upon the sensitiveness of the galvanometer. In using the bridge the following directions should be observed: (1) Always close the circuit at T before closing the bridge at S, and in breaking the circuit reverse this order. (2) Introduce between A and D and A and G resistance as nearly equal to the resistance sought (x) as practicable. If you have no conception what the unknown resistance is, it is best to begin by using high resistances. (3) Use a sensitive galvanometer, *e g.* a mirror galvanometer, or the galvanometer shown in Fig. 243, substituting the astatic needle for the tangent needle.

329. Measurement of Galvanometer Resistance. Lord Kelvin's Method.

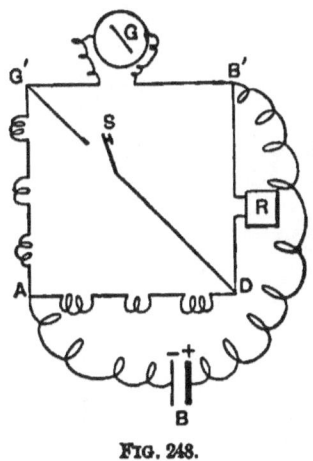

FIG. 248.

The bridge may be used for measuring the resistance of the galvanometer actually in use. The bridge is arranged as in Fig. 248. The resistance in the resistance box R is then varied until the deflection in G does not change when the key S is closed; then

$$r = R\frac{a}{b},$$

in which r is the resistance of the galvanometer, R is the resistance in the resistance box, and a and b are the resistances in the arms A G' and A D, respectively. If $a = b$, then $r = R$.

SECTION VII.

E.M.F. OF DIFFERENT CELLS. DIVIDED CIRCUITS. METHODS OF COMBINING VOLTAIC CELLS.

330. Electro-motive Force of Different Cells. If a galvanometer be introduced into a circuit with different battery cells, *e.g.* Bunsen, Daniell, Grenet, etc., very different deflections will be obtained, showing that the different cells yield currents of different strengths. This may be in some measure due to a difference in their internal resistances, but it is chiefly due to the difference in their electro-motive forces. We have learned that difference of electro-motive force is due to the difference of the chemical action on the two plates used, and this depends upon the nature of the substances used. It is wholly independent of the size of the plates; hence, the electro-motive force of a large cell is no greater than that of a small one of the same kind. Consequently, any difference in strength of current yielded by cells of the same kind, but of

LORD KELVIN.

different sizes, is due wholly to a difference in their internal resistances.

The electro-motive forces of the Bunsen, Daniell, and Grenet cells are, respectively, about 1.8, 1, and 2 volts.

331. Divided Circuits; Shunts.

Experiment. Make a divided circuit as in Fig. 249 (using double connectors a and b). Insert a galvanometer, G, in one branch and a resistance box, R, in the other. When the current reaches a, it divides, a portion traversing one branch through the galvanometer, and the remainder passing through the other branch and the resistance box. The branch a R b is called a *shunt* or *derived circuit*. Increase gradually the resistance in the resistance box. The result is that it throws more of the current through the galvanometer, as shown by the increase of deflection.

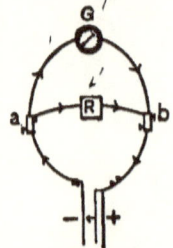

FIG. 249.

In a divided circuit the current is distributed between the paths in amounts inversely as their resistances. For example, if the resistance of the resistance box above be 4 ohms, and the resistance in the galvanometer be 1 ohm, then four fifths of the current will traverse the latter and one fifth the former.

Suppose that the resistance box and galvanometer be removed from the shunts, and that the shunts be of the same length, size, and kind of wire, and consequently have equal resistances, then using the two wires instead of one to connect a and b is equivalent to doubling the size of this portion of the conductor; consequently, the resistance of this portion is reduced one half.

Generally, *the joint resistance of two branches of a circuit is the product of their respective resistances divided by their sum.*

If any portion of a circuit be divided into three or more branches whose resistances are, respectively, r_1, r_2, r_3, etc., it may be demonstrated [1] that

$$\frac{1}{R} = \frac{1}{r_1} + \frac{1}{r_2} + \frac{1}{r_3} + \cdots,$$

[1] See the author's "Principles of Physics," p. 509.

in which R represents the joint resistance of the several branches. That is, *the reciprocal of the joint resistance of any number of branches is equal to the sum of the reciprocals of the resistances of the several branches.*

The reciprocal of the resistance R of a wire, *i.e.* $\frac{1}{R}$, is called its conductance (sometimes expressed in a unit called the *mho* [1]). We may say, therefore, in general, that when two points in a circuit are connected by a *multiple arc* (a term in common use to denote a divided circuit between any two points) consisting of n branches, *the conductance of the multiple arc is equal to the sum of the conductances of the* n *branches.*

332. Shunted Galvanometers. When it is necessary to measure a current that exceeds the capacity of a galvanometer, a wire may be connected across the terminals of the galvanometer, by means of which any fraction of the current may be deflected, and the galvanometer then measures a known fraction of the total current.

For example, if the resistance of the galvanometer be 3 ohms and the resistance of the shunt be ($\frac{1}{9}$ of $3 =$) .33 ohm, then the current in the galvanometer will be $\frac{1}{9}$ of the current in the shunt, or $\frac{1}{10}$ of the total current.

333. Combining Cells; Batteries.

A number of cells connected in such a manner that the currents generated by all have the same direction constitutes *a voltaic battery*. The object of combining cells is to get a stronger current than one cell will afford. We learn from Ohm's law that there are two, and only two, ways of increasing the strength of a current. It must be done either by increasing the E. M. F. or by decreasing the resistance. So we combine cells into batteries, either to secure greater E. M. F. or to diminish the internal resistance. Unfortunately, both purposes cannot be accomplished by the same method.

[1] A word formed by writing the word *ohm* in reverse order.

334. Batteries of Low Internal Resistance. Figure 250 represents three cells having all the carbon (+) plates electrically connected with one another, and all the zinc (−) plates connected with one another, and the triplet carbons connected with the triplet zincs by the leading-out wires through a galvanometer, G.

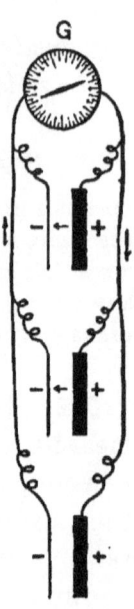

FIG. 250.

It is easy to see that through the battery the circuit is divided into three parts, and consequently the conductance in this part of the circuit, according to the principle stated in § 331, must be increased threefold; in other words, the internal resistance of the three cells is one third of that of a single cell. This is called connecting cells "in multiple arc," and the battery is called a "battery of low internal resistance." The resistance of the battery is decreased as many times as there are cells connected in multiple arc, but the E. M. F. is that of one cell only.

The formula for the current strength in this case is written thus:

$$C = \frac{E}{R + \dfrac{r}{n}},$$

in which n represents the number of cells. It is evident from this formula that when R is so great that $\dfrac{r}{n}$ is a small part of the whole resistance of the circuit, little is added to the value of C by increasing the number of cells in multiple arc.

335. Batteries of High Internal Resistance and Great E. M. F.

Fig. 251 represents four cells having the carbon or − plate of one connected with the zinc or + plate of the next, and the + plate at one end of the series connected by leading-out wires through a galvanometer with the — plate at the other end of the series. It is evident that the current in this series traverses the liquid four times, which is equivalent to lengthening the liquid conductor four times, and of course increasing the internal resistance fourfold. But, while the internal resistance is increased, *the E. M. F. of the battery is increased as many times as there are cells in series.* In many cases (always when the internal resistance is a small part of the whole resistance of the circuit) the gain by increasing the E. M. F. more than offsets the loss occasioned by increased resistance.

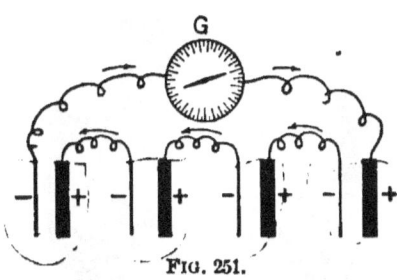

Fig. 251.

The formula for current strength in this case becomes

$$C = \frac{nE}{R + nr}.$$

It is evident that C is increased most by adding cells in series when nr is smallest in comparison with R.

336. Rule for Combining Cells.

When the external resistance is large, connect cells in series; when the external is less than the internal resistance, connect cells in multiple arc.

EXERCISES.

1. What E. M. F. is required to maintain a current of .2 ampere through a resistance of .8 ohm?
2. Through what resistance will an E. M. F. of 10 volts maintain a current of 9 amperes?
3. What current ought an E. M. F. of 85 volts to maintain through a resistance of 8 ohms?
4. A voltmeter (Fig. 252) applied each side of an electric lamp shows a difference of potential of 10 volts. What current flows through the lamp, if it have a resistance of 50 ohms?

5. The resistance between two points in a circuit is 70 ohms. An ammeter shows that there is a current strength in the circuit of 0.5 ampere. What is the difference in potential between the points?

6. What current will a Bunsen cell furnish when $r = 0.9$ ohm (about the resistance of a quart cell), $E = 1.8$ volts, and $R = 0.01$ ohm (about the resistance of 3 feet of No. 16 wire)?

[In the following exercises, whenever a Bunsen cell is mentioned it may be understood to be a quart cell, having a resistance of about 0.9 ohm. Its E. M. F. is about 1.8 volts.]

7. (a) When is a large cell considerably better than a small one?

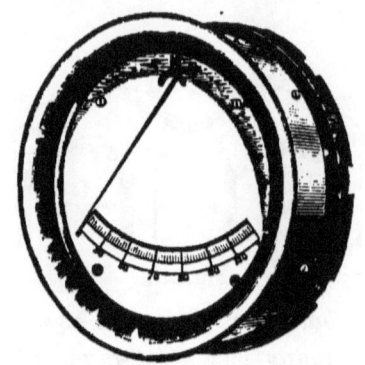

FIG. 252.

(b) When does the size of the cell make little difference in the current?

8. If you have a dozen quart cells, how can you make them equivalent to one 3-gallon cell?

9. If a battery of 10 cells have an E.M.F. 10 times greater than that of a single cell, why will not the battery yield a current 10 times as strong?

10. (a) The internal resistance of 10 cells connected in multiple arc is what part of that of a single cell? (b) If the cells were connected in series, how would the resistance of the battery compare with that of one of its cells? (c) How would the E.M.F. of the latter battery compare with that of a single cell?

11. What current will a single Bunsen cell furnish against an external resistance of 10 ohms?

12. What current will 8 Bunsen cells, in series, furnish against the same resistance?

SOLUTION: $\dfrac{E}{R + r} = \dfrac{1.8 \times 8}{10 + (0.9 \times 8)} = 0.83 +$ ampere.

13. What current will 8 Bunsen cells in multiple arc furnish against the same external resistance?

SOLUTION: $\dfrac{E}{R + r} = \dfrac{1.8}{10 + (0.9 \div 8)} = 0.17 +$ ampere.

14. What current will a Bunsen cell furnish against an external resistance of 0.4 ohm?

15. What current will a battery of two Bunsen cells, in series, furnish against the same resistance as the last?

16. What current will 2 cells in multiple arc furnish against the same resistance?

17. A coil of wire having a resistance of 10 ohms carries a current of 1.5 amperes. Required the difference of potential at its ends.

18. (a) The resistance between two points, A and B, of a conductor is 2.5 ohms; the resistance of a shunt between the same points is 1.5 ohms. What is the joint resistance between these points? (b) If a current of 10 amperes be maintained between these points, what will be the strength of current in each branch? (c) How will the strength of current between these points be affected if the shunt be removed and the same fall of potential be preserved? Why?

19. Three conductors have conductances of 4, 8, 10 mhos; if they be joined in multiple arc, what will be the conductance of the combination?

20. The resistances offered by three conductors are, respectively, 2, 5, and 7 ohms. What will be their joint resistance (a) if they be joined in series? (b) if they be joined in multiple?

21. Assume that the electrical dimensions of a Daniell's cell are as follows: E. M. F. $= 1$ volt and $r = 2$ ohms. Let a complex battery be formed of 10 Daniell's cells arranged in two groups, each group consisting of 5 cells connected in series, the two groups being connected in multiple. What current[1] will the battery furnish against an external resistance of 3 ohms? *Ans.* $\frac{5}{8}$ ampere.

22. Suggest some device by means of which the strength of a current flowing through any instrument, *e.g.* a motor on an electric car, may be easily and conveniently changed at will.

23. The E. M. F. of a Grenet cell $= 2$ volts; of a Daniell cell $= 1$ volt. (a) On what condition will the former yield a current twice as great as the latter against the same external resistance? (b) How may this condition be attained?

24. How many cells whose dimensions are E. M. F. $= 1.8$ volts and $r = 1.1$ ohms will be required in series to send a current of .5 ampere against an external resistance of 50 ohms?

25. What E. M. F. is required to maintain a current of .75 ampere in a lamp whose resistance is 80 ohms?

26. If a circuit have a large resistance, which would give the larger deflection, a high-resistance or a low-resistance coil galvanometer? Why?

27. If a circuit have a large resistance, should the helix of electro-magnets included in the circuit be wound with short large wire, or with long small wire?

28. What should be the resistance of a shunt, that a galvanometer may measure $\frac{1}{100}$ of the total current?

LAW OF MAGNETS.

SECTION VIII.

MAGNETS AND MAGNETISM.

337. Law of Magnets. Suspend by fine threads in a horizontal position two stout darning needles which have been drawn in the same direction (*e.g.* from eye to point) several times over the same pole of a powerful electro-magnet. These needles, separated a few feet from each other, take positions parallel with each other, and both lie in a northerly and southerly direction with the points of each turned in the same direction.

That point in the Arctic zone of the earth toward which magnetic needles point is called the north magnetic pole of the earth. That end of the needle which points toward the north magnetic pole of the earth is called the *north-seeking*, *marked*, or $+ pole$; this is the end that is always *marked* for the purpose of distinguishing one from the other. That end of the needle which points southward is called the *south-seeking*, *unmarked*, or $- pole$.

Experiment 1. Bring both points near each other; there is a mutual repulsion. Bring both eyes near each other; there is a mutual repulsion. Bring a point and an eye near each other; there is a mutual attraction.

Like poles of magnets repel, unlike poles attract each other.

338. Magnetic Transparency and Induction.

Experiment 2. Interpose a piece of glass, paper, or wood-shaving between the two magnets. These substances are not themselves perceptibly affected by the magnets, nor do they in the least affect the attraction or repulsion between the two magnets.

Substances that are not susceptible to magnetism are said to be *magnetically transparent*. When a magnet causes another body, in contact with it or in its neighborhood, to become a magnet, it is said to *induce* magnetism in that body. As attraction, and never repulsion, occurs between a magnet and an unmagnetized piece of iron or steel, it must

FIG. 253.

be that the magnetism induced in the latter is such that opposite poles are adjacent; that is, an N or + pole induces an S or − pole next itself, as shown in Fig. 253.

339. Polarity.

Experiment 3. Strew iron filings on a flat surface, and lay a bar magnet on them. On raising the magnet it is found that large tufts of filings cling to the poles, as in Fig. 254, especially to the edges; but the tufts diminish regularly in size from each pole towards the center, where none are found.

Magnetic attraction is greatest at the poles, and diminishes towards the center, where it is nothing; i.e. the center of the bar is neutral. This dual character of the magnet, as exhibited at its opposite extremities, is called *polarity*. If a magnet be broken, each piece becomes a magnet with two poles and a neutral line of its own.

340. Retentivity and Resistance.

FIG. 254. It is more difficult to magnetize steel than iron; on the other hand, it is difficult to demagnetize steel, while soft iron loses nearly all its magnetism as soon as it is removed from the influence of the inducing body. That property of steel in virtue of which it resists the escape of magnetism which it has once acquired is called *retentivity*. The greater the retentivity of a magnetizable body, the greater is the *resistance* which it offers to becoming magnetized. *The harder steel is, the greater is its retentivity.* Hence, highly tempered steel is used for *permanent* magnets. Hardened iron possesses considerable retentivity; hence, the cores of electro-magnets should be made of the *softest* iron, that they may acquire and part with magnetism instantaneously.

341. Forms of Artificial Magnets. Artificial magnets, including permanent magnets and electro-magnets, are usually made in the shape either of a straight *bar* or of the letter U, according to the use to be made of them. If we wish, as in the experiments already described, to use but a single pole, it is desirable to have the other as far away as possible; then, obviously, the bar magnet is more convenient. But if the magnet is to be used for lifting or holding weights, the U-form (see Fig. 258) is far better, because the attraction of both poles is conveniently available.

SECTION IX.

LINES OF MAGNETIC FORCE. THE MAGNETIC CIRCUIT.

342. Lines of Magnetic Force. These lines are easily studied by the use of iron filings. The field of force around

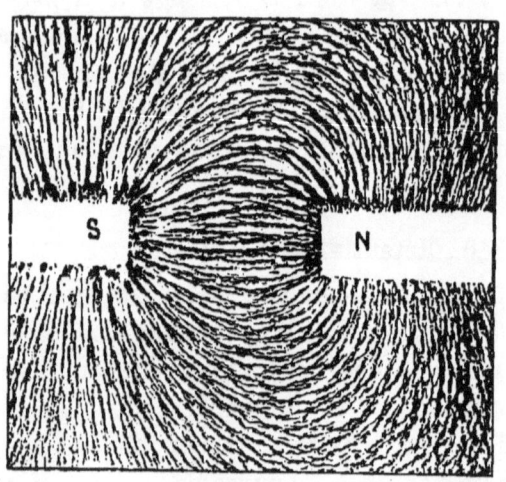

FIG. 255.

a magnet is shown by placing a paper over it, dusting filings upon the paper, and tapping it. The filings take symmetrical positions, form curves between the poles of the magnet or magnets, and show that *the lines of force connect the opposite poles of the magnet.* The fact is that each filing, when

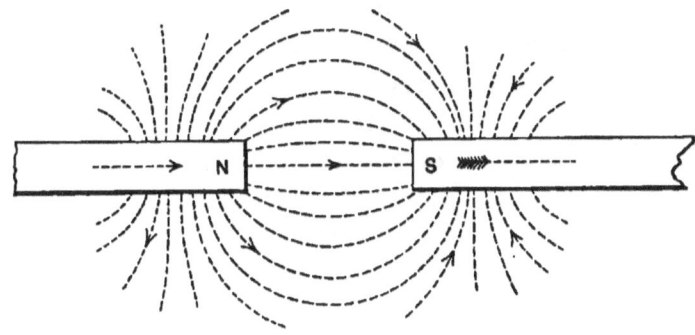

Fig. 256.

brought within the influence of the magnetic field,[1] becomes a magnet by induction, and of necessity tends to take a definite position which represents the resultant of the forces

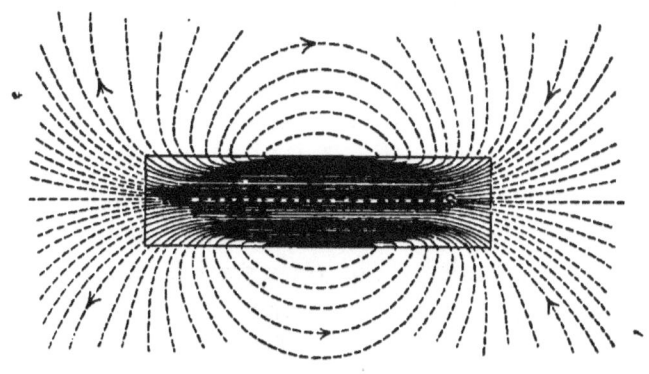

Fig. 257.

acting upon it from each pole of the system. *A line of magnetic force* is a line drawn in such a manner that the tangent to it at any point indicates the direction of the resultant magnetic force at that point; or it is a line at every

[1] Surrounding every magnet there is a region of magnetic influence, technically known as the *magnetic field.*

point of which the axis [1] of a magnetic needle is tangent. Fig. 255 represents a magnetic field photographed from a specimen paper, and Fig. 256 is a graphical representation of the same. In this illustration the unlike poles of two magnets are placed opposite each other. Fig. 257 is a diagram of paths of lines of force of a bar magnet when its axis coincides with the magnetic meridian of the earth (§ 248) and its N-pole points north; and Fig. 258, of those of a U-shaped magnet. A *magnetic pole* is a region within a magnet towards which the lines of force converge or from which they diverge.

343. Magnetic Circuit. The field of a magnet is permeated with lines of force, which may be more conveniently called *magnetic flux paths*, or lines of flow. A line of force is assumed arbitrarily to start from the N-pole and to pass through the sur-

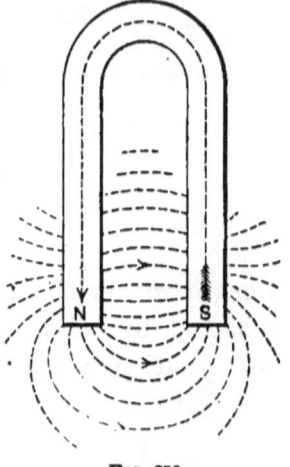

FIG. 258.

rounding medium (*e.g.* the air), entering the magnet by the S-pole, and completing its path through the magnet itself to its starting point (the N-pole), thus forming a complete circuit (Fig. 257). These lines do not all emerge, however, from the extremities. A multitude of lines start from all parts of the magnet and enter at corresponding points on the other side of its central or neutral line. Every line of magnetic force makes a complete circuit.

The exact nature of magnetic flux is not understood, but it appears to be attended by a strain in the ether. It possesses several peculiar characteristics.

In air and most other mediums the lines of force tend to separate from one another, but at the same time tend to become as short as possible.

[1] The axis of a needle is a straight line connecting its poles.

The strain is as if these lines were stretched elastic threads endowed with the property of repelling one another as well as of shortening themselves; in other words, there is tension along the lines and repulsion at right angles to them. They may be likened to the fibers of a muscle which contracts and at the same time thickens when exerting force.

If the N-pole of one magnet be placed opposite the S-pole of another (Fig. 256), the lines of force issuing from the former enter the latter, and, tending to shorten, produce *attraction*. If the similar ends be opposed

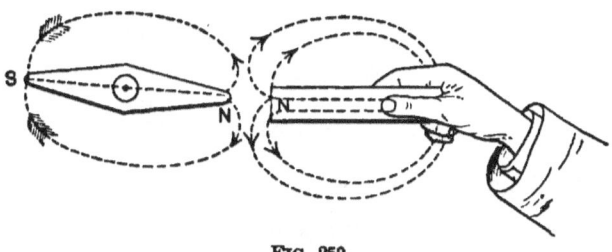

FIG. 259.

Fig. 259), the lines of force are turned away from each pole in all directions, and complete their circuits independently. Thus becoming parallel, they repel one another; thus, like magnetic poles repel each other.

It would seem that air is a poor conductor of lines of force, or, to use a technical term, its *permeability* is low; on the other hand, iron is highly permeable to lines of force. If a piece of iron be brought within a magnetic field, many lines of force will leave their normal paths through the air and crowd together in this medium of greater permeability.

Magnetic flux produces the following important effects: (1) A bar of iron, when introduced into the flux, becomes magnetized. (2) A freely suspended magnetic needle brought into the flux comes to rest in a definite position. (3) An E. M. F. is developed in an electric conductor when it is moved across the flux paths.[1]

It may not be amiss to state in this connection that, as a result of comparatively recent experimental and mathe-

---

[1] It is possible that all electric, magnetic, and electro-magnetic phenomena are referable to two conditions of stress in the ether, one of which is electric flux and the other magnetic flux. So intimately are the electric and magnetic fluxes correlated that any disturbance in one immediately calls the other into existence.

matical researches, scientists are becoming convinced that electric currents are transmitted as electric waves through the ether surrounding a conductor, being guided by the conductor, but not transmitted through it.

344. Law of Inverse Squares. It may be demonstrated experimentally [1] that *the force exerted between two magnetic poles varies inversely as the square of the distance between them.*

SECTION X.

TERRESTRIAL MAGNETISM.

345. The Earth a Magnet.

Experiment. Place a dipping needle [2] over the + pole of a bar magnet (Fig. 260). The needle takes a vertical position with its — pole down. Slide the supporting stand along the bar; the — pole gradually rises until the stand reaches the middle of the bar, where the needle becomes horizontal. Continue moving the

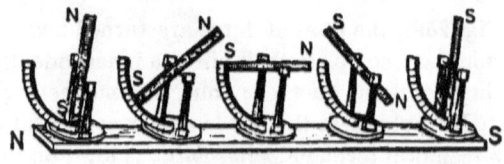

FIG. 260.

stand toward the — pole of the bar; after passing the middle of the bar the + pole begins to dip, and the dip increases until the needle reaches the end of the bar, when the needle is again vertical with its + pole down.

If the same needle be carried northward or southward along the earth's surface, it will dip in the same way as it approaches the polar regions, and be horizontal only at or near the equator.

The experiment presents within a small compass a series of phenomena precisely similar to those caused by the earth's influence upon the dipping needle, and this leads to the conclusion that *the earth is a magnet.* In other words, these phenomena are just what we should expect if a huge magnet

---

[1] See the author's "Principles of Physics," p. 528.

[2] A magnetic needle supported on a horizontal axle so that it can rotate in a vertical plane is called a *dipping needle.*

320   ETHER DYNAMICS.

were thrust through the earth, as represented in Fig. 261 — having its N-pole near the S geographical pole, and its S-pole near the N geographical pole;[1] or if the earth itself were a magnet.

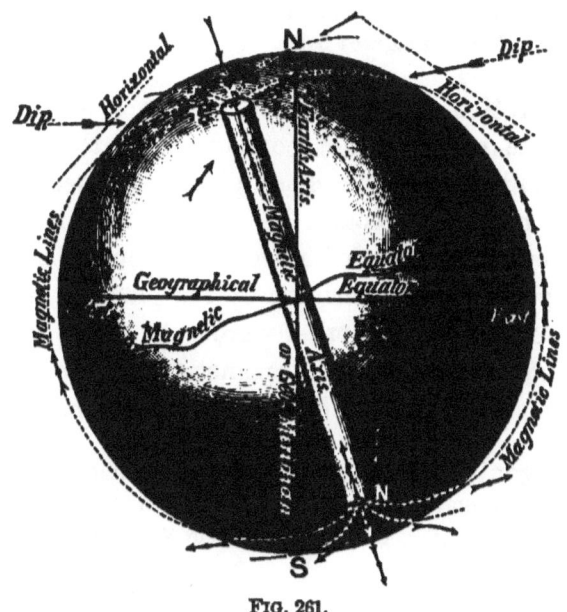

FIG. 261.

346. Magnetic Poles of the Earth. Places where the dipping needle assumes a vertical position are called the *magnetic poles of the earth*. A point was found on the western coast of Boothia by Sir James Ross, in the year 1831, where the dipping needle lacked only one sixtieth of a degree of pointing directly to the earth's center. The same voyager subsequently reached a point in Victoria Land where the opposite pole of the needle lacked only 1° 20' of pointing to the earth's center.

[1] In common parlance the magnetic poles of the earth are positions on the earth's surface where a dipping needle points vertically downward. The direction in which such a needle points would meet the direction in which, for example, a needle at Boston points, at some thousand miles down in the bowels of the earth, which shows that the poles or centers of magnetic action are really deep-seated; hence, the phrase *magnetic poles on the earth's surface* is somewhat misleading.

VARIATION OF THE NEEDLE. 321

It will be seen that, if we call that end of a magnetic needle which points north the N-pole, we must call that magnetic pole of the earth which is in the northern hemisphere the S-pole. (See Fig. 261.) To avoid confusion, many careful writers abstain from the use of the terms *north* and *south poles,* and substitute for them the terms *positive* and *negative,* or *marked* and *unmarked poles.*

347. Variation of the Needle. Inasmuch as the magnetic poles of the earth do not coincide with the geographical poles, it follows that in most places the needle does not point due north and south. The angle which the vertical plane through the axis of a freely suspended needle makes with the geographical meridian of the place is known as the *angle of declination.* In other words, the angle of declination is the angle formed by the magnetic and the geographical meridians. This angle differs at different places.

348. Isogonic Curves. These are lines connecting all points of equal declination on the earth's surface. The line of no declination, or isogonic of 0° (Fig. 262), commences at the N magnetic pole about lat.

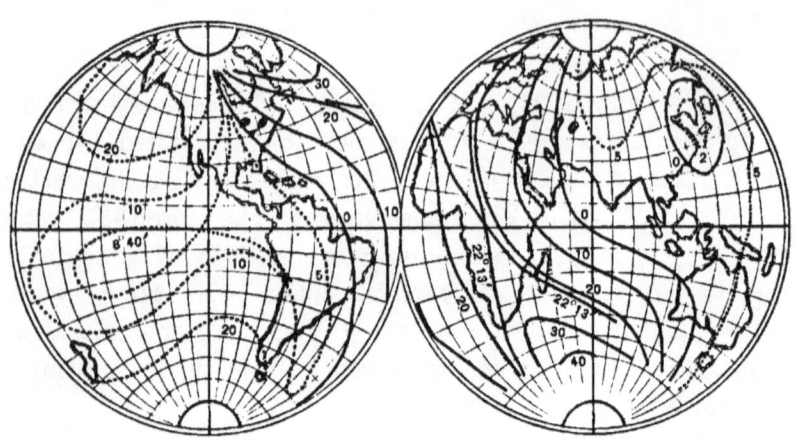

FIG. 262.

70°, long. 96°, passes in a southeasterly direction across Lake Erie and Western Pennsylvania, and enters the Atlantic Ocean near the boundary between the Carolinas. Pursuing its course through the south polar

322   ETHER DYNAMICS.

regions, it reappears in the eastern hemisphere and crosses Western Australia and the Caspian Sea, and thence passes to the Arctic Ocean. There is also a detached line of no declination enclosing an oval area in Eastern Asia and the Pacific Ocean. In all the New England states and in the states of Pennsylvania and New York the declination is westward. In all the states west of these states the declination is eastward.

The magnetic poles are not fixed objects that can be located like an island or cape, but are constantly changing. They appear to swing, something like a pendulum, in an easterly and westerly direction, each swing requiring centuries to complete it. The north magnetic pole is now on its westerly swing, and consequently the line of no declination is slowly moving westward.

SECTION XI.

MAGNETIC RELATIONS OF THE CURRENT. ELECTRO-MAGNETS.

349. Magnetic Field due to a Circular Current. If a wire be bent into the form of a circle of about 10 in. diameter, and placed vertically in the magnetic meridian, and a cardboard be placed at right angles to the circle so that its hori-

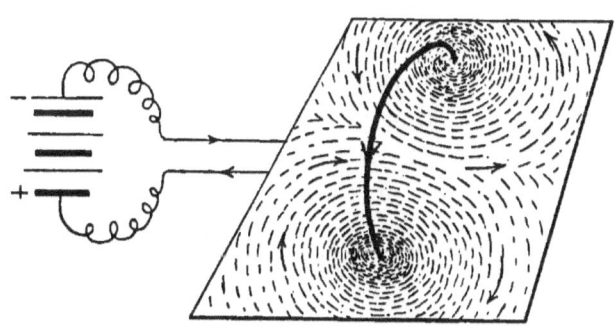

FIG. 263.

zontal diameter is coincident with the upper surface of the cardboard, and a very strong current be sent through the wire in the direction indicated by the arrowhead in the wire, iron filings sifted upon the card will arrange themselves as shown in Fig. 263. And if a small compass be carried inside and outside the circle, the several positions taken by the needle, as indicated in the figure by arrows, corroborate the directions of the lines of force as indicated by the filings. If the direction of the current be reversed, the direction of the needle will be reversed wherever it may be placed. The electric current and its encircling lines of force [1] always coexist,[2] and one varies directly as the other; that is, the greater the strength of the current, the greater is the number of lines of force that occupy the field.

350. Solenoid. If instead of a single circle of wire an insulated wire be wound into a helix of several turns, it is called a *solenoid*. The intensity of the magnetic field is greatly increased by the joint action of the many current turns. The passage of an electric current through a solenoid gives it all the properties

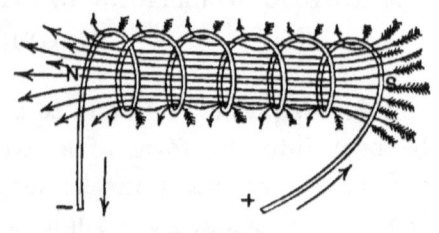

Fig. 264.

of a magnet. To within a short distance of its ends the lines of force are parallel with its axis, as shown in Fig. 264.

A solenoid encircling an iron core constitutes an *electromagnet*. By reason of its permeability the iron core greatly increases the number of lines of force which pass through the solenoid. Hence, the magnetic strength of a solenoid is greatly increased by the presence of an iron core.

[1] "Every conducting wire is surrounded by a sort of magnetic whirl. A great part of the energy of the so-called electric current in the wire consists in these external magnetic whirls. To set them up requires an expenditure of energy; and to maintain them requires a constant expenditure of energy. It is these magnetic whirls which act on magnets, and cause them to set, as galvanometer needles do, at right angles to the conducting wire." — S. P. THOMPSON.

[2] "Electricity in motion produces magnetic force, and magnetism in motion produces electric force." — HERTZ.

351. Polarity of an Electro-magnetic Solenoid.

Experiment. Fig. 265 represents a small cell floating on water. The leading out wire of the cell is wound into a horizontal solenoid.

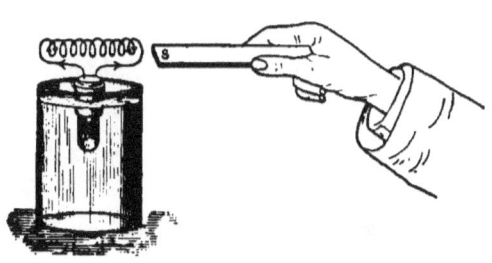

FIG. 265.

Slowly, after the cell is floated, it will take a position so that the axis of the solenoid will point north and south like a magnetic needle. Hold the S-pole of a bar magnet near that end of the solenoid which points north; the solenoid is attracted by the magnet. Hold the N-pole of the magnet near the north-pointing end of the solenoid; the magnet repels the solenoid.

Repeat the above, using in place of the bar magnet another current-bearing solenoid (Fig. 266); there will be a repetition of the phenomena obtained with the bar magnet.

Place the wire of another battery over and parallel with the coil (Fig. 267), so that the two currents will flow in planes at right angles to each other.

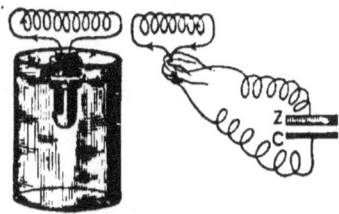

FIG. 266.

The coil is deflected like a magnetic needle (Fig. 268). Reverse the direction of the current above and the deflection is reversed.

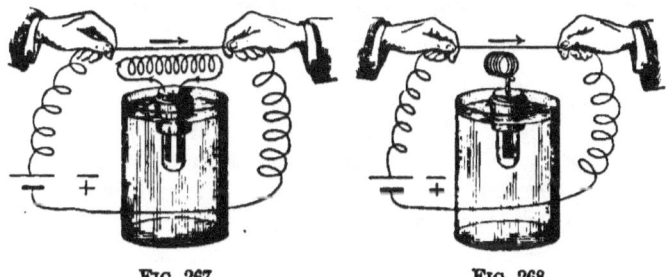

FIG. 267.   FIG. 268.

We thus prove that a solenoid bearing a current possesses polarity as if it were a magnet, and that there can be produced by a current-bearing solenoid a magnetic field of the same character as that produced by a permanent magnet.

RULES TO FIND POLES OF THE SOLENOID.

There is no essential difference between a permanent magnet, a current-bearing solenoid, and an electro-magnet, except that the last may be made much stronger than either of the others.

352. Given the Direction of the Current in a Solenoid, to Find the N- and S-poles of the Solenoid, and *vice versa*.

RULE 1. *Place the palm of the right hand against the side of the solenoid so that the fingers will point in the direction of the current passing through the windings* (as shown in Fig. 269); *the thumb will point in the direction of the N-pole of the solenoid or electro-magnet.*[1]

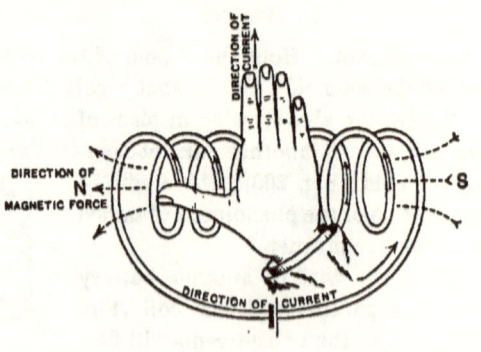

FIG. 269.

RULE 2. *Ascertain the N-pole of the solenoid or electro-magnet with a magnetic needle, and place the palm of the right hand upon the solenoid so that the outstretched thumb points in the direction of the N-pole; the fingers will point in the direction in which the current passes in the windings.*

Fig. 270 represents the two poles of a U-shaped electro-magnet, and shows the method in which the wire is wound in the two helices and the relative direction of the current in the same. It will be seen that that is the S-pole about which the current flows in the direction in which the hands of a clock move, while that is the N-pole about which the current has an anti-clockwise motion. Evidently, if the current be reversed the polarity will be reversed.

FIG. 270.

---

[1] The following suggestion will often prove of practical value: that is the south pole of a helix where the current corresponds to the motion of the hands of a watch, S, and that is the north pole where the current is in the reverse direction, N.

SECTION XII.

ELECTRODYNAMICS.   AMPERE'S THEORY OF MAGNETISM.

353. Mutual Action of Currents on One Another. If we suppose that a test-needle be moved up or down just back of

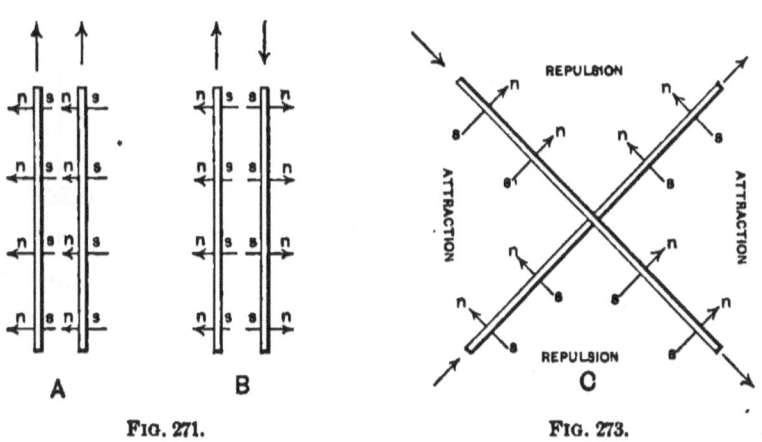

FIG. 271.   FIG. 273.

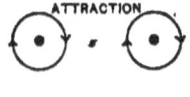

FIG. 272.

the current-bearing wires (Fig. 271), the N- and S-poles will take the positions indicated by n and s. From inspection of the polarity developed, we may readily predict that if the wires were so suspended as to be free to move either toward or from each other, the pair of wires in which the currents flow parallel to each other and in the same direction, A, would attract each other, and the pair of wires in which the currents flow in opposite directions, B, would repel each other;[1] but if the currents be inclined to each other, as in Fig. 273, they will tend to move into a position in which they will be parallel and in the same direction. That this actually takes place may be shown by the following experiments:

---

[1] Fig. 272 shows the cross section of the wires in the two cases and the directions of the lines of force encircling the wires. Compare with Fig. 258.

Experiment 1. Fig. 274 represents a portion of a divided circuit. The lower ends of the wires dip into mercury about one sixteenth of an inch, and the wires are so suspended that they are free to move toward or from each other. Send the current of a battery of three or four Bunsen cells, in multiple arc, through this divided circuit. The two portions of the current travel in the same direction and parallel with each other, and the two wires at the lower extremities move toward each other, showing an attraction.

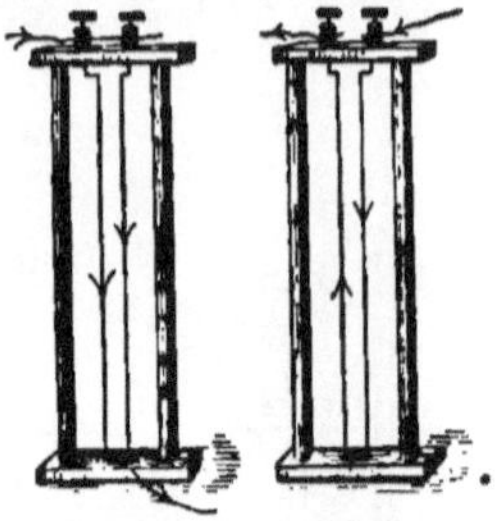

Experiment 2. Make the connections (Fig. 275) so that the current will go down one wire and up the other. They repel each other.

In the experiment with the floating cell and current-bearing wire placed over and parallel to the solenoid (Fig. 267), a careful examination will disclose the fact not only that the planes in which the current flows in the coil tend to become parallel to the current above, but that the current in the upper half of the coil, where the influence due to proximity is greatest, tends to place itself so as to flow in the same direction as that of the current above.

FIG. 274. FIG. 275.

354. Ampere's Laws. LAW 1. *Parallel currents in the same direction attract one another; parallel currents in opposite directions repel one another.*

LAW 2. *Currents that are not parallel tend to become parallel and flow in the same direction.*

355. Ampere's Theory of Magnetism. This celebrated theory, briefly stated, is that *magnets and solenoid systems are fundamentally the same;* that magnetism is simply electricity in rotation, and that a magnetic field is a sort of whirlpool of electricity. Not, of course, that a steel magnet contains an electric current circulating round and round it, as does an electro-magnet, but that every molecule of iron, steel, or other magnetizable substance, is the seat of a separate current circulating round it continuously and without resistance, and thus that every molecule is a magnet.

According to this theory, in an unmagnetized bar these currents lie in all possible planes, and, having no unity of direction, they neutralize one another, and so their effect as a system is zero. But if a current of electricity or a magnet be brought near, the effect of the induction is to turn the currents into parallel planes, and in the same direction, in conformity to Ampere's Second Law. If the retentivity be strong enough, this parallelism will be maintained after the removal of the inducing cause, and a permanent magnet is the result.

FIG. 276.

Intensity of magnetization depends on the degree of parallelism, and the latter depends on the strength of the influencing magnet. When these currents have become quite parallel, the body has received all the magnetism that it is capable of receiving, and is said to be *saturated*.

The hypothetical currents that circulate round a magnetic molecule we shall call *amperian currents*, to distinguish them from the known current that traverses the solenoid. In strict accordance with this theory, the poles of the electro-magnet are determined by the direction of the current in the helix. The inductive influence of the electric current causes the amperian currents to take the same direction with itself, as represented in Fig. 276.

SECTION XIII.

ELECTRO-MAGNETIC INDUCTION.

356. Description of Apparatus. A (Fig. 277) is a "short coil" of coarse wire (*i.e.* the wire which it contains is comparatively short), and has, of course, little resistance. B is a "long coil" of fine wire having many turns. Coil A is in circuit with a voltaic cell. This circuit we call the *primary circuit*, the current in this circuit the *primary* or *inducing current*, and the coil the *primary coil*. Another circuit, having in it no cell or other means of generating a current,

ELECTRO-MAGNETIC INDUCTION.

contains coil B and a galvanoscope with an astatic needle.[1] This circuit is called the *secondary circuit,* the coil the *secondary coil,* and the currents which circulate through this circuit are called *secondary* or *induced currents.*

Experiment 1. Lower the primary coil quickly into the secondary coil, watching at the same time the needle of the galvanoscope to see whether it moves, and, if so, in what direction. Simultaneously with this movement there is a movement of the needle, showing that a current must have passed through the secondary circuit. Let the primary coil rest within the secondary until the needle comes to rest.

FIG. 277.

After a few vibrations the needle settles at zero, showing that the secondary current was a temporary one. Now, watching the needle, quickly pull the primary coil out; another deflection in the opposite direction occurs, showing that a current in the opposite direction is caused by withdrawing the coil.

It is evident that in this case the current does not by its mere presence cause an induced current, but that a *change* in the relative positions of the two circuits, one of which bears a current, is necessary.

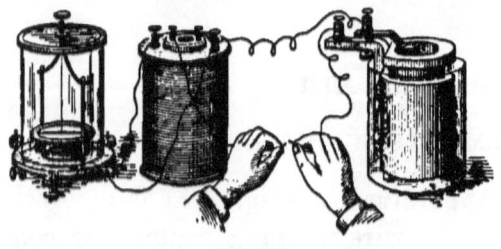

FIG. 278.

Instead of a current-bearing coil a bar magnet may be introduced into the secondary coil, and afterwards withdrawn from it. The needle is deflected at each act, as before.

[1] This needle consists of two needles of about the same intensity with their poles reversed, fixed parallel with each other. Though the needles nearly neutralize each other, and are therefore little affected by the field of the earth's magnetism, they are especially sensitive to the influence of the electric current.

Experiment 2. Place the primary coil within the secondary. Open the primary wire at some point and then close the circuit (Fig. 278) by bringing into contact the extremities of the wires. A deflection is produced. As soon as the needle becomes quiet, break the circuit by separating the wires; a deflection in the opposite direction occurs.

The same phenomena occur when the primary current is by any means suddenly strengthened or weakened.

An examination of the direction of these currents enables us to state the facts as follows: Starting a current in a primary, increasing the strength of the primary current, or moving the primary nearer (Fig. 279) produces in the secondary a transitory current in the *opposite* direction. Stopping the primary, diminishing the strength of the primary, or moving the primary away (Fig. 279), causes in the secondary a transitory current in the *same* direction.

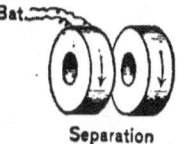
Fig. 279.

It is evident, therefore, that the conditions under which a current in the primary coil can cause a current in a neighboring secondary depend upon some *change* either in the strength of the primary current or in the relative positions of the primary and secondary circuits.

The act by which the primary, or a magnet, causes a current in a neighboring secondary is called *magneto-electric induction*.

357. Faraday's Law of Induction. *If any conducting circuit be placed in the magnetic field, then, if a change of relative position or change of strength of the primary current cause a change in the number of lines of force passing through the secondary, an electro-motive force is set up in the secondary proportional to the rate at which the number of lines of force included by the secondary is varying.*

MICHAEL FARADAY.

INDUCTION. 331

Consider the case of induction by a magnet. Let S (Fig. 280) be a secondary circuit and N a magnet projecting a certain number of lines of force through the circuit. If S be moved nearer to the magnet, say to S', a much greater number of lines of force of the magnet pass through the circuit than when in its former position, owing to the divergence of the lines as they recede from the pole.

358. Earth Induction. Call to mind that the earth itself is a great magnet, and that its lines of force pass through our atmosphere from pole to pole, and it will be easy to conceive that the mere motion of a coil of wire about an axis properly placed[1] is all that is necessary to produce a current. Such a coil with a galvanometer, G,

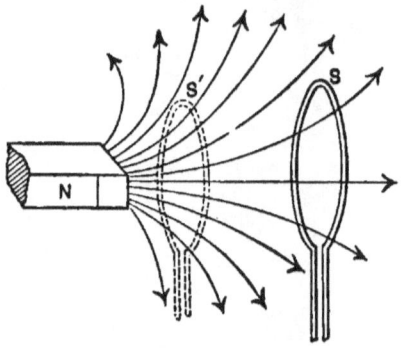

FIG. 280.

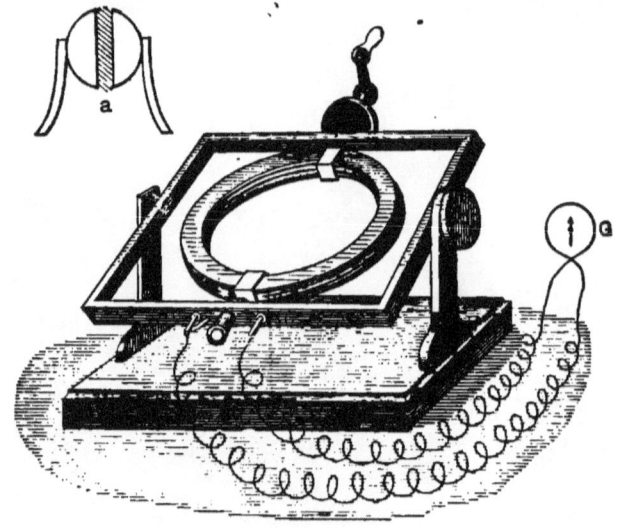

FIG. 281.

in circuit is represented in Fig. 281. The rotation of the coil across the magnetic flux encircling the earth causes the coil to be alternately filled with and emptied of this flux, and thereby an E. M. F. is generated in the coil, and this causes currents to flow through the galvanometer.

[1] The coil should be placed at right angles to the direction of the dip at the locality.

359. Self-induction. At the instant that a current enters a circuit, and while the magnetic flux is enveloping a conductor, an E.M.F. opposite to that of the advancing (or inducing) current is developed, thus opposing the current which produces it. But when the magnetic flux decreases (*e.g.* when the circuit is broken) an E.M.F. having a direction the same as that of the retiring current is established. This action is called *self-induction*. Thus it appears that when the current starts, self-induction prevents it from rising instantly to its full strength; when the current stops, self-induction tends to prolong the flow. In both cases the effect is only momentary.

The momentary current at breaking the circuit is called the *extra current*. This current has a high E.M.F., and is the cause of the spark seen whenever a strong current is interrupted. If the circuit contain an electro-magnet the spark is much intensified. Between every pair of turns of any coil there is a mutual inductance which is a part of the self-inductance of the coil. It is this principle that is utilized in the "spark-coils" employed in voltaic circuits for lighting gas.

360. Induction Coils. If a core of iron, or, still better, a bundle of wires (A A, Fig. 282), be inserted in the primary coil, it is evident that it will be magnetized and demagnetized every time the primary is made and broken. The starting and cessation of amperian currents in the core in conjunction with the commencement and ending of the primary current greatly intensifies the secondary currents. To save the trouble of making and breaking by hand, the core is also utilized in the construction of an automatic make-and-break piece. A soft iron hammer, b, is connected with the steel spring c, which is in turn connected with one of the terminals of the primary wire. The hammer presses against the point of a screw, d, and thus, through the screw, closes the circuit. But when a current passes through the primary wire, the core A A becomes magnetized, draws the hammer away from the screw, and breaks the circuit. The circuit broken, the core loses its magnetism, and the hammer springs back

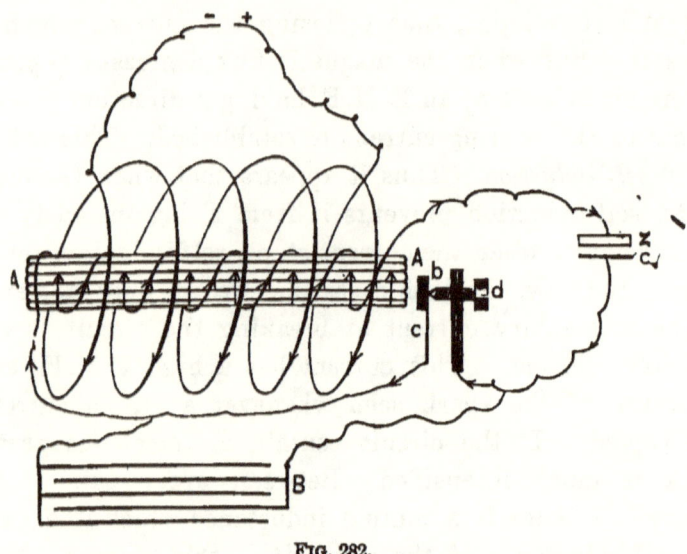

Fig. 282.

and closes the circuit again. Thus the spring and hammer vibrate, and open and close the primary circuit with great rapidity. Such an instrument is called an *induction coil.*

361. Ruhmkorff's Coil. This instrument has the important addition to the parts already explained of a *condenser*, B B (Fig. 282). This consists of two sets of layers of tin foil separated by paraffined paper; the layers are connected alternately with one and the other electrode of the battery, as the figure shows, so that they serve as a sort of expansion of the primary wire.

The effect of the condenser seems to be to prevent sparking at the make-and-break piece, and thereby to render the interruption of the primary current more abrupt, and hence the E. M. F. of the secondary is much increased.

Secondary currents developed in high resistance coils are, as we ought to expect, distinguished from primary, or voltaic currents, by their vastly greater E. M. F., or power to overcome resistances. A coil constructed for Mr. Spottiswoode, of London, has 280 miles of wire in its secondary coil, and 0.7 mile of wire in the primary. With five voltaic cells this coil gives a secondary spark forty-two inches long, and can perforate glass three inches thick. Many brilliant experiments may be performed with these coils.

Experiment 3. Connect a battery of two Bunsen cells, in multiple arc, with a Ruhmkorff coil (Fig. 283). Bring the electrodes of the secondary coil within from one fourth of an inch to one inch of each other, according to the capacity of the instrument. A series of sparks in rapid succession pass from pole to pole.

Fig. 283.

Experiment 4. Introduce a *Geissler tube*, A, into the secondary circuit. These tubes contain highly rarefied gases of different kinds. Platinum wires are sealed into the glass at each end to conduct the electric current through the glass. The sparks become diffused in these tubes so as to illuminate the entire tubes with an almost continuous glow. Observe that the electrodes are separated from each other much more widely than would be admissible in air of ordinary density, showing that rarefied gases offer less resistance than dense gases. Gases have been so highly rarefied, however, that an electric current would not pass.

In vacuum tube discharges the negative electrode can be distinguished from the positive electrode by the dark space which surrounds it, and by a patch of deep blue light which intervenes.

362. The Induction Coil Reversible. An induction coil is in a certain sense a reversible machine. If a current of considerable strength circulate under small E. M. F. in the primary, then variations in its strength give rise to very weak currents of exceedingly high E. M. F. in the secondary.

Conversely, if we cause to circulate in the *secondary* weak currents under very high E. M. F., by their fluctuations there will be generated in the primary strong currents of small E. M. F. We do not in either case create electric energy. Electric power is the product of two factors, current and electro-motive force. The induction coil enables us to increase one of these factors at the expense of the other, and to transform electric energy in form much as a mechanical power (*e.g.* a lever) enables us to convert a quantity of work which consists of small stress exerted through a great distance into a large stress exerted through a small distance.

363. The Transformer. The *transformer*—sometimes called a converter—is merely an induction coil used to change the relation of the number of volts to the number of amperes of any current. In a *perfect* transformer the number of watts in the primary equals the number of watts in the secondary.

The Ruhmkorff coil as ordinarily used may be regarded as a "step up" transformer from low potential to high potential. But if the coil of long thin wire be used as the primary, it becomes a "step down" transformer from high potential to low potential.

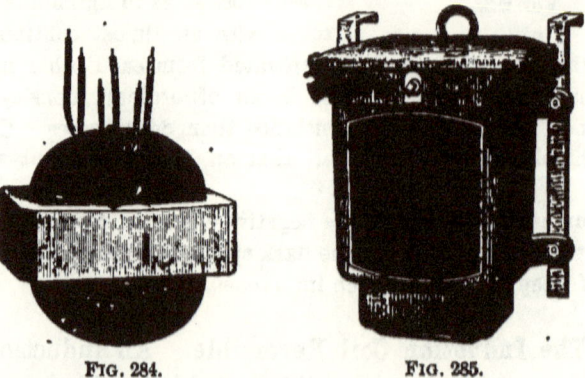

FIG. 284.   FIG. 285.

Fig. 284 represents the coils of a transformer used in the incandescent lamp service, and Fig. 285 represents the same enclosed in a case.

The transformer is applied in the welding of metals, *i.e.* to fuse the ends of metals that are to be joined together, where many hundred or even thousand amperes of current, and only a fraction of a volt, would be required for an instant.

A still wider application of transformers is in the transmission of electric power. The power of a current equals $C^2 R$. That is, when the current strength is doubled there will be four times as much energy transformed per second. We see, then, that to transfer electric energy to a great distance it may be desirable to have a high E. M. F. with a small current passing through the mains, and then to reduce the E. M. F. and increase the current by a transformer at the place where the energy is to be used. By this means the expense involved in the copper conductors is much reduced.

For electric lighting in private houses transformers are used to bring down the high potential of the mains to the safe limit of about 100 volts. These transformers are usually supported on the street poles.

SECTION XIV.

DYNAMO-ELECTRIC MACHINES.

364. Principles of the Dynamo. The *dynamo* is a device for transforming mechanical energy into electric energy. In the most improved types of dynamos this is done with a loss of less than five per cent of the energy.

The action of the dynamo is based on the principle of current induction. It embraces a system of coils which revolve in a magnetic field in such a way that the number of lines of force passing through them varies continuously. As we have previously learned, this creates a difference of potential in the system, so that if the points of different potential be connected by a wire, a current will be established in the circuit.

Experiment. Connect a flat coil of about two inches in diameter having several turns of wire, with a delicate galvanometer, and rotate the coil at one of the poles of a strong magnet on an axis at right angles to the axis of the magnet and the lines of force, as illustrated in Fig. 286.

The horizontal arrow a indicates the direction of the magnetic lines of force, the horizontal arrow b the direction of motion of the end of the coil of wire, and the vertical arrow c

PRINCIPLES OF THE DYNAMO. 337

the direction of the current induced in the coil of wire from the movement of the coil across the field of magnetic force in such a manner that the number of lines of force threading through the coil changes.

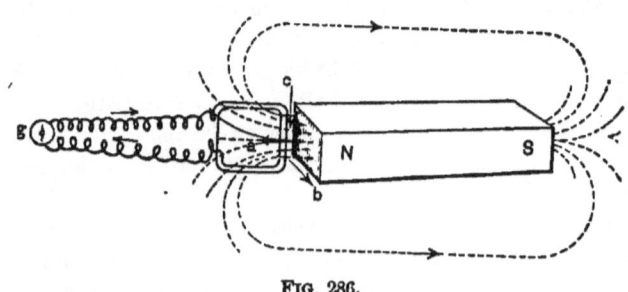

FIG. 286.

If the coil be moved rapidly in front of the magnet, the current is stronger, and hence the E. M. F. must be greater than if it be moved slowly. Also, if the number of turns of wire be increased, the E. M. F. will be increased, as will be shown by the increased strength of current.

We may continue our experiment still further by inserting a bar or disk of soft iron into the coil and again moving the end of the coil through the field of force in front of the north pole of the magnet. A very decided increase in the strength of current is observed. If, further, another bar magnet be placed so that its south end faces the other end of the coil, and the coil be fixed at its center while its two ends are made to rotate past the two poles, more lines of force are cut and greater E. M. F. is developed, as is seen from the increased strength of current. A powerful electro-magnet is preferable to a permanent magnet as an inducer, since its strength or *magnetic density* can be made much greater.

We have now found that E. M. F. and strength of current depend upon (1) *the rapidity of motion of the wire through the field;* (2) *the number of turns of the wire;* and (3) *the number of lines of force cut, or the strength of the field.*

365. Rule for Determining the Direction of the Induced Current. *Place the right hand* (Fig. 287) *so that the direction of the forefinger coincides with the direction of the lines of force* (as indicated by a test-needle), *and the thumb points in the direction of motion of the part of the conductor under consideration; the middle finger will indicate the direction of the induced current.*

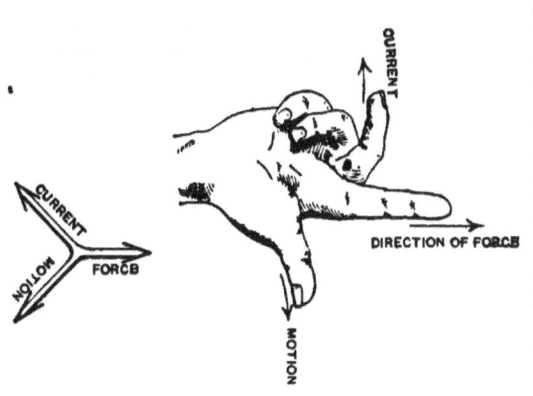

Fig. 287.

366. The Dynamo. We are now prepared to study the action of the dynamo. The inducing magnet, which is commonly an electro-magnet, is called the *field magnet*, and the coil or series of coils of wire, which is generally made to move in front of the poles of the field magnet, is called the *armature*. The armature is that part of the electric circuit in which the E. M. F. is generated. Like the battery, it may be considered as the source of the current. The number of

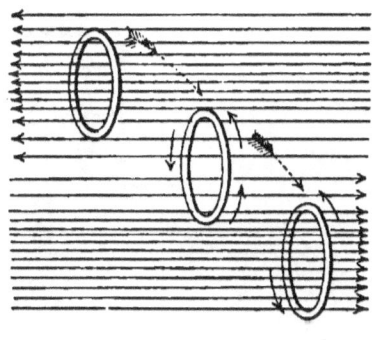

Fig. 288.

lines of force passing through a circuit may in general be changed in two ways: either (1) by moving the circuit through a field in which the density of the lines of force varies, as represented in Fig. 288; or (2) by rotating the plane of the circuit so as to change the angle which it makes with the line of force, thus increasing or decreasing the num-

THE DYNAMO. 339

ber which the circuit encloses (Fig. 289). A simple form of dynamo is illustrated in Fig. 290. A large core of soft iron of the U-form, surrounded with a coil of insulated wire, and terminating in the *pole pieces* N and S, forms the *field magnet*.

The armature consists of a single rectangular loop of wire fixed to a horizontal axis and terminating in two rings of metal, a and b, which are fixed to the axle, but insulated from it.

When a current passes through the field coils, and the core becomes magnetized, lines of

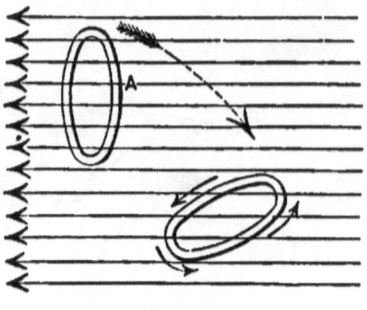

FIG. 289.

force cross and fill the space between the pole pieces of the field magnet. As these lines are cut by the horizontal parts of the rotating wire, an E. M. F. is generated in these parts, and a current flows in the direction indicated by the arrows.

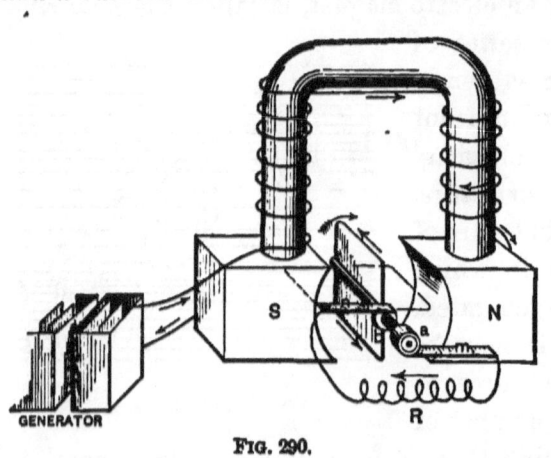

FIG. 290.

It is evident that the armature loop is successively filled and emptied with the flux — filled when its plane is vertical, empty when its plane is horizontal.

A metallic or carbon brush, m, touches and carries off the current from the lower horizontal segment of the rectangular coil. This current flows through the external resistance R, and completes the circuit through the brush n to the ring b, and the upper half of the loop. The current will continue to flow in this direction while the loop moves through one half of a revolution. Since the lines of force are cut in the opposite direction in the next half revolution, the current will be reversed in the armature wire and also through the external circuit. Thus with each half revolution of the armature a reversal of the current takes place. This, then, would be called an *alternating current dynamo*.

367. The Commutator. The alternating current is not adapted to all uses, and for many purposes it is necessary to have the current continuously flowing in the same direction. To accomplish this a *commutator* is attached to the axis of the armature.

In Fig. 291 the two brass rings (a and b, Fig. 290) are replaced by a single brass tube divided into two parts by cutting it lengthwise. These two segments are attached to but insulated from the axis, and are connected with the separate ends of the armature wire. When the plane of the armature coil is perpendicular to the line of force passing from N to S, as in Fig. 291, no lines of force are being cut, and hence no E. M. F. is developed and no current flows through the loop. But the instant it moves out of the vertical in the direction of the arrow, lines of force will be cut, and as the lower segment of the loop is moving upward past the pole S, and the other segment is moving downward in front of the pole N, a positive current flows from the loop through the segment a, the brush m, the resistance R, the brush n, and the segment of the commutator b. During the next half of a revolution the lines of force will be cut from an opposite direction by each of the horizontal segments of the armature loop, and hence the current will be reversed. But the segment b of the commutator will now be in contact with the brush m; and although the current is reversed in the armature it will flow off at the brush m as before. That is, one of the brushes is kept always positive and the other negative, and the current flows from the positive brush through the external circuit to the negative

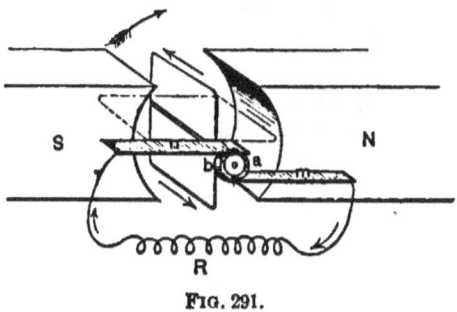

FIG. 291.

brush. Inasmuch as no E. M. F. is developed when the plane of the loop is perpendicular to the lines of force, it is at this point that the brushes pass from one segment of the commutator to the other.

Thus, although the current in the armature is reversed with each half revolution, by providing that the connections of the armature coil shall be shifted at the moment when the current in the coils is reversed, a reversal of the current in the external circuit is prevented. This arrangement constitutes a *direct-current* dynamo. Instead of one coil we may have two or any number of coils, each separate from the others, and terminating in strips or segments which are on opposite sides of the commutator. Generally the coils are connected in series, thus making any segment a terminal of one coil and the beginning of the next.

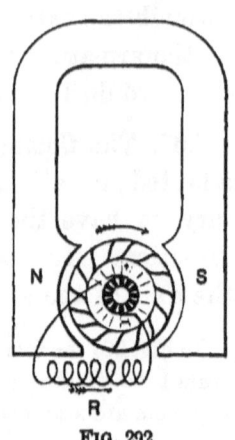

FIG. 292.

368. Classes of Dynamos.[1] Dynamos may be divided into different classes according to the method by which their field magnets are excited. Figure 292 illustrates a *magneto-electric* machine, where the field magnet is a permanent steel magnet. This form of machine is seldom used, since a permanent steel magnet cannot be made as powerful as an electro-magnet having a soft iron core of equal mass.

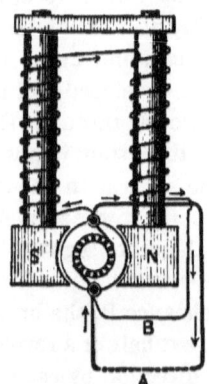

FIG. 293.

Figure 290 illustrates a *separately excited dynamo*, where the field-magnet coils receive their currents from a separate generator, *e.g.* a battery, and not from the armature coils.

---

[1] For the characteristics of the various classes of dynamos, as well as for a most lucid and comprehensive treatment of dynamos generally, see "Dynamo-Electric Machinery," by S. P. Thompson.

342   ETHER DYNAMICS.

Since an alternating current dynamo does not produce a constant magnetic field, alternating dynamos, in general, are separately excited. Small direct-current dynamos are commonly used to energize the field magnets of alternating dynamos. Such, for example, are the Westinghouse and the Thompson-Houston incandescent-lighting dynamos.

In a *series dynamo* the coils of the field magnet are joined in series with the armature so that the entire current passes through these coils.

Figure 293 illustrates a *shunt machine*, where the field coil serves as a shunt to the external circuit. A is the main wire, and B is the shunt wire. In the shunt machine only a part of the current generated in the armature passes through

FIG. 294.

the field coils. Sometimes, for purposes of regulation, the field magnet is encircled by both series and shunt coils. Such dynamos are said to be *compound wound*. A dynamo is said to be "self-exciting" when the whole or any part (Fig. 293) of the current which is produced is used to magnetize the field magnets.

The cores of the field magnets of a self-exciting dynamo, after being once excited from any source, *e.g.* another dynamo, always retain a little residual magnetism, so that when the armature begins to rotate, a slight current is at once induced in it. This current flows around the magnets and intensifies their power. This strengthens the field, and the stronger field reacts to increase the current, so that the current soon rises to its normal strength.

All figures hitherto have been diagrammatic representations of dynamos. Fig. 294 represents one of the most common forms of the Edison dynamo. It is a shunt-wound dynamo.

SECTION XV.

ELECTRIC MOTOR.

369. Reversibility of the Dynamo. If a current from an external source, *e.g.* a battery or another dynamo, be passed through the armature and field magnet of a direct-current dynamo, it will excite the armature and make of it an electro-magnet, and will also excite the fields. The current will enter at the terminals and will pass through the commutator into the armature. The relation of parts is such that in doing this it will develop N- and S-poles in parts of the periphery of the armature distant from the N- and S-poles of the fields. Hence, there will be set up between the armature and the poles of the field magnet a stress tending to move the former a little in the opposite direction to that in which it is compelled to move when generating a current. But as soon as it has turned a short distance the action of the commutator shifts the current, and new poles are established in the armature back of the first and in the same relative positions which they at first occupied. The armature continues to rotate as the new poles are attracted and repelled, and the action goes on so long as a current is supplied. Obviously, if there were no commutator the poles of the armature would be fixed, and it never could rotate through a greater angle than 180°.

It is evident, then, that if two dynamos be connected by wires in the same circuit and the armature of one be rotated, the armature of the other will rotate in a reverse direction as soon as the current transmitted from the first attains a certain intensity.

Let A and B (Fig. 295) represent in diagram two dynamos constructed exactly alike. Mechanical power is supplied to the dynamo A by the

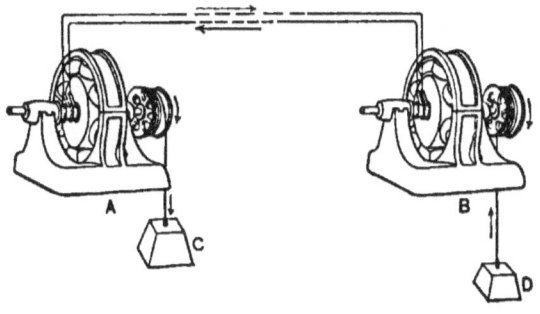

FIG. 295.

falling weight C. In this dynamo mechanical energy is transformed into the energy of an electric current, which in passing through B (now acting as a motor) becomes again transformed into mechanical energy and the weight D is raised thereby. The energy stored in D, after it is raised, plus some additional energy to compensate for that lost in the transmission from A to B and in the several transformations, may be used again by a reversal of transformations to raise the weight C.

The dynamo, then, is a reversible machine, in which mechanical energy can be changed directly into electrical energy, or electrical energy into mechanical energy. When the dynamo is used for the latter transformation, it is commonly known as an *electric motor*. In other words, a modern motor is a dynamo reversed. The discovery of the reversibility of the dynamo is considered to be one of high importance.

370. The Electric Railway. The system of electric car propulsion consists in the generation of an electric current at some power station by means of dynamos, its transmission over conductors to the electric motors on the cars, and its transformation into mechanical energy, which gives motion to the car.

The current, as shown by the arrows (Fig. 296), passes from the dynamo D through a switch board, S, with which are connected a current-indicator, voltmeter, fuses, etc.; thence over the trolley wire or

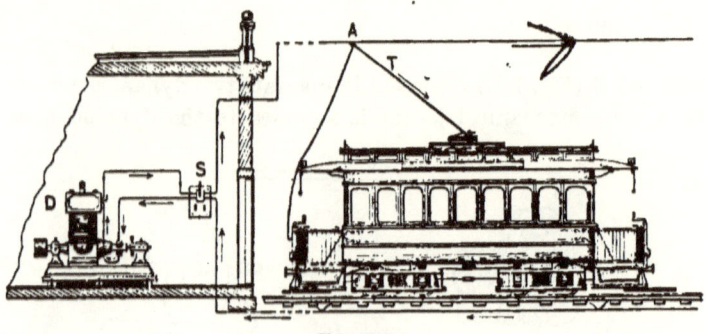

FIG. 296.

overhead conductor A, a portion passing down through the trolley arm T (the remainder going on to supply other cars), along wires concealed in the car to the motor, through the motor, and thence through the wheels to the rails, and back to the switch board and dynamo.

In some cases the rails connected by copper tie wires soldered to each rail at the joints serve for the return circuit; in other cases a "supplementary wire" laid between the rails is used. In the latter case each joint on both rails is connected with the supplementary wire. Much study is now being devoted to the problem of substituting a feed conductor laid in a conduit buried beneath the road surface for the overhead trolley wire.

At each end of the car is a "controller stand" provided with a handle, by means of which the resistance through a rheostat may be varied, and thereby the strength of the current through the motor and the speed of the car is regulated. There is also a reversing switch by means of which the direction of the rotation of the armature may be changed and thereby the motion of the car quickly checked in the case of emergency, or the motion may be reversed.

Fig. 297.

Each car usually has two motors connected in multiple, and each motor is geared to its own car axle. Fig. 298 represents one half of a car truck, with its motor, gearing, etc.

SECTION XVI.

SECONDARY OR STORAGE BATTERIES.

371. Reversibility of Electrolysis. If water be decomposed for a time between neutral electrodes such as platinum plates, and then the battery or other generator be withdrawn from the circuit and replaced by a sensitive galvanometer, a deflection of the needle shows that a transitory current flows in the opposite direction to the primary or electrolyzing current. It is evident that the electrolyzing current polarizes the electrodes in the electrolyte, and that energy is thus stored in the cell. Polarization is of the nature of a counter E. M. F. It is precisely this polarization which we have to contend with in nearly all voltaic cells, and which we seek to neutralize by means of depolarizing substances.

Devices for thus storing up energy by electrolysis are called *storage* or *secondary batteries,* and sometimes *accumulators*. Note that the process is an *electrical storage of energy,* not a storage of electricity. The energy of the charging current is transformed into the potential energy of chemical separation in the storage cell. When the circuit of the storage cell is closed this energy is reconverted into the energy of an electric current in precisely the same way as with an ordinary voltaic cell.

A common form of storage cell consists of a series of lead plates cast in the form of a framework of bars at right angles to one another, as shown in Fig. 298. The spaces between the bars of the positive plates are filled with a paste of red lead and sulphuric acid; the spaces in the negative plates are filled with a paste of litharge and sulphuric acid. The liquid surrounding the plates is dilute sulphuric acid. The positive plates are all connected in multiple at one end of the cell, and the negative plates are connected at the other end. The cell is charged by con-

FIG. 298.

necting a dynamo with its terminals, when the positive plates become peroxidized by electrolysis and the negative plates deoxidized.

The storage battery offers a means of accumulating energy at one time or place, and using it at some other time or place. For example, energy of a dynamo current may be stored during the daytime when the current is not needed for illuminating purposes; and this energy, reconverted into electric energy, may feed incandescent lamps at night at any convenient place; or the charged cells may be transported to lecture halls, workshops, electric cars, etc., where powerful currents may be needed.

A storage cell has an E. M. F. of 2.2 volts or more. Its resistance depends on the size and number of plates, but is not usually greater than .005 ohm.

348 ETHER DYNAMICS.

SECTION XVII.

THERMO-ELECTRIC CURRENTS.

372. Heat Energy Transformed Directly into Electric Energy.

Experiment. Let G (Fig. 299) be a low-resistance, astatic-needle galvanometer. Form two junctions between its copper-wire terminals and an iron wire by tightly twisting the wires together near their extremities. Let A and B be the two junctions. Immerse both junctions in separate beakers of water. (1) Raise the water in one of the beakers to the boiling point; a current passes through the galvanometer G, causing a deflection (say) to the right. (2) Reverse the position of the two beakers of water; the current now causes a reversed deflection. (3) Bring the cold water to the boiling point; the deflection diminishes steadily to zero as the difference in temperature in the two waters diminishes.

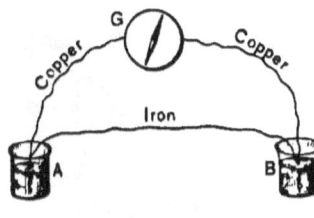
FIG. 299.

Thus, the thermo-electric current depends on the difference of temperature of the junctions, vanishing when that difference vanishes. Suppose the junctions to be at 10° C. and 100° C., the E. M. F. will be about .0015 volt. It would require about 1000 pairs of such junctions to give an E. M. F. comparable to that of an ordinary voltaic cell.

373. Thermo-electric Batteries and Thermopiles. The E. M. F. may be built up by combining a large number of pairs with one another in series. This is done on the same principle and in the same manner that voltaic pairs are united, viz. by joining the + metal of one part to the − metal of another. Fig. 300 represents such an arrangement. The light bars are bismuth and the dark ones antimony. If the source of heat be strong and near, one face may be

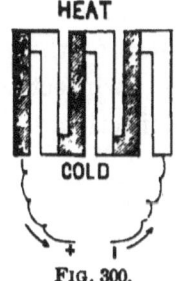
FIG. 300.

heated much hotter than the other, and a current equal to that from an ordinary galvanic cell is often obtained. Such contrivances for generating electric currents are called *thermo-electric batteries*. They are seldom used, inasmuch as the best of them transform less than one per cent of the heat energy given out by the source of heat. Furthermore, the E. M. F. of thermic piles is generally so small that any considerable external resistance makes the current extremely weak. If the source of heat be feeble or distant, the feeble current may serve to measure the difference of temperature between the ends of the bars turned toward the heat and the other ends, which are at the temperature of the air. The apparatus, when used for this purpose, is called a *thermopile* or *thermo-multiplier*. A combination (Fig. 301) of as many as thirty-six pairs of antimony and bismuth bars, connected with a very sensitive galvanometer, constitutes an exceedingly delicate *thermoscope* and *thermometer*. Changes of temperature that would not produce a perceptible variation in an ordinary thermometer can, by the use of a thermo-electric pile, be made to produce large deflections of the galvanometer needle. Radiations from the body of an insect several inches from the pile may cause a sensible deflection.

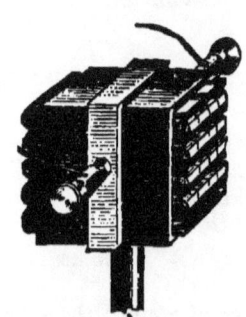

Fig. 301.

SECTION XVIII.

THE ELECTRIC LIGHT.

374. Electric Light; Voltaic Arc. If the terminals of wires from a powerful dynamo or galvanic battery be brought together, and then separated 1 or 2 mm., the current does not cease to flow, but volatilizes a portion of the terminals. The vapor formed becomes a conductor of high resistance, and, remaining at a very high temperature, produces intense light. The heat is so great that it fuses the most refractory substances. Metal terminals quickly melt and drop off like tallow, and thereby become so far separated that the electro-motive force is no longer sufficient for the increased resistance, and the light is extinguished. Hence, pencils of carbon (prepared from the coke deposited in the distillation of coal inside of gas retorts), being less fusible, are used for terminals.

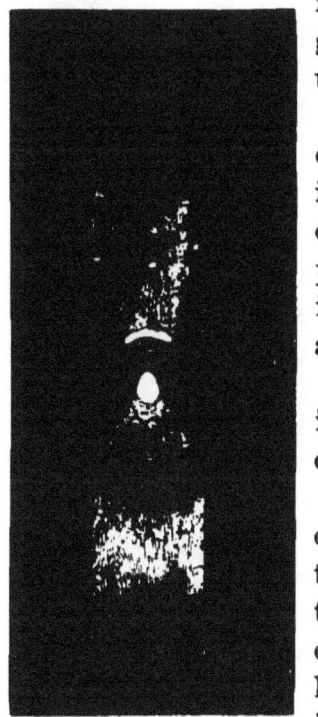

Fig. 302.

The light is too intense to admit of examination with the naked eye; but if an image of the terminals be thrown on a screen by means of a lens or a pin hole in a card, an arch-shaped light is seen extending from pole to pole, as shown in Fig. 302.

The heated air containing the glowing particles of carbon forms what is called the *electric arc*.

The larger portion of the light, however, emanates from the tips of the two carbon terminals, which are heated to an intense whiteness, although some emanates from the arc. The + pole is hotter than the − pole, as is shown by its glowing longer after the current is stopped. The carbon of the + pole becomes volatilized, and the light-giving particles are transported from the + pole to the − pole, forming a bridge of luminous vapor between the poles. What we see is not electricity, but *luminous matter*.

375. Electric Lamp. It is apparent that the + pole is subject to a wasting away; so also the − pole wastes away, but not so fast. At the point of the former a conical-shaped cavity is formed, while around the point of the latter warty protuberances appear. When, in consequence of the wearing away of the + pole, the distance between the two pencils becomes too great for the electric current to span, the light goes out. Numerous self-acting regulators for maintaining a uniform distance between the poles have been devised. Such an arrangement (Fig. 303) is called an *electric lamp*. The movements of the carbons are accomplished automatically by the action of the current itself.

376. Incandescent Electric Lamps. The incandescent (or "glow") light is produced by the heating of some refractory body to a state of incandescence by the passage of an electric current. Carbon filaments are now almost exclusively used in incandescent lamps. The filament of the Edison lamp is carbonized bamboo. It is essential that the oxygen of the air be removed from these bulbs, otherwise the carbons would be quickly burned out; hence, very high vacua are produced in the bulbs with a mercury pump.

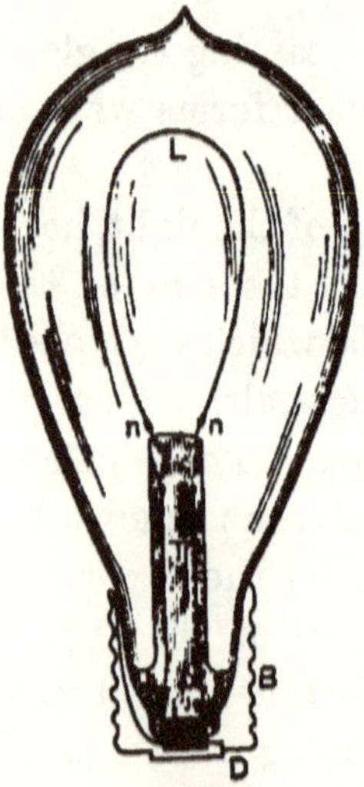

FIG. 303.

Fig. 304 represents an Edison lamp. The loop or filament of carbon L is joined at n n to two platinum wires, which pass through the closed end of the glass tube T. One of these wires is connected with the brass ring B, and the other with the brass button D, at the bottom of the lamp. When the lamp is screwed into its socket, connection is made with the line through pieces of brass in the socket, which are insulated from each other.

FIG. 304.

An Edison 16-candle-power lamp has a resistance (when hot) of about 140 ohms, the difference of potential at its terminals is about 110 volts, and it requires a current of 0.75 ampere. Each lamp consumes about one tenth of a horse-power, or about 4 watts per candle-power. One kilo-watt (1 k.w. = 1.34 h.p.) hour will supply sixteen 16-candle-power lamps for one hour, and give as much light as 100 cubic feet of gas.

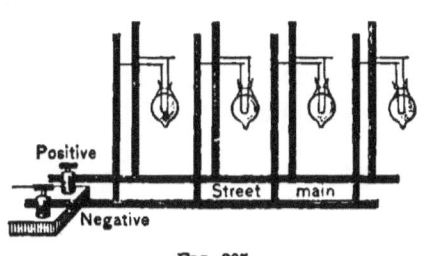

Fig. 305.

Incandescent lamps are usually introduced into the circuit in multiple arc (Fig. 305), the current being equally divided by properly regulating the resistance between all the lamps in the circuit.

SECTION XIX.

ELECTROTYPING AND ELECTROPLATING.

377. Electrotyping. This book is printed from electrotype plates. A molding-case of brass, in the shape of a shallow pan, is filled to the depth of about one quarter of an inch with melted wax. A few pages are set up in common type, and an impression or mold is made by pressing these into the wax. The type is then distributed, and again used to set up other pages. Powdered plumbago is applied by brushes to the surface of the wax mold to render it a conductor. The case is then suspended in a bath of copper sulphate dissolved in dilute sulphuric acid. The — electrode of a galvanic battery or dynamo machine is applied to it, and from the + electrode is suspended in the bath a copper plate opposite and near to the wax face. The salt of copper is decomposed by the electric current, and the copper is deposited on the surface of the mold. The sulphuric acid appears at the + electrode, and, combining with the copper of this electrode, forms new molecules of copper sulphate. When the copper film has acquired about the thickness of an ordinary visiting card, it is removed from the mold. This shell shows distinctly every line of the types or engraving. It is then backed, or filled in, with melted type-metal, to give firmness to the plate. The plate is next fastened on a block of wood, and thus built up type-high, and is now ready for the printer. (For full directions which will enable a pupil to electrotype in a small way, see the author's "Physical Technics.")

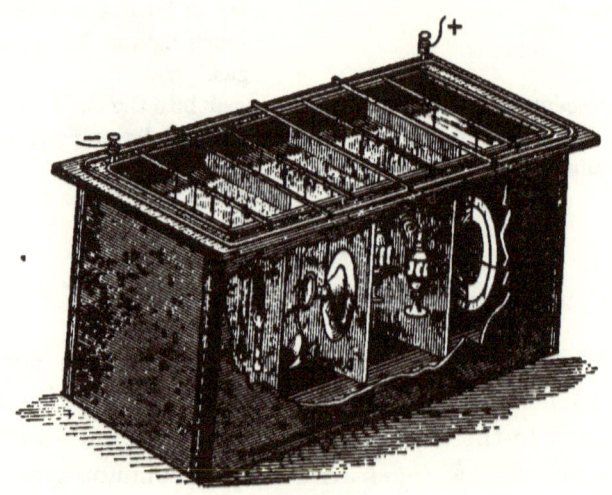

FIG. 306.

378. Electroplating. The distinction between electroplating and electrotyping is that with the former the metallic coat remains permanently on the object on which it is deposited, while with the latter it is intended to be removed. The processes are, in the main, the same. The articles to be plated are first thoroughly cleaned, and suspended on the — electrode of a battery, and then a plate of the same kind of metal that is to be deposited on the given articles is suspended from the + electrode (Fig. 306). The bath used is a solution of a salt of the metal to be deposited. The cyanides of gold and silver are generally used for gilding and silvering.

SECTION XX.

THE ELECTRIC TELEGRAPH.

379. Morse Telegraph. First, it should be understood that, instead of two lines of wires (one to convey the electric current far away from the battery, and another to return it to the battery), if the distant pole be connected with a large metallic plate buried in moist earth, or, still better, with a gas or water pipe that leads to the earth, and the other pole near the battery be connected in like manner with the earth, so that the earth forms about one half of the circuit, there will be needed *only one wire* to connect telegraphically two places that are distant from each other.

Let B (Fig. 307, Plate II) represent the message sender, or operator's key ; Y, the message receiver. It may be seen that the circuit is broken at B. Let the operator press his finger on the knob of the key. He closes the circuit, and the electric current instantly fills the wire from Boston to New York. It magnetizes a ; a draws down the lever b, and presses the point of a style on a strip of paper, c, that is drawn over a roller. The operator ceases to press upon the key, the circuit is broken, and instantly b is raised from the paper by a spiral spring, d. Let the operator press upon the key only for an instant, or long enough to count one ; a simple *dot* or indentation will be made in the paper. But if he press upon the key long enough to count three, the point of the style will remain in contact with the paper the same length of time ; and, as the paper is drawn along beneath the point, a short straight line is produced. This short line is called a *dash*. These dots and dashes constitute the *alphabet of telegraphy*.

380. The Sounder. If the strip of paper be removed, and the style be allowed to strike the metallic roller, a sharp click is heard. Again, when the lever is drawn up by the spiral spring it strikes a screw point above (not represented in the figure), and another click, differing slightly in sound from the first, is heard. A listener is able to distinguish dots from dashes by the length of the intervals of time that elapse between these two sounds. Operators generally read by ear, giving heed to the clicking sounds produced by the strokes of a little hammer. A receiver so used is called a *sounder*, a common form of which is represented in the lower central part of Plate II.

355

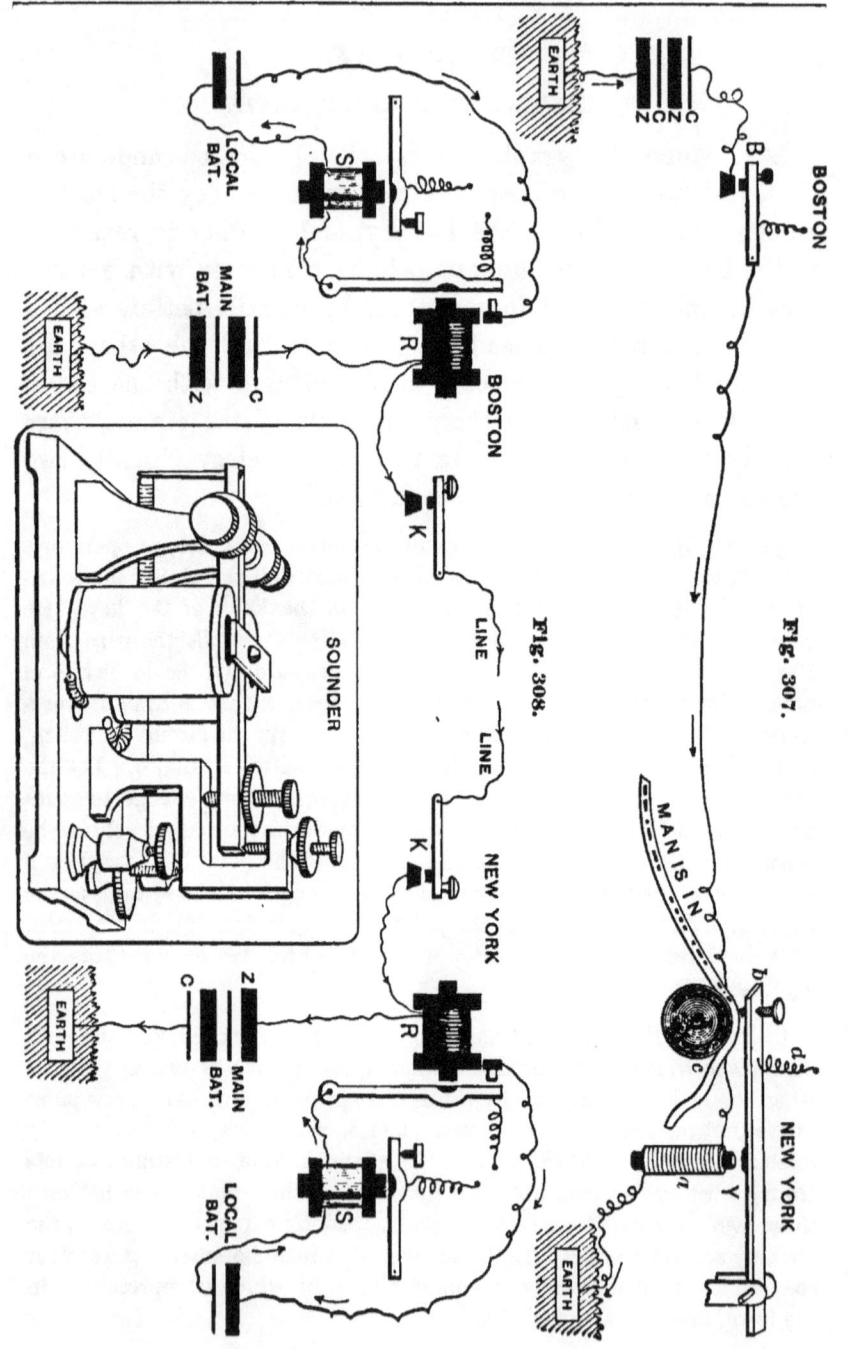

Fig. 307.

Fig. 308.

381. The Relay. The strength of the current is diminished, of course, as the line is extended and the number of instruments in the circuit is increased. Hence, a battery that would give a current sufficient to move the parts of a single sounder audibly on a short line would not move the same parts of many sounders on a long line with sufficient force to render the message audible. Resort is had to *relays*.

In Fig. 308, Plate II, R represents a relay and S a sounder. Suppose a weak current arrives at New York from Boston, and has sufficient strength to attract the armature of the relay at that station. This, as may be seen by examination of the diagram, will close another short circuit, called the *local circuit*, and send a current from a *local battery* located in the same office through the sounder at that station. The sounder, being operated by a battery in a circuit of only a few feet in length, delivers the message audibly.

SECTION XXI.

THE TELEPHONE.

382. Bell Telephone. Fig. 309 represents a sectional and a perspective view of this instrument. It consists of a steel magnet, A, encircled at one extremity by a spool of very fine insulated wire, B, the ends of which are connected with the binding screws D D. Immediately in front of the magnet is a thin circular iron disk, E E. The whole is enclosed in a wooden or rubber case, F. The conical-shaped cavity G serves the purpose of either a mouthpiece or an ear-trumpet. There is no difference between the transmitting and the receiving telephone; consequently, either instrument may be employed as a transmitter, while the other serves as a receiver. Two magneto telephones in a circuit are virtually in the relation of a dynamo and a motor. The transmitter being in itself a diminutive dynamo, no battery is required in the circuit. Connect in circuit two such telephones, and the apparatus is ready for use.

A person talking near the disk of the transmitter throws it into rapid vibration. The disk, being quite close to the magnet, is magnetized by induction; and as it vibrates its magnetic power is constantly changing, being strengthened as it approaches the magnet and enfeebled as it recedes. This fluctuating magnetic force will of course induce currents in alternate directions in the neighboring coil of wire. These currents traverse the whole length of the wire, and so pass through the coil of the distant instrument. When the direction of the arriving current is such as to increase the intensity of the magnetic field of the receiver, the magnet attracts the iron disk in front of it more strongly than before. When the current is in the opposite direction, the disk is less attracted, and flies back. Hence, the disk of the receiving telephone is forced to repeat whatever movement is imparted to the disk of the transmitting

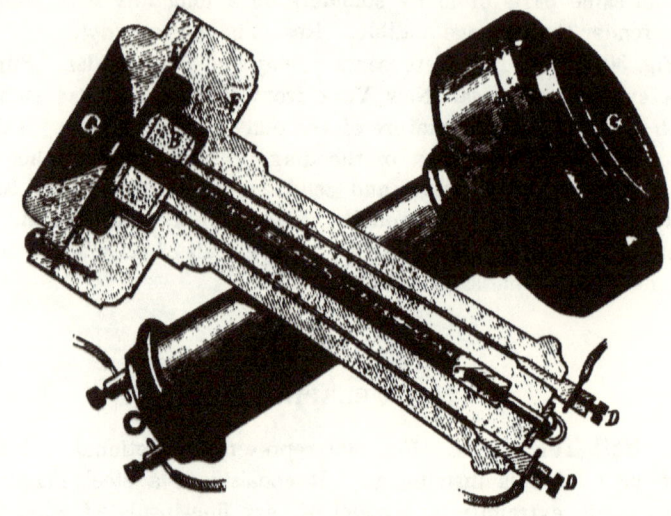

FIG. 309.

telephone. The vibrations of the former disk generate sound-waves in the same manner as the vibrations of a tuning fork or of the head of a drum.

The above is a description of the original and simplest form of the Bell telephone. It is apparent that the original energy (*i.e.* that of the

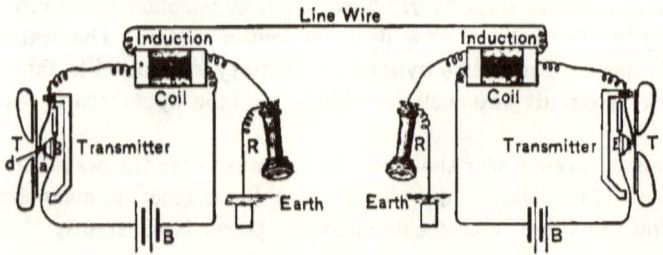

FIG. 310.

voice) applied at the transmitter must, during its successive transformations, and especially during its transmission in the form of electric energy through large resistances, become very much enfeebled, so that when it reappears as sound, the sound is quite feeble and frequently inaudible. The first grand improvement on the original consists in introducing a battery into the circuit, and so arranging that the voice, instead of being obliged to generate currents, shall be required only to render a current, already generated by a voltaic cell, fluctuating or *undulating*.

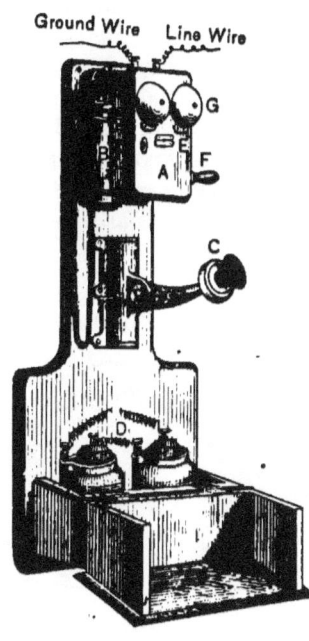

FIG. 311.

The fluctuations are caused by a varying resistance in the circuit. The pupil must have learned by experience ere this that the effect of a loose contact between any two parts of a circuit is to increase the resistance and thereby weaken the current; but the effect of a slight variation in pressure is especially noticeable when either or both of the parts are carbon. Fig. 310 illustrates a simple telephonic circuit in which are included two variable resistance transmitters, T T, and two batteries, B B. One of the electrodes, a platinum point, touches the center of the transmitter disk d; the other electrode, a carbon button, a, is pressed by a spring gently against the platinum point. Every vibration of the disk, however minute, causes a variation in the pressure between the two electrodes and a corresponding variation in the circuit resistance. As the resistance changes, so changes the current strength, and thus the current is rendered undulatory.

The next improvement of considerable importance consists in the adoption of an induction coil (Fig. 310), which, we have learned, may produce a current of much greater electromotive force than is possessed by the original battery current. Since the battery current traverses only a local circuit, as may be seen by reference to Fig. 310, a single Leclanché cell is generally sufficient to operate it. The currents induced by the fluctuating primary current traverse the line wire and generate sonorous vibrations in the disk of the receiver R in the same manner as in the original telephone.

FIG. 312.

Fig. 311 represents the entire telephonic apparatus required at any single station. The box A contains a small hand dynamo, such as is represented in Fig. 312. A person turning the crank F generates a current which rings two pairs of electric bells, one pair (G) at his own and another pair at a distant station, and thus attracts attention. He next takes the receiver B off the supporting hook and places it at his ear. When the weight is removed from the hook, the hook rises a little and throws the dynamo and bells out of the circuit, and at the same time introduces the receiver B, the transmitter C, and the battery D, so that the circuit stands as represented in Fig. 310. E is a "lightning arrester."

In Fig. 313, A and B are buttons of carbon; the former is attached to a thin iron disk, the latter to a steel spring, C, and both are connected in circuit with a battery and a telephone used as a receiver. The spring presses B against A, and any slight jar will cause a variation in the pressure and corresponding variations in the current strength. D is an induction coil.

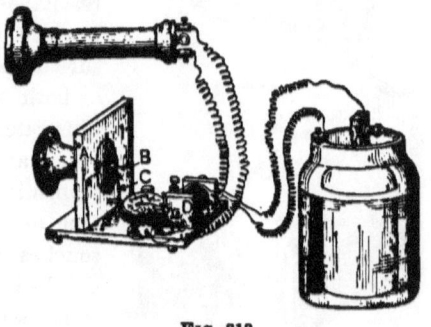

FIG. 313.

By means of this instrument, called the *microphone*, any *little sounds*, as its name indicates, such as the ticking of a watch or the footfall of an insect, may be reproduced at a considerable distance, and be as audible as though the original sounds were made close to the ear.

SECTION XXII.

ELECTRO-MAGNETIC THEORY OF LIGHT. ELECTRIC RADIATION.

383. Maxwell's Theory of Light. In 1865 Maxwell propounded the theory that light is the result of electro-magnetic disturbances of rapidly alternating character in the ether, such as would result from its being set in local strains and being released from them; that the vibrations which constitute light are electrical vibrations, and that light-waves are electro-magnetic waves.

ETHER DYNAMICS.

The Maxwellian theory of light may now be considered as completely verified by the wonderful experimental researches made by the late Dr. Hertz.[1] "So that we have now a real undulatory theory of light, no longer based on an analogy with sound. The whole domain of Optics is now annexed to Electricity, which has thus become an imperial science." — LODGE.

The importance of these experiments in a scientific sense can scarcely be overestimated, in so far as they teach us to refer electrostatic and electro-magnetic phenomena to the intervention of the same all-pervading medium. This medium forms the vehicle by which energy passes through space from one body to another, and the source to which we now must probably look for a knowledge of facts concerning the ultimate constitution of matter.[2] The term radiant energy is continually acquiring new scope in physics.[3]

---

[1] See Hertz's Researches on Electrical Oscillations, "Smithsonian Report," 1889.
Hertz showed experimentally that the electrical ether is wonderfully like, if not identical with, the ether which transmits light waves. By rapidly charging and discharging a conductor he causes the ether upon it to surge to and fro. This agitates the surrounding dielectric ether, and the disturbance travels in waves. The speed of these waves he determines to be the same as the speed of light-waves. He finds that these waves may interfere, may be reflected from metal mirrors, may be refracted by lenses and prisms, and are susceptible of diffraction effects. He has shown that many optical experiments can be electrically performed by substituting dielectrics for transparent bodies, and conductors for opaque bodies.

[2] "Matter is the rotating parts of an inert perfect fluid which fills all space." — LORD KELVIN.

[3] *A prophecy.* "The conclusions at which we have arrived, that light is an electrical disturbance, and that light-waves are excited by electric oscillations, must ultimately, and may shortly, have a practical import. Our present systems of making light artificially are wasteful and ineffective." (It is estimated that not more than 5 per cent of the energy put into an incandescent lamp is useful for illumination.) "We want a certain range of oscillation between 400 and 700 trillion vibrations per second; no other is useful to us, because no other has any effect on our retina; but we do not know how to produce vibrations at this rate. . . . We want a small range of rapid vibrations, and we know no better than to make the whole series leading up to them. It is as though, in order to sound some little shrill octave of pipes in an organ, we were obliged to depress every key and every pedal, and to blow a young hurricane."— LODGE. The production of light by very rapidly alternating currents seems to give promise of success in this direction. (See § 386.)

SECTION XXIII.

THE RÖNTGEN PHENOMENA.

384. Radiography. If a glass bulb be exhausted to approximately one one-millionth of an atmosphere, a condition of things is attained such as was investigated and described by Crookes in 1879. The mean free path of the residual gas molecules in this bulb is greater than the dimensions of a bulb of ordinary size. Gas in this state was called by Crookes *radiant matter*, since, when under the influence of an electric discharge, it presents properties quite different from those of ordinary gas.

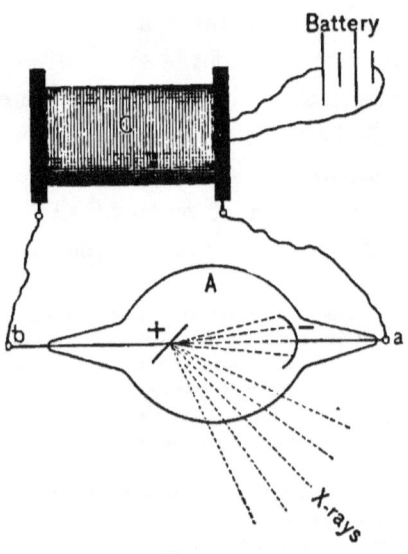

Fig. 314.

The molecules of gas are strongly attracted to the *cathode* or negative terminal of the bulb, where, becoming negatively electrified, they are repelled with great force. They travel in straight lines till they strike the walls of the bulb or other surface placed athwart their path. This molecular bombardment causes the surface which receives the blows to glow more or less brilliantly with a color depending on the substance which receives the impact. These molecular streams are known as the *cathode rays* of a Crookes tube.

The usual method of exciting these bulbs is to connect them by wires to the terminals of the secondary of an ordinary induction coil, or to the conductors of a Holtz machine.

Let A (Fig. 314) represent an exhausted glass bulb. Sealed into this bulb are the electrodes a and b; the former usually consists of a concave disk of aluminum, and the latter of a flat disk of platinum. Let a be connected to the negative and b to the positive pole of an induction coil. The discharge consists of faint streamers which proceed in straight lines from the negative electrode, as indicated in the figure. These streamers are the cathode rays.

Röntgen (1895) accidentally discovered that a non-luminous radiation emanates from an excited Crookes bulb, and after passing through certain kinds of matter quite opaque to light, is capable of producing photographic effects beyond. He with others discovered that numerous substances, such as metals, glass, and bone, are rather opaque to the new form of radiation; while many other substances, notably wood, paper, flesh, and rubber, are relatively transparent. The eye is quite insensitive to these rays. It soon became possible to produce shadow pictures of the bones of the body, as shown in Plate III. Röntgen called the new radiation the X-rays, since he was unable to state its nature. No satisfactory theory of Röntgen radiance has yet been given. His discovery, however, opens up a vast new field for experimental research, and is likely to lead to more definite views concerning the nature of the ether. In the bulb the cathode ray energy probably is transformed into X-ray energy at the points of impact on the platinum disk.

The principal effects of X-rays by which we are made aware of their existence are as follows: (1) They affect a silver salt on a photographic dry plate. (2) They make a fluorescent[1] substance shine. (3) They cause an electrical charge to leak away, however well insulated it may be. The last property affords one of the most delicate tests of their presence. The discharge occurs whether the body be positively or negatively electrified. The leakage takes place not only when the body is surrounded by air, but also when it is imbedded in a solid non-conductor. Hence, any substance when exposed to their action becomes a conductor.

[1] Fluorescence is a property which certain substances possess of becoming self-luminous after exposure to the action of light-rays.

Plate III.

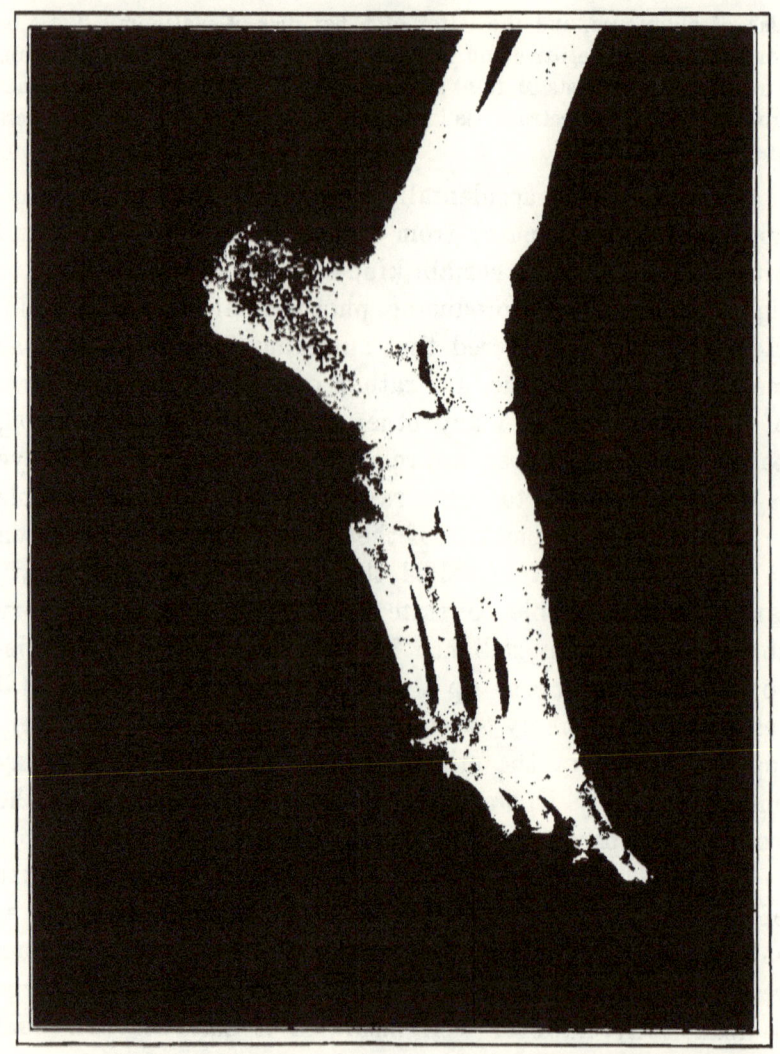

RADIOGRAPH OF THE NORMAL HUMAN FOOT.

[Taken by Dr. A. W. Goodspeed.]

385. Fluoroscope. A very important accessory to the apparatus described above is an instrument called a *fluoroscope*. It consists of a box (Fig. 315), dark in the interior, with an opening at one end, A, into which to look. At the opposite and larger end B is spread some fluorescent material, preferably barium platinum cyanide. Any body which is opaque to X-rays, if placed outside the fluorescent screen, and between it and the Crookes bulb, casts upon the screen a shadow which may be viewed by looking in at the opening; for example, one can see a shadow picture of the bones of his own hand upon the screen, which is elsewhere illuminated by the X-rays.

FIG. 315.

SECTION XXIV.

ALTERNATING CURRENTS.

386. Tesla's Investigations.[1] Of the various branches of electrical investigation now in progress perhaps the most interesting and most promising is that relating to alternating currents. In this connection a few words concerning the remarkable experiments of Tesla may not be out of place.

Tesla's work has been especially with alternating currents of very high frequency and enormously high potential. In his experiments he operates an induction coil either with a specially constructed alternating dynamo capable of giving many thousands of reversals of current per second, or by disruptively discharging a condenser through the primary. By the latter means may be produced a vibration in the secondary

---

[1] For details on this subject, see "Experiments with Alternating Currents of High Potential and High Frequency," by Tesla.

circuit of a frequency of many hundred thousands or even millions per second. The apparatus is usually enclosed in a wooden box and completely immersed in oil, that the insulation may be more nearly perfect. The discharge produced by this means is quite different from the series of sparks produced by the ordinary induction coil. Herein is opened to the experimenter a field as yet quite unexplored.

If two straight wires terminate the secondary coil and run parallel to each other for a short distance, the discharge appears in the form of powerful brushes and luminous streams issuing from all points of the wires. When an ordinary low-frequency discharge is passed through moderately rarefied air, the air assumes a purplish hue; but if by some means the intensity of the molecular or atomic disturbance be increased, the gas changes to a white color. A similar change occurs in air at ordinary pressure when agitated by electric impulses of the high frequency obtained by Tesla.

The chief interest in investigations along this line seems to lie in the *possibilities* they offer for the production of an efficient illuminating device. The present means of producing artificial light is woefully inefficient and wasteful. Who can say that it is not to be by means of alternating currents of high frequency and high potential that we shall soon vie with the firefly in economical illumination?

REVIEW EXERCISES.

1. How is a body charged by conduction? How by induction?
2. In a circuit of large resistance which would be more sensitive, a long-coil or a short-coil galvanometer? Why?
3. How does the condition of a wire and its surroundings when traversed by a current differ from that of a wire when not traversed by a current?
4. What conditions are the prerequisite of a current of electricity?
5. If a current be sent through the armature of a dynamo, what happens? Why?

REVIEW EXERCISES. 365

6. Upon what conditions does the strength of a current furnished by a dynamo depend?

7. You wish to make of an iron rod an electro-magnet with a certain end as an N-pole. Explain the method by a diagram.

8. What is the strength of a current which, falling 15 volts, yields .002 horse-power?

9. When would you wind an electro-magnet with fine wire?

10. If the difference of potential between the terminals of an arc lamp supplied with a 10-ampere current be 50 volts, what is the power consumed in the lamp?

11. A difference of potential of 5.5 volts is maintained at the terminals of a wire of 0.1 ohm resistance. (a) What current flows? (b) What is the power of the current?

12. To which electrode must an article to be electroplated be attached? Why?

13. Explain, in accordance with Ampère's theory of magnetism, the deflection of a magnetic needle by an electric current.

14. What length of copper wire .012 inch in diameter will offer a resistance of 1 ohm?

15. What currents are difficult to insulate? Why?

16. Upon what does the E. M. F. of a dynamo depend?

17. (a) How does a storage cell differ from a voltaic cell? (b) What can you say as to the direction of the current produced by each?

18. Which pole of an electro-magnet is that where the direction of the current in the coil is anti-clockwise?

19. What must be the E. M. F. of a generator which maintains a current of 5 amperes and works at the rate of 60 watts?

20. If the resistance of 40 feet of wire $\frac{1}{16}$ inch in diameter be 4 ohms, what is the resistance of 80 feet of like wire $\frac{1}{8}$ inch in diameter?

21. If the resistance of an electric lamp be 90 ohms, what would be the resistance of 10 such lamps connected (a) in series? (b) in multiple?

22. The resistance between two points in a conductor is 40 ohms, but on shunting these points it falls to 10 ohms. What is the resistance of the shunt?

23. How many watts are consumed in a 32 candle-power lamp requiring an E. M. F. of 54 volts and a current of 1.75 amperes?

24. How many cells in series each having an E. M. F. of 1.5 volts and an internal resistance of 2.5 ohms will produce a current of $\frac{1}{4}$ ampere through an external resistance of 10 ohms?

25. Why is a battery composed of cells no larger than percussion gun caps practically as efficient for sending a message through an ocean cable as a battery of an equal number of cells as large as flour barrels?

Milliliter

Cubic Centimeter

The area of this figure is a square decimeter. A cube of water, one of whose sides has this area, is a cubic decimeter or a liter of water, and at the temperature of 4° C. has a mass of a kilogram. The same volume of air at 0° C., and under a pressure of one atmosphere, has a mass of 1.293 grams. The gram is the mass of 1 cc of pure water at 4° C.

Square Inch

Square Centimeter

10 Centimeters

100 Millimeters

APPENDIX.

TABLES OF METRIC MEASURES.

Measure of Length.

1 Millimeter (mm.) = .001 meter (m.) = .03937 inch.
1 Centimeter (cm.) = .01 meter = .39371 inch.
1 Decimeter (dm.) = .1 meter = about 4 inches.
1 *Meter* = 39.37079 inches = about 3 ft. 3⅜ in.
1 Kilometer (km.) 1000 meters = about ⅝ mile.

Measure of Surface.

1 Square millimeter (mm.^{2}) = .000001 square meter (m.^{2}) = .0015 sq. in.
1 Square centimeter (cm.^{2}) = .0001 square meter = .1550 sq. in.
1 Square decimeter (dm.^{2}) = .01 square meter.
1 Are = 100 square meters.

Measure of Volume.

1 Cubic millimeter (mm.^{3}) = .000000001 cubic meter (m.^{3}).
1 Cubic centimeter (cm.^{3} or cc.) = .000001 m.^{3} = .061 cu. in.
1 Cubic decimeter (dm.^{3}) = .001 m.^{3} = 1000 cm.^{3}.
1 Cubic meter = about 1.308 cu. yds.

Measure of Capacity.

1 Milliliter (ml.) = .001 liter (l.) = 1 cc. = .061 cu. in.
1 Centiliter (cl.) = .01 liter = 10 cc.
1 Deciliter = .1 liter = 100 cc.
1 *Liter* = 1000 cm.^{3} = 61.027 cu. in. = 1.0567 qts.
(liquid measure).

APPENDIX.

Measure of Mass and Weight.

1 Milligram (mg.) = .001 gram (g.) = .0154 grain.
1 Centigram (cg.) = .01 gram = .1543 grain.
1 Decigram (dg.) = .1 gram = 1.5432 grains.
             1 Gram = 1.5432 grains = .03527 av. oz.
1 Kilogram (kg. or k.) = 1000 grams = 2.2046 av. lbs.

Table of Equivalent Values.

1 in. = .0254 m. = 2.53995 cm. = about $2\frac{1}{2}$ cm.
1 ft. = .3048 m. = 30.48 cm. = about $30\frac{1}{2}$ cm.
1 yd. = .9144 m. = about $1\frac{9}{10}$ m.
1 mile = 1609 m. = about 1.609315 km.

1 sq. in. = 6.4514 cm.^{2}.
1 sq. ft. = 929.01 cm.^{2}.
1 sq. yd. = 8361.1 cm.^{2} = .83611 m.^{2}.

1 cu. in. = 16.38618 cm.^{3}.
1 cu. ft. = 28,316 cm.^{3}.
1 cu. yd. = 764,526 cm.^{3} = about .76 m.^{3}.

1 U. S. quart = 946 cc. = .946 l.
1 av. oz. = 28.3494 g.
1 av. lb. = 453.59 g. = .45359 k. = about $\frac{5}{11}$ k.

Reduction of Measures to and from the C. G. S. System.

1 gram weight = 980 dynes (where g. = 980 cm. per sec.).
1 av. lb. weight = 4.445×10^5 dynes   "   "
1 k. weight = 980,000 dynes   "   "
1 gram-centimeter = 980 ergs.
1 erg = 1 dyne-centimeter = .0000001 joule = $\frac{1}{980}$ gram-centimeter.
1 kilogrammeter = 98,000,000 ergs = 7.23314 ft. lbs.
1 ft. lb. = 13,550,000 ergs.
1 foot-poundal = 421,402 ergs.[1]
1 joule = 10,000,000 [1] ergs.
1 horse-power $\begin{cases} 33000 \text{ ft. lbs. per min.} = \text{about } 7.452 \times 10^9 \text{ ergs per sec.} \\ 75 \text{ kilogrammeters per sec.} = 7.35 \times 10^9 \text{ ergs per sec.} \end{cases}$
1 watt = 1 joule per sec. = 10^7 ergs per sec. = 44.2394 ft. lbs. per min.
1 gram-degree = 4.17×10^7 ergs.
1 calorie = 4.17×10^{10} ergs.

[1] Independent of acceleration of gravity (*g.*).

APPENDIX.

Properties of Solids.

	Specific Density 17° C.	Hardness.	Expansion Coefficient 0°–100°.	Melting Point.	Specific Heat.
Agate	2.6	7			
Alum	1.7	2+			
Aluminum	2.7	3	.00002	700°	.21
Antimony	6.7	3	.00001	432°	.05
Arsenic	5.7	3	.000007		.08
Beech	.8				
Bismuth	9.8	2	.000013	266°	.03
Boxwood	.9				
Brass (cast)	8.3	3	.000019	900°?	.09
" (hard drawn)	8.5				
Bricks	1.7				.2
Bronze	8.8	3			
Canada Balsam	1.07				
Cherry	.7				
Copper	8.8	3	.000017	1100°	.09
Cork	.24				
Diamond	3.5	10			.14
Emerald	2.7	8			
Feldspar	2.5	6			
German-silver	8.5	3	.00002		
Glass (crown)	2.5		.000007	400°	.19
" (flint)	3.6				
Gold	19.3		.000012	1050°	.03
Granite	2.7				
Graphite	2.3				.20
Ice	.9	1.5		0°	.5
Iceland spar	2.7				
Iron (cast)	7.2	6?	.000012	1500°	.11
" (wrought)	7.7	4			
Ivory	1.9				
Lead	11.3	2	.000028	326°	.03
Marble	2.7	3			.21
Mica	2.8				
Paraffine	.9			55°	
Platinum (wire)	21.4		.000008	1800°	.03
Quartz	2.6	7			
Rock salt	2.2	2		800°	.21
Selenite	2.3	2			
Silver	10.4		.000019	1000°	.05
Slate	2.8				
Spermaceti	.9			44°	.08
Steel (tempered)	7.8	9?	.000013		
Sulphur (native)	2.			115°	.17
Talc	2.6	1			
Tallow	.9			40°	
Tin	7.3	2	.000019	232°	.05
Zinc	7.1	3	.00003	360°	.09

APPENDIX.

Properties of Liquids.

	Specific Density.	Coefficient Expansion at 0°.	Freezing Point.	Boiling Point 760 mm.	Specific Heat.	Refractive Index.
Acid, nitric, 0°	1.5	.00111	—47°			
" sulphuric, 0°	1.84	.00059		330°?	.34	1.43
Alcohol (grain), 0°	.81	.00100		78.2°	.59	1.36
Benzine, 20°	.87	.00118	4°	80°	.39	1.49
Carbon dioxide, 20°	1.37					
" disulphide, 15°	1.20				.23	1.64
Ether, 0°	.73	.00148		35°	.54	1.35
Glycerine, 0°	1.26	.0005		290°		1.47
Mercury, 0° (Regnault)	13.596	.00018	—39°	350°	.034	
" 15°–20°	13.558					
" (solid) —40°	14.3					
Milk, 0°	1.032					
Oil of turpentine, 0°	.89	.00071	—10°	160°	.43	1.47
Olive oil, 0°	.92	.00080				1.47
Sea water, 0°	1.026					
Water, 0°	.999		0°	100°	1.00	
" 4.07°	1.000					1.33
" 20°	.998					
" 100°	.958					

Specific Density of Gases and Vapors.

(Standard: Air at 0° C.; barometer, 76 cm.)

Air	1.0000	Hydrogen	0.0693
Ammonia	0.5367	Nitrogen	0.9714
Carbonic acid	1.5290	Oxygen	1.1057
Chlorine	3.4400	Sulphuretted hydrogen	1.1912
Hydrochloric acid	1.2540	Sulphurous acid	2.2474

APPENDIX. 371

YOUNG'S MODULUS OF ELASTICITY.

A rod of metal or a wire may have its length increased by pulling. The important quantity known as *Young's Modulus*, M, *is the stretching force per unit area of cross section divided by the elongation produced per unit length.*

Thus, let a rod of length l, and cross section a, be stretched by the force f so that its length becomes $l + l'$; then the strain per unit of length is $\frac{l'}{l}$, and the stress is $\frac{f}{a}$; hence, Young's modulus is

$$M = \left(\frac{f}{a} \div \frac{l'}{l} = \right) \frac{lf}{l'a}.$$

The numerical value of M in any case depends upon the unit of force used and the unit employed in measuring the cross section. For example, if the unit of force be the pound, and if the cross section be measured in square inches, the value of E for iron is about 25,000,000. This means that a rod of iron 1 sq. in. in section will, under a load of (say) 10,000 lbs. (4.46 tons), have its length increased $\frac{1}{2500}$ part. If the rod be 6 ft. long, its length will be increased by fully $\frac{1}{35}$ of an inch.

TABLE OF RESISTANCE OF WIRE,

Chemically pure, one meter long, one millimeter in diameter, at 0° C. (Jenkin). *Also relative resistances* (Ayrton).

		Relative Resistances.
Silver, annealed	.01937 ohm	1.000
Copper, annealed	.02104 "	1.086
Zinc, pressed	.07244 "	3.741
Platinum	.11660 "	6.022
Iron, annealed	.12510 "	6.460
Lead, pressed	.25270 "	13.050
German silver	.26950 "	13.920

TABLE OF TANGENTS.

Arc.	Tangent.	Arc.	Tangent.	Arc.	Tangent.
1	.017	31	.601	61	1.80
2	.035	32	.625	62	1.88
3	.052	33	.649	63	1.96
4	.070	34	.675	64	2.05
5	.087	35	.700	65	2.14
6	.105	36	.727	66	2.25
7	.123	37	.754	67	2.36
8	.141	38	.781	68	2.48
9	.158	39	.810	69	2.61
10	.176	40	.839	70	2.75
11	.194	41	.869	71	2.90
12	.213	42	.900	72	3.08
13	.231	43	.933	73	3.27
14	.249	44	.966	74	3.49
15	.268	45	1.000	75	3.73
16	.287	46	1.036	76	4.01
17	.306	47	1.07	77	4.33
18	.325	48	1.11	78	4.70
19	.344	49	1.15	79	5.14
20	.364	50	1.19	80	5.67
21	.384	51	1.23	81	6.31
22	.404	52	1.28	82	7.12
23	.424	53	1.33	83	8.14
24	.445	54	1.38	84	9.51
25	.466	55	1.43	85	11.43
26	.488	56	1.48	86	14.30
27	.510	57	1.54	87	19.08
28	.632	58	1.60	88	28.64
29	.554	59	1.66	89	57.29
30	.577	60	1.73	90	Infinite

VALUE IN MILLIMETERS OF BROWN & SHARPE WIRE-GAUGE NUMBERS.

Number.	Diameter mm.	Number.	Diameter mm.
1	7.348	21	0.723
2	6.544	22	0.644
3	5.827	23	0.573
4	5.189	24	0.511
5	4.621	25	0.455
6	4.115	26	0.405
7	3.656	27	0.361
8	3.264	28	0.321
9	2.906	29	0.286
10	2.582	30	0.255
11	2.305	31	0.227
12	2.053	32	0.202
13	1.828	33	0.180
14	1.628	34	0.160
15	1.459	35	0.143
16	1.291	36	0.127
17	1.150	37	0.113
18	1.024	38	0.101
19	0.912	39	0.090
20	0.812	40	0.080

DEVELOPMENT OF PHYSICAL SCIENCE.

Modern scientific knowledge comprises a very large body of facts, grouped under a comparatively few comprehensive generalizations, and these form a starting point for further investigations. But for centuries scientific study was confined to the observation of the more obvious isolated facts only, and to the construction of crude, because purely speculative, hypotheses.

It is the glory of the early investigators that in apparently unrelated fields of observation, and with no aid from the conception of the unity of scientific processes, they yet were able to formulate certain laws of the utmost value to later investigators. Among these pioneers was

Archimedes (page 118), 287–212 B.C., a native of Syracuse, who was the greatest mathematician of his age, and was skilled in various branches in Natural Philosophy. His principal discovery, that of the *law of buoy-*

ancy of fluids (see § 113), is the foundation of the science of specific gravity. He also discovered the *principle of the lever*, and the application of the inclined plane in the *Archimedean screw*.

From the third to the sixteenth century, science made but little progress. Modern science, in which theory is verified by observation under selected conditions, may be said to date from

Galileo (page 38), A.D. 1564–1642, an Italian mathematician, astronomer, and physicist. He is credited with being the founder of experimental science. The discovery of the *isochronism of the vibrations of the pendulum*, attributed to him, made it possible to develop a science of the relations of the three basal quantities, time, space, and force, to which his demonstration of the *law of falling bodies* (§ 40) still further contributed; while his invention of the *astronomical telescope* (§ 272) made possible modern astronomy, and that of the *thermometer* made possible the study of heat.

Sir Isaac Newton (page 28), 1642–1727, is styled "Prince of Philosophers." His penetrating mind traced the principles which govern the motions of the planets in their orbits and preserve the order of the universe. Not only did he formulate the *Law of Gravitation* (§ 65), but he enunciated the *three universal laws of motion* (page 27), and by his discussion of mass, momentum, inertia, force, etc., went far towards constructing a complete science of molar dynamics; and the principles propounded in his works contain by implication the modern doctrine of energy. In the field of optics, he discovered the principle of the *different refrangibility of colors* and propounded the corpuscular theory of light (§ 214). His work in mathematics was as profound and comprehensive as that in physics. Pope said of his vast achievements:

> Nature and Nature's Laws lay hid in Night;
> God said, Let Newton be! and all was Light.

Yet at the close of life he declared, "I seem to have been only like a boy playing on the seashore, and diverting myself in now and then finding a smoother pebble, a prettier shell than ordinary, while the great ocean of truth lay all undiscovered before me."

Newton was followed by a very large number of earnest students of nature, some of whom founded the science of chemistry, while others investigated the subjects of light, heat, and electricity.

Benjamin Franklin (page 278), 1706–90, was an American statesman and a scientist of original powers. He demonstrated the identity of lightning with the electric spark, and investigated vitreous and resinous electricity (§ 282). The former he attributed to an excess, the latter to a deficiency of an hypothetical "electric fluid," and thus gave to electrical terminology the expressions "positive" and "negative" electricity. His discovery that the *boiling point of liquids varies with the pressure* (§ 150) was a distinct step in advance in the study of the three states of matter, and his investigations of atmospheric phenomena foreshadowed the modern science of meteorology.

The nineteenth century has far surpassed all previous centuries in the amount and the quality of its contributions to scientific knowledge and theory. Of the many names of scientists who have contributed in a marked degree to the advancement of modern science, we can mention only Faraday and Thomson.

Michael Faraday (page 330), 1791-1867: as a scientific investigator, he stands preëminent by his rare combination of the speculative and imaginative power in originating experiments with manipulative skill in conducting them. He established the identity of the forces manifested in the phenomena known as electrical, galvanic, and magnetic, and laid the foundation of the science of electricity. The whole language of the magnetic field and lines of force is Faraday's. Especially notable are his studies in *electrolysis*, in *electrical induction*, and in *electro-magnetism*.

Sir William Thomson (page 306), 1824-, was raised in 1892 to the peerage as Lord Kelvin. In the scientific world he is noted for his researches and his contributions to scientific literature, and for the invention of many electrical instruments of great scientific value.

To the general public he is best known by his work in connection with submarine telegraphy (1858-66), while among scientists he commands respect by his profound and exhaustive study of molecular motion, his masterly attempts to unify the sciences of light, heat, electricity, and magnetism under a general theory of molecular physics, and his subtle speculations in regard to the ultimate constitution of matter.

Helmholtz wrote of him as follows: "His particular merit, in my opinion, consists of his treatment of mathematical physics. He has striven with great consistency to purify the mathematical theory from hypothetical assumptions which are not pure expressions of facts. In this way he has done very much to destroy the old unnatural separation between the experimental and mathematical physics, and to reduce the latter to a precise and pure expression of the laws of phenomena."

5

INDEX.

[Numbers refer to pages.]

A

Aberration, Chromatic, 240; Spherical, 237.
Absolute temperature, 140; units, 35, 36; zero, 139.
Acceleration, 10; Laws of, 11, 38.
Action and reaction, 31.
Adhesion, 94.
Air-pump, 114.
Ammeter, 299.
Ampère, 292.
Ampère's laws, 327.
Ampère's theory of magnetism, 327.
Ampere-volt, 294.
Analysis of light, 239.
Astigmatism, 262.
Atmospheric pressure, 106.

B

Barometer, 109; Fortin, 110.
Batteries, 308.
Beam of light, 209.
Beats in music, 194.
Boyle's law, 112.
Buoyancy, 117.

C

Caloric, 132.
Calorimetry, 131.
Candle-power, 214.
Capillary phenomena, 95.
Cell, Voltaic, 279, 306; Storage, 347.

Center of mass, 48; of oscillation, 62.
Central force, 57.
Centrifugal force, 58.
Chord, Musical, 196.
Cohesion, 91, 94.
Cold, Artificial, 150.
Color, 249–56; blindness, 254.
Colors, Complementary, 254.
Commutator, 340.
Composition of forces, 41, 44, 45; of velocities, 17.
Compressibility, 3.
Concord, 196.
Conduction of heat, 154; of electricity, 271.
Conductivity, 294.
Conjugate foci, 233.
Conservation of energy, 159.
Convection of heat, 155.
Correlation of energy, 159.
Coulomb, 292.
Couple, Dynamical, 45.
Critical angle, 228.
Curvilinear motion, 57.

D

Declination of needle, 321.
Deflection of needle, 288.
Densimeter, 121.
Density, 6, 119.
Dew-point, 153.
Diathermancy, 263.

INDEX.

Dielectrics, 272.
Diffused light, 218.
Discord, 195, 196.
Dispersion, 241.
Distillation, 148.
Divided circuit, 307.
Divisibility of matter, 2.
Ductility, 94.
Dynamics, 25.
Dynamos, 336.
Dyne, 35.

E

Ear, 203.
Earth a magnet, 319.
Ebullition, 145.
Echo, 179.
Elasticity, 92; of gases, 112; Modulus of, 372.
Electric lighting, 349.
Electrification, 268.
Electrolysis, 286, 346.
Electro-magnets, 288, 325.
Electro-motive force, 292.
Electroscopes, 270.
Electrotyping and electroplating, 352.
Energy, Electrical, 294; Kinetic and potential, 68; of chemical separation, 69; of sound-waves, 176; Units of, 70.
Engine, Steam, 162.
Equilibrant, 42.
Equilibrium, 41, 50.
Erg, 71.
Ether, The, 206.
Evaporation, 145.
Expansibility, 3.
Expansion, Anomalous, 137; Coefficient of, 136; factors, 136; by heat, 135.
Eye, 260.

F

Fluidity, 7.
Fluids, 7, 26.
Fluoroscope, 363.
Foci, Conjugate, 233.
Focus, Principal, 221.
Force, 25, 27, 34; Balanced, 41; Central, 57; Gravitation units of, 34; Leverage of, 46; Measurement of, 35, 37; Moment of, 45; Resolution of, 55; Unbalanced, 42.
Forces, Composition of, 41, 44, 55; Equilibrium of, 41; Parallelogram of, 53.
Formula for concave mirrors, 22; for lenses, 233.
Fraunhofer's lines, 247.
Freezing machines, 151.
Fusion, 142.

G

Galileo's experiment, 38.
Galvanometer, 298, 306, 308.
Galvanoscope, 290.
Gaseous masses, Laws of, 140.
Gravitation, 65.
Gravity, Specific, 119.

H

Hardness, 93.
Harmonic motion, 166.
Harmonics, 192.
Harmony, 195.
Heat capacity, 132; Consumption of, 150; defined, 125; Kinetic theory of, 125; Mechanical theory of, 161; of fusion, 143; of vaporization, 147; Sources of, 126; Specific, 132.
Horse-power, 77.

INDEX. 379

Hydrokinetics, 98.
Hydrometer, 121.
Hydrostatic press, 100.
Hydrostatics, 98.
Hygrometry, 152.

I

Illumination, 214.
Images, 210, 217, 219, 222, 234.
Impenetrability, 2.
Inclined plane, 85.
Induction, 273; coils, 332; Electromagnetic, 328; Faraday's law of, 330.
Inertia, 27.
Interference of ether-waves, 255; of sound-waves, 183.
Isogonic curves, 321.

J

Joule, 294.
Joule's experiment, 160.

K

Kinematics, 8.
Kinetics, 42.

L

Lamps, Electric, 349.
Law of Charles, 140; of gaseous masses, 140.
Lenses, 231.
Leverage, 46.
Levers, 82.
Light, 207; Maxwell's theory of, 359.
Lightning, 277.
Lines of magnetic force, 315.
Luminous bodies, 208.

M

Machines, 78–91; Efficiency of, 82.
Magnetic circuit, 317; field, 322; flux, 317; poles of the earth, 320.
Magnetism, Ampère's theory of, 327.
Magnets, 313.
Malleability, 94.
Mass, 5, 37; Center of, 48.
Matter, Kinetic theory of, 138; Theory of the constitution of, 3, 138; Three states of, 7.
Megohm, 294.
Metric system, 4.
Microphone, 359.
Microscope, Compound, 257; Simple, 236.
Mirrors, 217, 219.
Molecule, 2.
Moment of a couple, 47; of a force, 45.
Moments, Equilibrium of, 46.
Momentum, 27, 73.
Motion, 8; Composition and resolution of, 17; Curvilinear, 23, 57; Harmonic, 166; Kinds of, 22; Newton's laws of, 27, 28, 31.
Motor, Electric, 343.
Musical instruments, 199; scale, 186; sound, 186.

N

Nodes, 191.
Notes, 193.

O

Ohm, 293.
Ohm's law, 295, 296.
Opalescence, 250.
Optical prisms, 231.
Oscillation, Center of, 62.
Overtones, 192.

P

Parallelogram of forces, 53.
Pascal's principle, 101.
Pendulum, 60–65; Simple, 62.
Phenomenon, 1.
Phonograph, 197.
Photometry, 214.
Physical measurements, 4.
Physics, 1.
Pitch, Musical, 185.
Plasticity, 92.
Plates, Sounding, 201.
Pneumatics, 98.
Polarity, 314, 324.
Polarization, Electrical, 282.
Porosity, 3.
Potential, Electrical, 275, 296.
Poundal, 36.
Power, 76; Electrical, 294.
Pressure, 43; of gases explained, 138.
Prévost's theory of exchanges, 266.
Principle of Archimedes, 117.
Prisms, Optical, 231.
Pumps, 114–16.

Q

Quality of sound, 196.

R

Radiant energy, 206; Effects of, 207.
Radiation, 159, 206; Electric, 359; Thermal effects of, 263.
Radiography, 361.
Railway, Electric, 345.
Ray, 209.
Rays, Infra-red and ultra-violet, 248.
Reënforcement of sound-waves, 180.
Reflection of sound-waves, 178.

Refraction, 224; Cause of, 226; Indices of, 228.
Relay, 356.
Resistance, Electrical, 293, 296, 300, 303.
Resonance, 179.
Resonators, 181.
Retentivity, 314.
Rigidity, 26.
Ruhmkorff's coil, 333.

S

Screw, 86.
Self-induction, 332.
Shadows, 211.
Shunts, 307.
Siphon, 116.
Solenoid, 323.
Sonometer, 190.
Sound and sound-waves, 173, 196.
Speaking-tubes, 177.
Specific density, 119; gravity, 119; heat, 132.
Spectroscope, 243.
Spectrum analysis, 245.
Spectrums, 243.
Spherical aberration, 237.
Stability of bodies, 51.
Statics, 42.
Steam engine, 162.
Stereopticon, 262.
Storage cells, 347.
Strain, 26.
Strength of current, 291, 296.
Stress, 26.

T

Telegraph, 354.
Telephone, 356.
Telescope, 259.
Temperature, 128.

Tenacity, 91.
Tension, 43; Surface, 95.
Tesla's investigations, 363.
Thermal units, 132.
Thermo-dynamics, 159.
Thermo-electric currents, 348.
Thermometer, 128.
Tones, 193.
Transformer, 335.
Transparency, 209; Magnetic, 313.

U

Units, Absolute, 35, 36; Physical, 4.

V

Vaporization, 147.
Variation of the needle, 321.
Velocity, 10; Composition and resolution of, 17; of light-waves, 212; of sound-waves, 175.
Ventilation, 156.

Vibration, 167, 169; Sympathetic, 184.
Vibrations, Complex, 192; Stationary, 191.
Vibrograph, 169.
Viscosity, 92, 95.
Visual angle, 216.
Vocal organs, 202.
Volt, 293.
Voltaic arc, 350.
Volt-ampere, 294.
Volt-coulomb, 294.
Volume, 5.

W

Watt, 294.
Wave-motion, 167–72.
Weight, 6, 34, 36. 66.
Wheatstone bridge, 304.
Wheel and axle, 84.
Work, 67, 71; Electrical, 294.

www.ingramcontent.com/pod-product-compliance
Lightning Source LLC
Chambersburg PA
CBHW020740020526
44115CB00030B/689